Paul Otter

W9-CRB-728

Massive Neutrinos In Physics And Astrophysics

World Scientific Lecture Notes in Physics

World Scientific Lecture Notes in Physics – Vol. 60

Massive Neutrinos In Physics And Astrophysics

Second Edition

Rabindra N Mohapatra
University of Maryland, USA

Palash B Pal
Saha Institute of Nuclear Physics, India

World Scientific
Singapore • New Jersey • London • Hong Kong

Published by

World Scientific Publishing Co. Pte. Ltd.

P O Box 128, Farrer Road, Singapore 912805

USA office: Suite 1B, 1060 Main Street, River Edge, NJ 07661

UK office: 57 Shelton Street, Covent Garden, London WC2H 9HE

British Library Cataloguing-in-Publication Data
A catalogue record for this book is available from the British Library.

MASSIVE NEUTRIONS IN PHYSICS AND ASTROPHYSICS

Copyright © 1998 by World Scientific Publishing Co. Pte. Ltd.

All rights reserved. This book, or parts thereof, may not be reproduced in any form or by any means, electronic or mechanical, including photocopying, recording or any information storage and retrieval system now known or to be invented, without written permission from the Publisher.

For photocopying of material in this volume, please pay a copying fee through the Copyright Clearance Center, Inc., 222 Rosewood Drive, Danvers, MA 01923, USA. In this case permission to photocopy is not required from the publisher.

ISBN 981-02-3373-6

This book is printed on acid-free paper.

Printed in Singapore by Uto-Print

QC
793
.5
N426
M64
1998
PHYS

— To our parents —

Preface to the second edition

Neutrino physics is a field full of hopes and challenges, sometimes mysterious and occasionally frustrating but never devoid of excitement and always rich in its implications for new physics. Just as the postulate of the very existence of the neutrino that led to the birth and the understanding of the field weak interactions and just as it was with the discovery of the neutrino neutral current phenomena that formed the first cornerstone of the standard model, so it has been dreamed that the confirmation of the various terrestrial and extraterrestrial hints of neutrino oscillations will be the first beacon of physics beyond the standard model. At long last as this millennium comes to a close, this dream is looking more and more like reality and the outline of a long hoped for $B - L$ symmetry as new symmetry of nature is beginning to emerge. Whether this will come in association with its twin — the right handed weak iso-spin symmetry, we do not know — but there is a strong likelihood that it would and particle physics would have taken a giant leap forward.

Since the first edition of the book was completed in 1990, a temporary storm cloud over the field raised by the 17 keV neutrino has died down; the solar neutrino physics has crossed several important milestones; the hints of oscillations by the cosmic ray neutrinos is receiving its confirmation and the first hints of neutrino oscillations in the laboratory have appeared. All these have made the field more vibrant than it was in 1990. In this new edition, we have tried to capture some of this excitement by updating and improving most of the chapters. Two new chapters, one giving an introduction to supersymmetry and another discussing the propagation of neutrinos in a medium have been included. We have also added a final chapter, which contains a summary of different constraints from different chapters and provides an outlook for neutrino physics for the coming years. The exercises, which appeared at the end of the first edition, have been moved to positions within the

vii

chapters where they would be most helpful to a beginner. The number of exercises have also been increased considerably. In making all these changes, we have always kept in mind our original goal for the book i.e. this is a *book* and not a collection of the latest facts and references and is meant for beginning graduate students as well as advanced researchers in the field of neutrino physics. We are extremely grateful to our colleagues who have recommended the book as a textbook in their courses in various universities around the world.

We are extremely grateful to many colleagues who have generously shared their knowledge about the neutrinos and related subjects with us — in particular, we wish to thank K. S. Babu, P. Bhattacharjee, C. Burgess, D. Caldwell, R. Cowsik, D. A. Dicus, J. C. D'Olivo, R. Gandhi, E. Golowich, A. Joshipura, H. Klapdor-Kleingrothaus, J. F. Nieves, S. Nussinov, A. D. Patel, J. C. Pati, S. T. Petcov, A. Smirnov, S. Mishra, U. Sarkar, G. Senjanović, V. L. Teplitz, L. Wolfenstein. In addition, U. Sarkar has taught us how to edit postscript files, which has helped enormously in the preparation of the figures for this edition.

Rabindra N. Mohapatra
Palash B. Pal

August, 1997

From the preface to the first edition

Neutrinos have played a key role in the evolution of our understanding of particles, forces and the universe and most likely the next step in this process of exploration of physics beyond the standard model will come from new properties of the neutrino revealed in many on going experiments such as the solar neutrino experiments, neutrino mass measurements, double beta decay searches etc. In the absence of any solid experimental clues regarding this new direction, many interesting theoretical speculations have been advanced with implications for collider as well as non-collider experiments. The central theme of most of these speculations in the possibility of a massive neutrino which in turn impacts on our thinking about the universe past and present. Most of these ideas will be put to test by the same kind of experiments discussed.

We therefore felt that it may be an appropriate time to summarize the theoretical, phenomenological and astrophysical implications of the massive neutrino so that first, a person starting out in the field will have a ready reference to the major existing ideas and, secondly, an expert in the field may be spared frequent trips to the library to clarify simple points in his or her thinking. In this spirit, we have not attempted a complete, exhaustive survey of all the details in various areas of neutrino physics but rather an introduction to the important major ideas in the field. We have tried to restrain our personal prejudices in the presentation to the extent humanly possible. We have most certainly left out some ideas as we must have missed citing some important works. While some of it is perhaps unavoidable because of the size of the book, mostly it is inadvertent and will be included in subsequent editions if brought to our attention with sufficient conviction. We have, however, given references to review articles where additional references can be obtained. For articles originally published in languages other than English, we have given reference to the English translation if we knew of one. The reader can easily find out the original reference by looking at the translated material.

A suggestion for new comers to the area trying to read this book : a necessary pre-requisite is a course on field theory, group theory and basic concepts in particle physics, and a right attitude. We have added some exercises at the end of the book to help the new comers.

Over the years, both the authors have learned a great deal about the subjects discussed in the book from many of their colleagues — either through collaboration or through discussions. Specifically we would like mention K. S. Babu, Riccardo Barbieri, Darwin Chang, Nilendra Deshpande, Boris Kayser, D. Krakauer, Ling-Fong Li, Robert Marshak, Shmuel Nussinov, José Nieves, Lev Okun, Sandip Pakvasa, Roberto Peccei, Peter Rosen, Goran Senjanović, Eichii Takasugi, José Valle, Lincoln Wolfenstein and Tsutomu Yanagida.

RNM wishes thank the National Science Foundation for support during the time this book was being written and the University of Maryland for a sabbatical leave. The work of PBP was supported by the Department of Energy, and this support is gratefully acknowledged. Our sincerest thanks go to Michèle Wilson Gibson, who has typed most of the plain text of the book and has drawn many of the diagrams. In addition, we thank Betty Krusberg for typing some parts of the manuscript. We are grateful to Fred Olness for helping with problems that arose at various stages of the use of the computer to produce a camera-ready manuscript, and to Tristan Hübsch for timely suggestions regarding the use of LaTeX.

Rabindra N. Mohapatra
Palash B. Pal

November 1990

Notations

Our notations regarding the metric and gamma matrices etc are the same as the standard text books on Quantum Field theory such as the book by Bjorken and Drell or by Itzykson and Zuber. Thus, for example, our metric tensor is

$$g_{\mu\nu} = \text{diag}\,(1, -1, -1, -1)\,.$$

The γ-matrices have been given explicitly in §4.3, and are not repeated here.

However, our notation differ from the textbooks mentioned above in some important aspects. These are listed below.

1. Our normalization of spinor solutions to Dirac equation is different. It has been described in §4.3 and has the advantage that it applies equally well to massive as well as massless fields.

2. To denote antiparticles, we do not use the customary bar since it is easy to confuse it with the operation of hermitian conjugation followed by multiplication of the Dirac matrix γ^0. Neither do we use the notation, which is used in some modern literature, of attaching a superscript c to the particle since it might lead to the misconception that the antiparticle is the C conjugate of the particle. For particles like the neutrinos whose interactions violate C by large amounts, the antiparticle can be defined only through the operation CPT. We use a hat to denote antiparticles and hope that this symbol will find acceptance in the community.

3. We have used the summation convention for repeated indices when the index denotes space-time, but not when it labels different particles or represents elements of internal symmetry groups.

Contents

Contents ———————————————————————— xix

Part I

Preliminaries

Part 1

Preliminaries

Chapter 1

Introduction

1.1 History

The phenomenon of radioactivity was discovered in 1892 by Becquerel. Bohr was the first to realize that beta decay is a process in which the electron is ejected from the nucleus. In 1914, Chadwick made the crucial discovery that the primary beta spectrum is continuous. Until well into the twenties, this result was believed to have other explanations. In 1929, in a letter to a conference [1, 2], Pauli wrote, "Dear Radioactive Ladies and Gentlemen, ... as a desperate remedy to save the principle of energy conservation in beta decay, ... I propose the idea of a neutral particle of spin half." So, the neutrino was born into the world of theoretical physics. Chadwick discovered the neutron in 1932, which was realized to be the particle in the nucleus which emitted the electron and the neutrino in the process of beta decay. Fermi wrote the four-Fermi Hamiltonian for beta decay using the neutrino, electron, neutron and the proton. A new field of theoretical physics came into existence — the field of Weak Interactions. In 1956, Reines and Cowan discovered the neutrino. As more and more particles were discovered and found to participate in weak processes, weak interactions acquired legitimacy as a new force of nature and neutrino became an integral part of this interaction. With the invention of new accelerators and sophisticated detectors and after the work of many dedicated experimentalists, we now know that neutrinos come in three varieties. Each neutrino comes with its own family of quarks and leptons. Observations involving neutrinos have played a key role in our understanding of weak interactions in the form of the so-called standard model and it is believed that further observation involving them may hold the key to our understanding of physics beyond the standard model.

In the past few decades it has also been realized that our understanding of the Universe we live in depends critically on our understanding of the neutrino. For instance the neutrino is the most abundant form of matter in the Universe next to radiation, and through its role in nucleosynthesis, it is responsible for the origin of heavy elements that form the basis of life. There are also speculations that neutrinos may have played a role in the formation of galaxies that provided the stage for the evolution of life. Thus the neutrino is clearly an important particle. This book is an attempt to summarize the present status of our knowledge of the neutrino and some popular speculations about its yet undiscovered properties.

Let us start with a brief overview of the various properties of the neutrino. It is known to have a spin of half in units of $\hbar$ and is believed to have no electric charge. Whether it has a mass or not is one of the key questions of present day particle physics and will constitute one of the chief obsessions of some of the latter chapters in this book. It is however known from various laboratory experiments that, if it has a mass, it is very much smaller than any of the masses of other known particles such as quarks or leptons in the same family.

As mentioned earlier, now there are known to be three kinds of neutrinos. The existence of the muon and tau neutrino was suspected from the properties of the corresponding charged partners: for instance it was found that the muon does not decay to an electron and a photon which is a priori allowed and neither does a muon decay to $e^+e^-e^-$. The simplest way to understand this is to postulate that there exists a separate neutrino corresponding to the muon. The muon neutrino was discovered in 1962 by the group of Lederman, Schwarz and Steinberger. The third charged lepton, the tau, was discovered by the group of Perl in the mid-seventies and by similar reasoning, a tau neutrino is believed to exist, although no direct detection of it has yet been made. The families of fermions can therefore be written as follows:

$$(u, d, e, \nu_e)$$
$$(c, s, \mu, \nu_\mu)$$
$$(t, b, \tau, \nu_\tau).$$

(1.1)

A prime focus of experiments now is to ascertain whether there exists a fourth type of neutrino in nature. Recent measurements of the width of the Z-boson have definitely ruled out a fourth neutrino if it is lighter than 40 GeV or so. If there exists a fourth type of neutrino, by analogy it would imply the existence of a fourth type of matter, which of course will be a profound discovery.

1.2 Four-Fermi interaction

Using the neutrino hypothesis of Pauli, Fermi [3] was the first to write down the field theoretical form of an interaction involving the neutron, the proton, electron and neutrino fields that would describe weak interaction. If we denote these fermion fields by ψ_i, where $i = n$, p, e and ν_e, the so-called *four-Fermi interaction* Hamiltonian density is written as

$$\mathcal{H}_{\text{weak}} = \frac{G_F}{\sqrt{2}} \overline{\psi}_p \gamma_\mu \psi_n \, \overline{\psi}_e \gamma^\mu \psi_{\nu_e} . \tag{1.2}$$

The form of the weak Hamiltonian in low energy weak decays has been confirmed over the year to basically have the same form as in Eq. (1.2) although a great deal more is known about its Lorentz structure than is given in Eq. (1.2). Also, many new four-Fermi terms describing other kinds of weak processes are now known to be part of $\mathcal{H}_{\text{weak}}$. Let us give a very brief and qualitative summary of these new features.

1.2.1 Modern form of four-Fermi interaction

The main observations that demand some modification of Fermi's original form of the four-Fermi interaction are as follows:

- The fundamental fields entering the Hamiltonian are not hadronic fields like the proton or the neutron, but rather are fields describing quarks.

- The space-time structure of the currents have not only the vector currents (V) as in Eq. (1.2), but have also axial vector currents (A). The fact that the currents are of the $V - A$ form was noted by Marshak and Sudarshan, and by Feynman and Gell-Mann [4].

- The weak interactions consist not only of *charged current processes* like the beta decay in which electric charge of leptons and quarks change, but also of processes where the electric charge remain unchanged among the leptons and hadrons. Such processes are called *neutral current processes*.

Using the above considerations, the weak interaction Hamiltonian can be written in a compact form as follows:

$$\mathcal{H}_{\text{weak}} = \frac{4G_F}{\sqrt{2}} \left[J^\mu(x) J^\dagger_\mu(x) + \rho K^\mu(x) K_\mu(x) \right], \tag{1.3}$$

where J^μ is a charged current, which in terms of quark and lepton fields can be written as

$$J_\mu(x) = \overline{\mathbf{u}} \gamma_\mu \mathsf{L} V \mathbf{d} + \overline{\boldsymbol{\nu}} \gamma_\mu \mathsf{L} \boldsymbol{\ell} \tag{1.4}$$

where the chirality projection operators are defined as

$$\mathsf{L} = \frac{1}{2}(1 - \gamma_5), \quad \mathsf{R} = \frac{1}{2}(1 + \gamma_5). \tag{1.5}$$

The fermion fields in Eq. (1.4) are

$$\mathbf{u} = \begin{pmatrix} u \\ c \\ t \end{pmatrix}, \ \mathbf{d} = \begin{pmatrix} d \\ s \\ b \end{pmatrix}, \ \boldsymbol{\nu} = \begin{pmatrix} \nu_e \\ \nu_\mu \\ \nu_\tau \end{pmatrix}, \ \boldsymbol{\ell} = \begin{pmatrix} e \\ \mu \\ \tau \end{pmatrix}, \tag{1.6}$$

and V is a matrix which denotes the mixing effect between different generations of quarks. Present information about the magnitudes of various elements of V is given below [5]:

$$V = \begin{pmatrix} 0.9745 \text{ to } 0.9757 & 0.219 \text{ to } 0.224 & 0.002 \text{ to } 0.005 \\ 0.218 \text{ to } 0.224 & 0.9736 \text{ to } 0.9750 & 0.036 \text{ to } 0.046 \\ 0.004 \text{ to } 0.014 & 0.034 \text{ to } 0.046 & 0.9989 \text{ to } 0.9993 \end{pmatrix}. \tag{1.7}$$

There is also a phase in the matrix which is responsible for CP violation. In Ch. 2, we will see that in the standard electroweak theory of Glashow, Weinberg and Salam, V is a unitary matrix.

The first term in the interaction of Eq. (1.3) gives rise to a host of new weak interaction phenomena such as:[1]

$$\mu^- \rightarrow e^- \widehat{\nu}_e \nu_\mu$$
$$c \rightarrow s e^+ \nu_e$$
$$s \rightarrow u e^- \widehat{\nu}_e. \tag{1.8}$$

---

[1] As explained in the prechapter called "Notations", we will use the hat to denote antiparticles.

Whereas we have described above weak processes in terms of quarks, the actual observation involves a color-singlet hadron containing the quark. For instance, the last process in Eq. (1.8) is really inferred from $\Lambda \to pe^-\hat{\nu}_e$.

In Eq. (1.3), K_μ is the neutral current, which has the following form:

$$
\begin{aligned}
K_\mu(x) &= \sum_q [\epsilon_L(q)\bar{q}\gamma_\mu Lq + \epsilon_R(q)\bar{q}\gamma_\mu Rq] \\
&+ \frac{1}{2}\sum_\nu \bar{\nu}\gamma_\mu L\nu + \frac{1}{2}\sum_\ell \bar{\ell}\gamma_\mu [g_V(\ell) - \gamma_5 g_A(\ell)]\ell.
\end{aligned} \quad (1.9)
$$

In Ch. 2, we will discuss the presently available information on various parameters appearing in the neutral currents. This involves the parameters in Eq. (1.9) as well as the ρ parameter defined in Eq. (1.3), which measures the relative strengths of neutral and charged current interactions.

Several characteristics of the Hamiltonian in Eq. (1.3) are worth noting. In the neutral current K_μ, there is no term which connects two different species of quarks or leptons. The quarks and leptons of a particular generation are often called *flavor*, and the possible currents that could connect two different flavors are called *flavor changing neutral currents* (FCNC). At present extensive search is going on for flavor changing hadronic as well as leptonic neutral currents. The experiments look for processes [6] such as $K_L \to \mu^+\mu^-$, $K_L \to e^+e^-$, $K_L \to \mu^\pm e^\mp$, $K^+ \to \pi^+\nu\hat{\nu}$, $\mu \to e\gamma$, $\mu^\pm \to e^-e^+e^\pm$, $\mu + \text{Nucl} \to e + \text{Nucl}$ etc. The present limits are very stringent and imply upper bounds on the strength of couplings expressed in the form of Eq. (1.3) to be $10^{-4}G_F$ for Kaon decays to $10^{-6}G_F$ for muon decays. These selection rules imply that in the leptonic sector, one can define new quantum numbers called *generational lepton numbers* L_e, L_μ and L_τ. The known weak interactions appear to obey the conservation of all these numbers. Note that due to the presence of the mixing matrix V, no analogous kind of selection rule can be written down for the hadronic (or quark) sector.

1.2.2   Fierz-Michel transformation

In the weak interaction Hamiltonian of Eq. (1.3), there are two kinds of terms, viz, the terms coming from charged-current interactions and those coming from neutral current interactions. However, four-Fermi interactions of these two kinds can be transformed into each other. This is achieved through the Fierz-Michel transformations, which we now discuss. Sometimes, this helps a lot in calculations, or in understanding the main results without getting into details.

Since any fermionic field ψ is a four-component object, there can be sixteen independent fermionic bilinears of the form $\overline{\psi} F \psi$. Thus, any F can be written in terms of 16 basis bilinears, which we choose in the following way:

$$
\begin{aligned}
\Gamma^1 &\equiv 1, \\
\Gamma^{2,3,4,5} &\equiv \gamma^\mu, \\
\Gamma^{6 \text{ to } 11} &\equiv \sigma^{\mu\nu}, \quad (\mu < \nu) \\
\Gamma^{12,13,14,15} &\equiv i\gamma^\mu\gamma_5, \\
\Gamma^{16} &\equiv \gamma_5.
\end{aligned}
\tag{1.10}
$$

The Γ matrices with lower indices can be defined by taking all the Lorentz indices down. Then it is easily verified that

$$
\text{Tr}\,(\Gamma^r \Gamma_q) = 4\delta^r_q.
\tag{1.11}
$$

The most general four-fermion interaction that is Lorentz invariant and parity conserving can be written as

$$
\sum_{r=1}^{16} C_r [\Gamma_r]_{1,2} [\Gamma^r]_{3,4},
\tag{1.12}
$$

where we have used the shorthand

$$
[\Gamma_r]_{1,2} \equiv \overline{\psi}_1 \Gamma_r \psi_2,
\tag{1.13}
$$

for two fermion fields ψ_1 and ψ_2. Notice that, in Eq. (1.12) as well as in the rest of this section, repeated indices of Γ are not assumed to be automatically summed over, unless there is an explicit summation sign.

However, in Eq. (1.12), fermions ψ_1 and ψ_3 are created, whereas ψ_2 and ψ_4 are annihilated, so there is no reason why it cannot be written in the form

$$
\sum_{q=1}^{16} C'_q [\Gamma_q]_{1,4} [\Gamma^q]_{3,2}.
\tag{1.14}
$$

The forms given in Eqs. (1.12) and (1.14) are therefore equivalent, and the task is to find the relation between the coefficients C_r and C'_q. Since this equivalence should be valid for arbitrary fermion fields, we can extract the coefficients of $(\overline{\psi}_1)_A(\psi_2)_B(\overline{\psi}_3)_C(\psi_4)_D$ from each expression, where A, B, C, D are Dirac indices. This enables us to write

$$\sum_{r=1}^{16} C_r(\Gamma_r)_{AB}(\Gamma^r)_{CD} = -\sum_{q=1}^{16} C'_q(\Gamma_q)_{AD}(\Gamma^q)_{CB}\,, \qquad (1.15)$$

where the minus sign comes because the fermion fields on the right side involve one interchange compared to those on the left side. Multiplying both sides now by $(\Gamma_{r'})_{DC}(\Gamma^{r''})_{BA}$ and summing over all the Dirac indices, we obtain

$$C_r = \sum_q \Lambda_{r,q} C'_q\,, \qquad (1.16)$$

where

$$\Lambda_{r,q} = -\frac{1}{16}\mathrm{Tr}\,(\Gamma_q\Gamma_r\Gamma^q\Gamma^r)\,. \qquad (1.17)$$

It is now straightforward to calculate the matrix Λ. However, a more useful form is obtained by noting that Eq. (1.15) can also be written as

$$\sum_i \tilde{C}_i(\mathcal{O}_i)_{AB}(\mathcal{O}^i)_{CD} = -\sum_j \tilde{C}'_j(\mathcal{O}_j)_{AD}(\mathcal{O}^j)_{CB}\,, \qquad (1.18)$$

where now the index i runs over the five different rows of Eq. (1.10), i.e., corresponding to the scalar (S), vector (V), tensor (T), axial vector (A) and pseudo-scalar (P) interactions. In this case, one can write

$$\tilde{C}_i = \sum_j \tilde{\Lambda}_{ij}\tilde{C}'_j\,, \qquad (1.19)$$

instead of Eq. (1.16). Obviously, the elements of the matrix $\tilde{\Lambda}$ are simply related to those of the matrix Λ. For example,

$$\tilde{\Lambda}_{SS} = \Lambda_{1,1} = -\frac{1}{16}\mathrm{Tr}\,1 = -\frac{1}{4}\,. \qquad (1.20)$$

Now consider the element $\tilde{\Lambda}_{VS}$. This means that we want to see the scalar contribution, with the ordering of Eq. (1.12), coming from all four components of the vector interaction with the ordering of Eq. (1.14). Thus,

$$\tilde{\Lambda}_{VS} = -\frac{1}{16}\text{Tr}\,(1\,\gamma_\mu\,1\,\gamma^\mu) = -1\,.$$ (1.21)

Similarly, one can proceed to see that the matrix $\tilde{\Lambda}$ is given by

$$\tilde{\Lambda} = -\frac{1}{4}\begin{pmatrix} 1 & 1 & 1 & 1 & 1 \\ 4 & -2 & 0 & 2 & -4 \\ 6 & 0 & -2 & 0 & 6 \\ 4 & 2 & 0 & -2 & -4 \\ 1 & -1 & 1 & -1 & 1 \end{pmatrix}\,.$$ (1.22)

This is the Fierz-Michel transformation matrix.

So far, we talked about parity conserving interactions only. In a discussion of weak interactions, parity violation will of course be present. This is possible if the terms shown in Eq. (1.12) appear in conjunction with terms of the form

$$\sum_{r=1}^{16} D_r [\Gamma_r]_{1,2} [\Gamma^r \gamma_5]_{3,4}\,.$$ (1.23)

One can similarly use a different ordering of the fermion fields in this case and find the relation between between the coefficients in the two cases. However, this can also be read off from Eq. (1.22) directly once we note that Eq. (1.23) can be written in the form of Eq. (1.12) with a different field ψ'_4 which is related to ψ_4 by the relation $\psi'_4 = \gamma_5 \psi_4$.

Exercise 1.1 *Show that the $V - A$ and $V + A$ interactions are both invariant under the Fierz-Michel transformations, i.e.,*

$$\begin{aligned}
[\gamma_\mu(1-\gamma_5)]_{1,2}[\gamma^\mu(1-\gamma_5)]_{3,4} &= [\gamma_\mu(1-\gamma_5)]_{1,4}[\gamma^\mu(1-\gamma_5)]_{3,2} \\
[\gamma_\mu(1+\gamma_5)]_{1,2}[\gamma^\mu(1+\gamma_5)]_{3,4} &= [\gamma_\mu(1+\gamma_5)]_{1,4}[\gamma^\mu(1+\gamma_5)]_{3,2}
\end{aligned}$$ (1.24)

1.2.3 Problems with the four-Fermi interaction

While the four-Fermi interaction $\mathcal{H}_{\text{weak}}$ given in Eq. (1.3) provides an excellent description of observed low energy weak interaction phenomena, it is fraught with difficulties when considered as a field theory. There are many ways to see it. For instance, if one calculates higher order weak interaction corrections to any lowest order weak process, the contributions are divergent and the divergence at the L-th loop goes as Λ^{2L}, Λ being the cutoff for the theory. Therefore, in a strict sense, the lowest order calculations are not reliable.

Even at the lowest order, the total cross section for neutrino-electron scattering, for example, comes out to be proportional to $G_F^2 s$, where s is the Mandelstam variable defined by $s = (p_{\nu_e} + p_e)^2$ in terms of the momenta of the incoming neutrino and electron. With increasing energy this cross section grows without limit. Since $\nu_e e$ scattering occurs in the s-wave, the amplitude for this process should obey the s-wave unitarity bound, viz.,

$$\sigma_{\text{tot}}^s \leq \frac{16\pi}{s} \,. \tag{1.25}$$

This leads to a contradiction, implying that the four-Fermi description of weak interaction must breakdown above a certain energy, which in old days was called the weak-interaction cutoff, Λ_{weak}. Examination of various weak processes produced values of Λ_{weak} between 300 GeV down to 4 GeV [7]. This presented a crisis for four-Fermi description of weak interaction.

The basic problem was connected with the fact that the four-Fermi Lagrangian is not renormalizable. So search for a renormalizable weak interaction Lagrangian began in the late sixties. It culminated in the discovery of gauge theories, whose basic ingredients we describe below.

1.3 Symmetries and forces

Symmetries have played a fundamental role in our understanding of particle physics. Starting with the Poincaré symmetry group of space-time transformations to the isospin invariance group of $SU(2)$ in nuclear physics and the $SU(3)$ symmetry group of Gell-Mann and Neéman for hadron physics, our understanding of physics has always deepened with the identification of an invariance group in the system.

1.3.1 Global symmetries

There are two distinct kinds of symmetries for physical systems: *global symmetries* where the same symmetry transformation is applied to a field at all space-time points, and *local symmetries* where the symmetry transformations at different space-time points are unrelated. The isospin symmetry of nuclear physics introduced by Heisenberg, which provided a lot of insight into nuclear energy levels and forces, is an example of global symmetry. With the discovery of many baryon and meson states, the global $SU(3)$ symmetry was proposed as a generalization of isospin to classify the various hadronic states and mass spectra.

Under the isospin symmetry the up (u) and the down (d) quarks form a doublet whereas all other quarks are invariant. The proton and the neutron are made out of u and d quarks alone: $p \equiv uud$ and $n \equiv udd$. The mesons $\pi^{\pm}$ and π^0 are made out of u and d in combination with their antiparticles.

Under $SU(3)$, the three quarks u, d and s form a triplet. Again, baryons made out of these three quarks in various combinations, and mesons made out of various combinations of u, d, s and $\overline{u}$, $\overline{d}$, $\overline{s}$, have been identified, thus confirming the validity of the $SU(3)$ symmetry in particle physics.

The existence of exact global symmetries always implies relations between masses and coupling constants among particles, provided the symmetry is realized in the so-called Wigner-Weyl mode. This just means that the ground state of the system is invariant under the symmetry transformations. If the isospin symmetry were realized in this way, this would imply the proton and the neutron have the same mass. Since that is not true ($m_n - m_p = 1.3\,\mathrm{MeV}$), the isospin symmetry must therefore be approximate. Similarly, the $SU(3)$ symmetry is also approximate. Study of approximate $SU(3)$ symmetry led, in fact, to the concept of quarks as constituents of hadrons, which was a major leap forward in our understanding of particle interactions.

1.3.2 Local symmetries

Let us now turn to local symmetries. Their importance was realized long ago in connection with electromagnetic theory, and it is the cornerstone of modern particle physics.

The demands of a local symmetry are much more stringent and can only be met provided some new spin-1 fields – to be called the *gauge fields*, are introduced into the theory. These gauge fields, to be denoted by A_μ, have interaction with all fields that transform non-trivially under the local symmetry and their exchange gives rise to the forces. Thus, requiring invariance under a local symmetry always leads to forces — a result which takes the mystery out of the forces and puts them back into the nature of invariances.

To see in detail how it works, consider a local symmetry group G and let some matter fields transform as a certain representation under this group. If we denote these matter fields by a column vector $\psi(x)$, then under a gauge transformation

$$\psi(x) \to U(x)\psi(x) , \tag{1.26}$$

where $U(x)$ is a matrix representation of an element of the group G. The ordinary kinetic energy term for $\psi(x)$ will not be invariant under this transformation. To write down a term that is invariant under Eq. (1.26), we replace the ordinary space-time derivative $\partial_\mu \psi$ in the Lagrangian by the covariant derivative

$$D_\mu \psi = \partial_\mu \psi - ig A_\mu \psi , \tag{1.27}$$

and write, for the fermions

$$\mathcal{L}_{\text{kin}} = \overline{\psi} i \gamma^\mu D_\mu \psi , \tag{1.28}$$

which is gauge invariant. Here, $A_\mu(x)$ contains spin-1 fields and is matrix-valued in the group space, i.e.,

$$A_\mu(x) = \sum_a T^a A_\mu^a(x) , \tag{1.29}$$

where T^a are the generators of the group G in the appropriate representation. From the requirement that Eq. (1.28) is invariant under the gauge transformation of Eq. (1.26), we obtain that $A_\mu(x)$ must transform like

$$A_\mu(x) \to U A_\mu(x) U^{-1} + \frac{1}{ig}(\partial_\mu U)U^{-1} . \tag{1.30}$$

Exercise 1.2 *Show that for scalar fields, the Lagrangian given by*

$$\mathcal{L}_{\text{kin}} = (D_\mu \phi)^\dagger (D^\mu \phi) \tag{1.31}$$

is invariant under the gauge transformation, where the covariant derivative is defined as in Eq. (1.27), with ϕ replacing ψ.

On substituting Eq. (1.27) to Eqs. (1.28) and (1.31), we see that the principle of gauge invariance has generated interaction terms between matter fields ψ, ϕ and the gauge fields A_μ. Exchange of gauge fields then generates forces, as mentioned earlier.

There is a good side and a bad side to this. The good side is that the coupling g is universal to all matter fields as long as G is a nonabelian group. Therefore, this has the potential to explain universality of couplings to different particles observed in weak interactions (see Eq. (1.3)). The problem however is that gauge invariance under Eq. (1.30) demands that the terms involving the gauge fields only are of the following form:

$$\mathcal{L}_{\text{(pure gauge)}} = -\frac{1}{4}\text{tr}\, F_{\mu\nu}F^{\mu\nu} \tag{1.32}$$

where

$$F_{\mu\nu} = \partial_\mu A_\nu - \partial_\nu A_\mu + ig[A_\mu, A_\nu]. \tag{1.33}$$

Notice that there is no mass term in Eq. (1.32), because it is forbidden by gauge invariance. So the gauge field must be massless. Therefore, its exchange can only lead to long range forces. But we know that weak and strong interactions are short range. Therefore, if we want to generate weak interaction via a gauge principle, we must somehow give mass to the gauge boson. This was achieved with the discovery of the Higgs mechanism described below. The short range nature of strong interactions is understood on the basis of a different principle called confinement.

1.3.3 Spontaneous breaking of symmetries

The time-evolution of a physical system depends on two things: (i) the Hamiltonian (or the Lagrangian) and (ii) the initial condition. In quantum field theories with symmetries, two analogous conditions determine the way in which a symmetry manifests itself. First, the Lagrangian must be invariant under the symmetry transformations; secondly, the vacuum state must be invariant under the same. When this happens, the symmetry is said to be realized in the Wigner-Weyl mode. In this case, natural expectations such as equality of masses of particles in a given irreducible representation, equality of S-matrix elements among the appropriate physical states — are satisfied.

It was discovered around 1960, through the work of Nambu and Goldstone, that a Lagrangian may be invariant under a symmetry transformation but the vacuum state may not be. If the symmetry is global, the spectrum of the theory contains a massless particle known as the Nambu-Goldstone boson. It was realized through the works of Higgs, Kibble, Guralnik, Hagen, Brout and Englert [8] that if a gauge symmetry is spontaneously broken, no such massless particle results. Rather, the gauge boson corresponding to broken generators pick up mass. Since this will be the principal technique in the construction of the standard model as well as its extensions, we outline the practical way of implementing this in gauge theories.

Let G be the local symmetry group and T^a be the generators of the group. Consider a set of scalar fields ϕ_a belonging to a representation of G. To discuss spontaneous breaking of G, consider the potential of the scalar fields, $V(\phi)$, which is invariant under G. This invariance implies that

$$\sum_{i,j} \frac{\partial V}{\partial \phi_i} (T^a)_{ij}\phi_j = 0. \tag{1.34}$$

Condition for the minimum is given by

$$\frac{\partial V}{\partial \phi_i}\bigg|_{\phi=\langle\phi\rangle} = 0. \tag{1.35}$$

Differentiating Eq. (1.34) with respect to ϕ_k and setting $\phi_k = \langle\phi_k\rangle$, we get

$$\sum_{i,j} \frac{\partial^2 V}{\partial \phi_k \partial \phi_i}\bigg|_{\phi=\langle\phi\rangle} (T^a)_{ij} \langle\phi_j\rangle = 0, \tag{1.36}$$

by using Eq. (1.35). Now, the mass matrix of the scalar fields is given by

$$\mathcal{M}_{ki}^2 = \frac{\partial^2 V}{\partial \phi_k \partial \phi_i}\bigg|_{\phi=\langle\phi\rangle}. \tag{1.37}$$

Thus, if $\sum_j (T^a)_{ij} \langle\phi_j\rangle$ is not a null vector, i.e., if the generator T^a does not annihilate the vacuum, then Eq. (1.36) shows that $\sum_j (T^a)_{ij} \langle\phi_j\rangle$ is

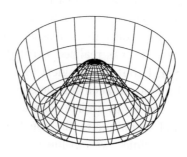

Figure 1.1: The potential which gives rise to broken-symmetry solutions.

a zero mass eigenvector of the mass matrix. This gives rise to a Nambu-Goldstone mode.

The G-invariant kinetic energy for ϕ is given by Eq. (1.31). On substituting $\phi = \langle \phi \rangle$ (and the fact that $\partial_\mu \langle \phi \rangle = 0$) in that equation, we obtain

$$
\begin{aligned}
\mathcal{L}_{\text{kin}}(\phi)|_{\phi=\langle\phi\rangle} &= \sum_{a,b}(T^a\,\langle\phi\rangle)^\dagger(T^b\,\langle\phi\rangle)A^a_\mu A^{b\mu} \\
&\equiv \sum_{a,b}M^2_{ab}A^a_\mu A^{b\mu}\,.
\end{aligned}
\tag{1.38}
$$

This is precisely the term that makes the gauge bosons massive after symmetry breaking. But note that if for a particular generator $T^a\,\langle\phi\rangle = 0$, the corresponding gauge field has no mass. Thus, only the gauge fields corresponding to spontaneously broken generators, i.e., the ones for which $T^a\,\langle\phi\rangle \neq 0$, pick up mass. This is the *Higgs mechanism*.

It may be useful for the reader to see a simple abelian example of Higgs mechanism. Consider a $U(1)$ local symmetry with the scalar field ϕ transforming non-trivially under the gauge group as follows:

$$
\phi \to e^{ie\lambda(x)}\phi\,.
\tag{1.39}
$$

If we demand that the spin-1 gauge field transforms under the $U(1)$

transformation as

$$A_\mu(x) \to A_\mu(x) + \partial_\mu\lambda(x)\,, \qquad (1.40)$$

then the gauge invariant Lagrangian for the system can be written as

$$\mathcal{L} = \mathcal{L}_0 - V(\phi)\,, \qquad (1.41)$$

where

$$\mathcal{L}_0 = (D_\mu\phi)^* D^\mu\phi - \frac{1}{4}F_{\mu\nu}F^{\mu\nu} \qquad (1.42)$$

with the covariant derivative defined by

$$D_\mu\phi = \partial_\mu\phi - ieA_\mu\phi\,. \qquad (1.43)$$

The potential $V(\phi)$ must be gauge invariant, i.e., it is a function of $\phi^*\phi$. The most general renormalizable potential then consists of two terms:

$$V(\phi) = -\mu^2\phi^*\phi + \lambda(\phi^*\phi)^2\,. \qquad (1.44)$$

Now, λ must be positive in order that the potential has a lower bound. But nothing restricts the sign of μ^2. If it happens to be positive as well, then the shape of the potential looks like Fig. 1.1. Minimization of this potential gives

$$\langle\phi\rangle = \sqrt{\frac{\mu^2}{2\lambda}}\,. \qquad (1.45)$$

Of course, the first derivative vanishes at $\phi = 0$ as well, but Fig. 1.1 clearly shows that the minimum corresponds to the nonzero value. Let us now redefine ϕ as

$$\phi(x) = \varphi(x) + \frac{v}{\sqrt{2}} \qquad (1.46)$$

where $v = \sqrt{\mu^2/\lambda}$ defines the minimum and φ is the fluctuation around this minimum. This way, we get $\langle\varphi\rangle = 0$, so that φ can be expanded in terms of creation and annihilation operators as is usual in quantum field theory. We then see that the kinetic energy term for ϕ leads to a term $\frac{1}{2}e^2v^2A_\mu A^\mu$ in the Lagrangian, which is a mass term for A^μ. Thus, the process of spontaneous symmetry breaking gives mass to a gauge boson.

Exercise 1.3 *Instead of Eq. (1.46), use*

$$\phi(x) = \frac{\eta(x) + v}{\sqrt{2}} \, e^{i\zeta(x)}, \tag{1.47}$$

where $\eta(x)$ and $\zeta(x)$ are real scalar fields which replace the complex field $\phi(x)$. Show that, in this representation, the field $\zeta(x)$ remains massless after the symmetry breaking. Moreover, by performing a gauge transformation $\phi(x) \to \phi(x)e^{-i\zeta(x)}$, show that the field $\zeta(x)$ vanishes from the Lagrangian altogether.

At this point, we have presented the three basic ingredients needed for gauge model building, which we now summarize [9]:

- Choice of the gauge group G.

- Assignment of the fermions to suitable representation of the gauge group.

- Choice of Higgs bosons and their vacuum expectation values to break the gauge symmetry down to $U(1)_Q$. This choice must also be good enough to reproduce the quark and lepton masses in a phenomenologically acceptable way.

Exercise 1.4 *Consider an SU(2) gauge theory with a scalar multiplet transforming like a doublet. In the vacuum, one component of the doublet scalar field acquires a non-zero value. Verify that all the three gauge bosons of SU(2) acquire equal masses.*

Exercise 1.5 *Repeat the same problem with the vev in one component of a triplet scalar multiplet. Do all the gauge bosons acquire mass in this case? If not, why not?*

1.4 Renormalizability and anomalies

One might wonder about the necessity of going through the elaborate machinery of Higgs mechanism to give mass to gauge fields instead of simply adding to a gauge invariant theory an explicit mass term for the gauge bosons that need to pick up mass to fit observations. In fact, historically, these kinds of attempts were made in the late sixties and early seventies. It was however realized that these theories are plagued with divergences, which proliferate as one goes to higher loops. Such theories therefore have no predictive power. It was shown by 't Hooft [10]

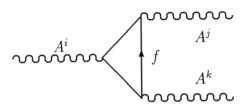

Figure 1.2: Triangle graphs giving rise to anomalies.

in 1971 that if gauge boson mass is generated by the Higgs mechanism, then the theory is renormalizable. It is therefore very important to remember in building gauge models that, the gauge symmetries that need to be broken (so that the corresponding gauge bosons pick up mass), must be broken by the Higgs mechanism. Furthermore, to keep the theory renormalizable, all terms in the Lagrangian must have mass dimensions less than or equal to four.

Another important criterion for renormalizability is the absence of triangle anomalies. To understand this, we recall that in proving renormalizability of a theory, gauge invariance is essential. Gauge invariance means that the currents corresponding to gauge symmetries must be conserved to all orders in perturbation theory. It was however pointed out by Adler, Bell and Jackiw [11] that if a theory involves chiral interactions (i.e., interactions involving γ_5 currents) of fermions, triangular one-loop graphs (see Fig. 1.2) in general destroy the current conservation which was true at the tree level. This is called the *axial anomaly*. If such anomalies are not canceled, then the theory loses its renormalizability. One must, therefore, impose the constraint of anomaly cancellation on gauge theories. This imposes restrictions on the nature of fermion spectrum. In general, if in the space of all fermions, T^a denotes the coupling matrix of fermions to the current J^a_μ, then the condition for anomaly cancellation is

$$\text{Tr}\left(T^a\{T^b, T^c\}\right)_L - \text{Tr}\left(T^a\{T^b, T^c\}\right)_R = 0, \qquad (1.48)$$

where the subscripts L and R denote left and right chirality states of fermions. This constraint plays an important role in understanding fermion spectrum, charge quantization etc and helps reduce the arbitrariness of gauge theories.

Chapter 2

The standard model and the neutrino

From the early days of particle physics, the universality of Fermi coupling led many people to suspect that there may exist an underlying symmetry in weak interaction which is not manifest in the mass spectrum of weakly interacting particles. From the nature of the charged current connecting two different particles in beta decay and other weak decays observed in the late fifties and early sixties, it was suspected that the underlying symmetry must have an $SU(2)$ part to it. In 1961, Glashow [1] proposed $SU(2) \times U(1)$ as a possible local symmetry of weak interactions. In 1964, Salam and Ward [2] used this symmetry group to construct a theory of both electrons and muons. Then, Weinberg in 1967 [3] and Salam in 1968 [4] proposed the spontaneously broken $SU(2)_L \times U(1)_Y$ theory of leptons in its modern version. The correct extension of this model to include quarks was done in the seventies using an earlier suggestion of Glashow, Iliopoulos and Maiani [5]. A crucial question regarding the model was its renormalizability, and this question remained until 1971, when 't Hooft [6] proved that all spontaneously broken gauge theories that include only interactions of mass dimension less than or equal to four are renormalizable. This not only solved the problems associated with the four-Fermi theory of weak interaction concerning its high energy behavior as well as reliability of the tree-level calculations but it also opened up the possibility of calculating radiative corrections to weak amplitudes and checking against experiments. Soon afterwards, in 1973, Gross, Wilczek and Politzer [7] showed that the unbroken $SU(3)_c$ group of strong interactions is asymptotically free, i.e., that strong interactions become weaker at higher energies, thus explaining Bjorken scaling observed in deep inelastic lepton-nucleon scattering. Thus was born the standard model of electro-weak and strong interactions based on the gauge group $SU(3)_c \times SU(2)_L \times U(1)_Y$. Since neutrinos have no strong interactions, we will not concern ourselves with the $SU(3)_c$ part of the group but instead will discuss the $SU(2)_L \times U(1)_Y$ model.

2.1 Gauge interactions in the standard model

Let us first discuss the spontaneous breaking of the $SU(2)_L \times U(1)_Y$ symmetry. The three gauge bosons of $SU(2)_L$ are denoted by $W_\mu^\pm$, W_μ^0, whereas the $U(1)_Y$ gauge boson is denoted by B_μ. To use the Higgs mechanism to break this group down to $U(1)_Q$, we use a doublet Higgs representation ϕ:

$$\phi = \begin{pmatrix} \phi_+ \\ \phi_0 \end{pmatrix}, \tag{2.1}$$

with the Higgs potential given by

$$V(\phi) = -\mu^2 \phi^\dagger \phi + \lambda(\phi^\dagger \phi)^2 . \tag{2.2}$$

For $\mu^2 > 0$, the minimum of $V(\phi)$ occurs at

$$\langle \phi_0 \rangle = \sqrt{\frac{\mu^2}{2\lambda}} \equiv \frac{v}{\sqrt{2}} . \tag{2.3}$$

This gives mass to the charged gauge bosons $W^\pm$:

$$M_W = \frac{1}{2}gv . \tag{2.4}$$

In the neutral gauge boson sector, only one particular combination of W^0 and B, acquires mass:

$$M_Z = \frac{M_W}{\cos\theta_W} . \tag{2.5}$$

This combination is given by

$$Z = \cos\theta_W W^0 - \sin\theta_W B , \tag{2.6}$$

where

$$\tan\theta_W = g'/g , \tag{2.7}$$

where this angle θ_W is known as the Weinberg angle. There is one gauge boson, the combination orthogonal to Z, which remains massless corresponding to an unbroken $U(1)$ symmetry. This will be identified with the electromagnetic potential of the photon. The associated gauge generator, the electric charge, is given by

$$Q = I_{3L} + \frac{Y}{2}. \tag{2.8}$$

In terms of the gauge coupling constants, the electric charge of the positron is given by

$$e = g \sin \theta_W. \tag{2.9}$$

In order to discuss the weak interaction of fermions, we assign quarks and leptons to the following representations of $SU(2)_L \times U(1)_Y$:

$$q_L = \begin{pmatrix} u_L \\ d_L \end{pmatrix} \quad : \quad (2, \frac{1}{3})$$

$$u_R \quad : \quad (1, \frac{4}{3})$$

$$d_R \quad : \quad (1, -\frac{2}{3})$$

$$\psi_L = \begin{pmatrix} \nu_{eL} \\ e_L \end{pmatrix} \quad : \quad (2, -1)$$

$$e_R \quad : \quad (1, -2). \tag{2.10}$$

Similar assignments are repeated for all three families of quarks and leptons with (u, d, e, ν_e) replaced by (c, s, μ, ν_μ) and (t, b, τ, ν_τ) respectively. Eq. (2.10) implies the following charged $W^\pm$ and neutral Z interaction with the quarks and leptons:

$$\mathcal{L}_{\text{weak}} = \frac{g}{\sqrt{2}} \left(J^\mu W_\mu^+ + J^{\mu\dagger} W_\mu^- \right) + \frac{g}{\cos \theta_W} K^\mu Z_\mu, \tag{2.11}$$

where

$$J^\mu = (\overline{u^0} \quad \overline{c^0} \quad \overline{t^0}) \gamma^\mu L \begin{pmatrix} d^0 \\ s^0 \\ b^0 \end{pmatrix} + (\overline{\nu_e} \quad \overline{\nu_\mu} \quad \overline{\nu_\tau}) \gamma^\mu L \begin{pmatrix} e \\ \mu \\ \tau \end{pmatrix} \tag{2.12}$$

The subscript zero on quarks implies that they are not mass eigenstates. The mass eigenstates are determined only after quarks acquire masses in the process of spontaneous symmetry breaking. The neutral current K_μ is given by

$$K^\mu = \sum_{f^0} \overline{f^0} \gamma^\mu \left[I_{3L} L - \sin^2 \theta_W Q \right] f^0, \qquad (2.13)$$

where $f^0 = \nu_e, e, u^0$ or d^0 or the corresponding objects in the higher generations, Q is the charge of f^0 and I_{3L} is the value of the neutral generator of $SU(2)_L$ for the left-chiral projection of the fermion field.

It is now clear that exchange of $W^\pm$ leads to charged current weak processes such as beta decay, muon decay etc. whereas exchange of Z leads to new type of interactions called neutral current weak interactions. Thus, the advent of gauge theories endowed the neutrino with new dynamical properties which were not known prior to the 1970's. The discovery of neutrino neutral currents was a major triumph of gauge theories. An important property of the neutral currents is that K_μ is flavor diagonal prior to mass generation. We will see that in the standard model, this property is preserved by symmetry breaking. The details of this will be discussed in the next section.

In order to complete the discussion of charged current weak interactions, let us discuss the origin of quark and lepton masses in the standard model. The gauge invariance prevents adding bare masses for them in the Lagrangian. They arise from the following Yukawa interactions allowed by gauge symmetry:

$$-\mathcal{L}_Y = \sum_{a,b} \left[h_{ab}^{(u)} \overline{q}_{aL} \hat{\phi} u_{bR} + h_{ab}^{(d)} \overline{q}_{aL} \phi d_{bR} + h_{ab}^{(\ell)} \overline{\psi}_{aL} \phi \ell_{bR} \right] + \text{h.c.}. \quad (2.14)$$

Here, a, b stand for generation indices and

$$\hat{\phi} = i\tau_2 \phi^*. \qquad (2.15)$$

Exercise 2.1 *Under an $SU(2)$ transformation, a doublet transforms as*

$$\phi \to \exp(i\frac{\tau}{2} \cdot \boldsymbol{\theta}) \, \phi,$$

where τ denote the Pauli matrices. Show that the components of ϕ^* do not transform like a doublet, but those of $i\tau_2 \phi^*$ do.

On substituting the non-zero vacuum expectation values for ϕ_0 given in Eq. (2.3), the following mass terms for up and down quarks as well as the charged leptons are generated:

$$-\mathcal{L}_{\text{mass}} = \sum_{a,b} \left[\overline{u}_{a\text{L}} M_{ab}^{(\text{u})} u_{b\text{R}} + \overline{d}_{a\text{L}} M_{ab}^{(\text{d})} d_{b\text{R}} + \overline{\ell}_{a\text{L}} M_{ab}^{(\ell)} \ell_{b\text{R}} \right] + \text{h.c.} . \quad (2.16)$$

where

$$M_{ab}^{(f)} = h_{ab}^{(f)} v/\sqrt{2} \qquad \text{with } f = \mathbf{u}, \mathbf{d} \text{ or } \ell. \quad (2.17)$$

By an appropriate choice of the quark and lepton basis, the coupling matrices $h^{(\mathbf{u})}$ and $h^{(\ell)}$ can be chosen diagonal so that we have $\mathbf{u}_a^{(0)} = \mathbf{u}_a$ and $\ell_a^{(0)} = \ell_a$. The $M^{\mathbf{d}}$ is however, a complex non-diagonal matrix in this basis and can be diagonalized by the following biunitary transformation:

$$V_L M^{(\mathbf{d})} V_R^\dagger = D^{(\mathbf{d})} . \quad (2.18)$$

The down-type quark mass eigenstates are then related to the $\mathbf{d}_a^{(0)}$ states as

$$(\mathbf{d}_{\text{L,R}})_a = \sum_b (V_{L,R})_{ab} (\mathbf{d}_{\text{L,R}}^{(0)})_b . \quad (2.19)$$

In terms of the mass eigenstates, i.e., the physical states, the weak charged current can be written as

$$J^\mu = (\overline{u} \quad \overline{c} \quad \overline{t}) \gamma^\mu \text{L} V_L^\dagger \begin{pmatrix} d \\ s \\ b \end{pmatrix} + (\overline{\nu}_e \quad \overline{\nu}_\mu \quad \overline{\nu}_\tau) \gamma^\mu \text{L} \begin{pmatrix} e \\ \mu \\ \tau \end{pmatrix} . \quad (2.20)$$

This V_L can be identified with the matrix V introduced in Eq. (1.4). Thus, a very important implication of the standard model is that the quark mixing matrix is unitary. This property of unitarity has been used in giving the values of the elements of this matrix in Eq. (1.7). It is also important to note here that the neutral current in Eq. (2.13) is not affected by this transition to physical basis and remains flavor diagonal in the mass eigenbasis.

The W and Z masses are predicted in terms of a single parameter $\sin^2 \theta_W$ at the tree level as follows:

$$M_W^0 = \left(\frac{\pi \alpha^0}{\sqrt{2} G_\mu^0 \sin^2 \theta_W} \right)^{1/2} \quad (2.21)$$

$$M_Z^0 = M_W^0 / \cos \theta_W , \quad (2.22)$$

where α is the fine structure constant, G_μ is the Fermi constant as evaluated from the muon decay, and the superscripts 0 on these and other quantities stand for the values at the tree level of calculation. Once the radiative corrections are included, new parameters such as the t-quark mass or the Higgs mass enter and we have the radiatively corrected value:

$$M_W = \frac{A_0}{\sin\theta_W (1 - \Delta r)^{1/2}}$$
$$M_Z = M_W / \cos\theta_W .$$
(2.23)

where $A_0 = \sqrt{\pi\alpha/\sqrt{2}G_\mu} \simeq 37.2802\,\text{GeV}$. The correction Δr includes the radiative corrections relating α, $\alpha(M_Z)$, G_F, M_W and M_Z and has been calculated by various authors [8]. It is known to depend quadratically on m_t and logarithmically on the Higgs mass. For $m_t = 180 \pm 7\,\text{GeV}$ and $M_H = 300\,\text{GeV}$ for instance, $\Delta r \simeq 0.0376 \pm 0.0025 \pm 0.0007$ [9]. Eq. (2.23) can then be used to obtain the value of $\sin^2\theta_W$ from precise measurements of M_W and M_Z.

Exercise 2.2 *Write down the tree-level amplitude for muon decay, $\mu^- \to e^- + \nu_\mu + \widehat{\nu}_e$, using the charged current interaction of Eq. (2.20). In the low-energy approximation where the momentum transfer carried by the W-boson can be neglected, compare this amplitude with Eq. (1.3) to show that*

$$\frac{G_F}{\sqrt{2}} = \frac{g^2}{8M_W^2} .$$
(2.24)

From this, using Eq. (2.9), deduce the tree-level expression for M_W given in Eq. (2.22).

2.2 Neutral current interactions of neutrinos

The weak neutral current interaction for the first generation of quarks and leptons following from Eq. (2.13) is:

$$\mathcal{L} = \frac{g}{\cos\theta_W} K_\mu(x) Z^\mu(x)$$
(2.25)

where, for the fermions in the first generation, we may use the standard notation, introduced in Eq. (1.9), and write

$$K_\mu(x) = \sum_q [\epsilon_L(q)\bar{q}\gamma_\mu Lq + \epsilon_R(q)\bar{q}\gamma_\mu Rq]$$
$$+ \frac{1}{2}\sum_\nu \bar{\nu}\gamma_\mu L\nu + \frac{1}{2}\sum_\ell \bar{\ell}\gamma_\mu(g_V^\ell - \gamma_5 g_A^\ell)\ell, \quad (2.26)$$

where standard model predicts,

$$\epsilon_L(u) = \frac{1}{2} - \frac{2}{3}\sin^2\theta_W , \qquad \epsilon_R(u) = -\frac{2}{3}\sin^2\theta_W$$
$$\epsilon_L(d) = -\frac{1}{2} + \frac{1}{3}\sin^2\theta_W , \qquad \epsilon_R(d) = \frac{1}{3}\sin^2\theta_W \quad (2.27)$$

and

$$g_V(e) = -\frac{1}{2} + 2\sin^2\theta_W \quad , \quad g_A(e) = -\frac{1}{2}. \quad (2.28)$$

Exercise 2.3 *The definition of the covariant derivative in the $SU(2)_L \times U(1)_Y$ model is given by*

$$D_\mu = \partial_\mu + ig\frac{\mathbf{T}}{2}\cdot\mathbf{W}_\mu + ig'\frac{Y}{2}B_\mu , \quad (2.29)$$

where $\mathbf{T}/2$ represents the three generators of $SU(2)_L$ in the appropriate representation. Use this and Eq. (2.10) to write down the neutral currents in terms of W_μ^0 and B_μ. Then use the definition of Z_μ in terms of W_μ^0 and B_μ, and the orthogonal combination for A_μ, to deduce the neutral current in Eq. (2.26).

For low momenta, Eq. (2.25) leads to the effective Hamiltonian:

$$\mathcal{H}_{\text{weak}} = \frac{4\rho G_F}{\sqrt{2}} K_\mu(x) K^\mu(x) , \quad (2.30)$$

where

$$\rho = \frac{M_W^2}{M_Z^2 \cos^2\theta_W} , \quad (2.31)$$

which can be set equal to unity in view of the mass relations given in Eq. (2.23). The Hamiltonian in Eq. (2.30) leads to neutral current processes such as $\nu_\mu e \to \nu_\mu e$ or $\nu q \to \nu q$.

Table 2.1: Experimental values of various neutral current parameters and their comparison with predictions of the standard model. (From Ref. [9]).

Quantity	Experimental value	Standard model prediction	
$\epsilon_L(u)$	$0.332\pm.016$	0.345	±0.0003
$\epsilon_L(d)$	$-0.438\pm.012$	-0.429	±0.0004
$\epsilon_R(u)$	$-0.178\pm.013$	-0.156	
$\epsilon_R(d)$	$-0.026\pm^{.075}_{.048}$	-0.078	
$g_A(e)$	$-0.507\pm.014$	-0.507	±0.0004
$g_V(e)$	$-0.041\pm.015$	-0.036	±0.0003

Exercise 2.4 *Consider the decay $Z \to f\widehat{f}$, where f is some fermion with $m_f \ll M_Z$. Let*

$$\left\langle f\widehat{f}\left|K_\mu\right|0\right\rangle = \overline{u}\gamma_\mu(a - b\gamma_5)v\,, \tag{2.32}$$

where K_μ is the neutral current. Show that

$$\Gamma(Z \to f\widehat{f}) = \frac{\sqrt{2}}{3\pi}(a^2 + b^2)G_F M_Z^3\,. \tag{2.33}$$

Using the proper values of a and b for neutrinos, show that the rate for $Z \to \nu\widehat{\nu}$ is 165 MeV.

The weak neutral currents were first observed experimentally in 1972 in an experiment at CERN [10] in $\nu_\mu e$ scattering and was confirmed shortly thereafter by experiments at Fermilab [11]. In the subsequent years a great variety of neutral current processes involving ν-quark, ν-lepton, e-quark have been observed and thoroughly studied [12]. On the theoretical front, the radiative corrections to the neutral current processes were carried out [8], thus enabling a detailed comparison between theory and experiment. Various authors [13, 14] have carried out a detailed comparison between theory and experiment and have obtained the following value of $\sin^2\theta_W$ which provides the best fit to all available neutral current data (see Table 2.1):

$$\begin{aligned} \sin^2\theta_W &= 0.229 \pm 0.0064 \\ \rho &= 0.998 \pm 0.0086\,. \end{aligned} \tag{2.34}$$

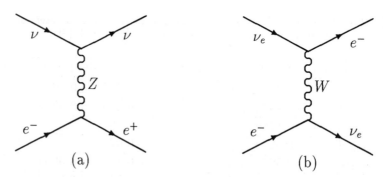

Figure 2.1: Tree-level diagrams that contribute to the $\nu_e e$ scattering in the standard model.

However recent measurements at LEP [15] and SLC [16] of the Z-pole observables have led to a much more precise determination of $\sin^2 \theta_W$. The global fit to all data in conjunction with the CDF/D0 [17] value for $m_t = 180 \pm 12 \pm$ GeV yields [9]

$$\sin^2 \theta_W = 0.2315 \pm 0.0002 \pm 0.0003 \,. \tag{2.35}$$

2.3 Neutrino-electron scattering in the standard model

In this section, we present the predictions of the standard model for neutrino scattering cross-sections off electrons. For the sake of simplicity, we will restrict ourselves to low energies, i.e., when the momentum q carried by the intermediate vector bosons satisfy $|q^2| \ll M_W^2$.

2.3.1 $\nu_e e$ and $\hat{\nu}_e e$ scattering

The $\nu_e e$ scattering receives contribution both from charged and neutral current interactions, as shown in Fig. 2.1. Since we will work in the low energy limit, we will neglect the momentum dependence of the W and Z-propagators and work in the four Fermi approximation.

For the neutral current diagram given in Fig. 2.1a, the Feynman amplitude is given by

$$\mathcal{M}_{\rm nc} = \sqrt{2} G_F \left[\bar{u}'(k') \gamma^\lambda {\sf L} u'(k) \right] \left[\bar{u}(p') \gamma_\lambda (g_L {\sf L} + g_R {\sf R}) u(p) \right] \,, \tag{2.36}$$

where the neutrino spinors have been denoted by a prime, and

$$g_L = 2\sin^2\theta_W - 1\,, \quad g_R = 2\sin^2\theta_W\,, \tag{2.37}$$

which can be read off from the neutral current interaction given in Eqs. (2.26) and (2.28). In writing down the matrix element, a factor of 2 comes in because each current can come from either factor of K_μ in the interaction Lagrangian.

The contribution of the charged current diagram of Fig. 2.1b, on the other hand, can be written as

$$\begin{aligned}
\mathcal{M}_{\rm cc} &= 2\sqrt{2}G_F \left[\overline{u}'(k')\gamma^\lambda L u(p)\right]\left[\overline{u}(p')\gamma_\lambda L u'(k')\right] \\
&= 2\sqrt{2}G_F \left[\overline{u}'(k')\gamma^\lambda L u'(k)\right]\left[\overline{u}(p')\gamma_\lambda L u(p)\right]\,, \tag{2.38}
\end{aligned}$$

using the Fierz transformation rules in the last step. Adding up the two contributions, we thus obtain

$$\mathcal{M} = \sqrt{2}G_F \left[\overline{u}'(k')\gamma^\lambda L u'(k)\right]\left[\overline{u}(p')\gamma_\lambda \left\{(g_L+2)L + g_R R\right\} u(p)\right]\,. \tag{2.39}$$

Averaging over the initial electron spin and summing over the final electron spin, we thus obtain

$$\frac{1}{2}\sum_{\rm spin}|\mathcal{M}|^2 = G_F^2 N^{\lambda\rho}L_{\lambda\rho}\,, \tag{2.40}$$

where

$$\begin{aligned}
N^{\lambda\rho} &= \operatorname{tr}\left[\slashed{k}\gamma^\rho L \slashed{k}'\gamma^\lambda L\right] \\
&= 2\left[k^\rho k'^\lambda + k'^\rho k^\lambda - g^{\lambda\rho}k\cdot k' + i\varepsilon^{\alpha\rho\beta\lambda}k_\alpha k'_\beta\right]\,, \tag{2.41}
\end{aligned}$$

and $L_{\lambda\rho}$ is the corresponding trace with the electron bilinears. Upon contraction, one obtains

$$N^{\lambda\rho}L_{\lambda\rho} = 16\left[(g_L+2)^2(k\cdot p)^2 + g_R^2(k'\cdot p)^2 - (g_L+2)g_R m_e^2 k\cdot k'\right]\,, \tag{2.42}$$

using momentum conservation, $k+p = k'+p'$, to ensure that $k\cdot p = k'\cdot p'$ and $k\cdot p' = k'\cdot p$.

Exercise 2.5 *Verify Eq. (2.42) by evaluating $L_{\lambda\rho}$ explicitly and using the form of $N^{\lambda\rho}$ given above.*

To proceed now, we decide to work in the frame in which the initial electron is at rest and choose the axes so that the incoming neutrino is traveling along the positive z-direction. Then

$$p^\lambda = (m_e, 0, 0, 0), \qquad k^\lambda = (\omega, 0, 0, \omega). \tag{2.43}$$

In this frame, let the energies of the final electron and the neutrino be E' and ω' respectively. Starting from the general formula for the cross-section given in Eq. (2.57) and performing the integration over the final electron momentum, we now obtain

$$\sigma_E(\nu_e e) = \frac{1}{16\pi^2 m_e \omega} \int \frac{d^3 k'}{4E'\omega'} \, \delta(m_e + \omega - E' - \omega') \cdot \frac{1}{2} \sum_{\text{spin}} |\mathcal{M}|^2, \tag{2.44}$$

where the subscript E on σ denotes the elastic cross section. It is straight forward to check that the argument of the δ-function of energy vanishes when

$$\omega' = \frac{m_e \omega}{m_e + \omega(1 - \zeta)}, \tag{2.45}$$

where ζ is the cosine of the angle the outgoing neutrino makes with the z-axis. Also, in the frame used by us,

$$N^{\lambda\rho} L_{\lambda\rho} \simeq 16 m_e^2 \left[(g_L + 2)^2 \omega^2 + g_R^2 \omega'^2 \right], \tag{2.46}$$

where we have dropped the last term of Eq. (2.42), which is negligible if we assume $\omega \gg m_e$.

The integration over the azimuthal angle as well as the magnitude of $\mathbf{k}'$ can now be performed and one obtains

$$\frac{d\sigma_E(\nu_e e)}{d\zeta} = \frac{G_F^2 m_e^2}{2\pi} \frac{(g_L + 2)^2 \omega^2 + g_R^2 \omega'^2}{[m_e + \omega(1 - \zeta)]^2}, \tag{2.47}$$

where ω' is given by Eq. (2.45).

This gives the differential cross-section, but it is customary to express it in terms of the variable

$$y \equiv \frac{E' - m_e}{\omega} = 1 - \frac{\omega'}{\omega}. \tag{2.48}$$

In terms of this variable, we get the following differential cross section when $m_e \ll \omega$:

$$\frac{d\sigma_E(\nu_e e)}{dy} = \frac{G_F^2 m_e \omega}{2\pi} \left[(2\sin^2\theta_W + 1)^2 + 4\sin^4\theta_W(1-y)^2 \right], \quad (2.49)$$

putting in the values for g_L and g_R at this stage. On integrating over y we get

$$\sigma_E(\nu_e e) = \frac{G_F^2 m_e \omega}{2\pi} \left[(2\sin^2\theta_W + 1)^2 + \frac{4}{3}\sin^4\theta_W \right]. \quad (2.50)$$

Using the value $G_F^2 = 5.29 \times 10^{-38}\,\mathrm{cm}^2/\mathrm{GeV}^2$, one obtains

$$\sigma_E(\nu_e e) = 0.9 \times 10^{-43} \left(\frac{\omega}{10\,\mathrm{MeV}} \right) \mathrm{cm}^2. \quad (2.51)$$

This cross section has recently been measured by a Maryland-LosAlamos-Irvine collaboration [18] at Los Alamos for neutrino energies up to 50 MeV. They measure total cross section

$$\langle\sigma\rangle = (3.01 \pm 0.46(\mathrm{stat}) \pm 0.38(\mathrm{syst})) \times 10^{-43}\,\mathrm{cm}^2 \quad (2.52)$$

for a mean neutrino energy of 31.7 MeV, in excellent agreement with the prediction of the standard model. Specially important is the fact that it provides evidence for the interference between the charged and neutral current amplitudes as predicted by the standard model.

Exercise 2.6 *In deriving the $\nu_e e$ elastic cross-section, we neglected the last term of Eq. (2.42). Show that, if this term is not neglected, one obtains*

$$\frac{d\sigma_E(\nu_e e)}{dy} = \frac{G_F^2 m_e \omega}{2\pi} \left[(g_L + 2)^2 + g_R^2(1-y)^2 - (g_L + 2)g_R \frac{y m_e}{\omega} \right] \quad (2.53)$$

Turning now to $\hat{\nu}_e e$ scattering, there are also both W and Z-contributions in this case, as in Fig. 2.2. The differential cross section is given here by

$$\frac{d\sigma_E(\hat{\nu}_e e)}{dy} = \frac{G_F^2 m_e \omega}{2\pi} \left[(2\sin^2\theta_W + 1)^2(1-y)^2 + 4\sin^4\theta_W \right] \quad (2.54)$$

and integration over y yields the total cross section

$$\sigma_E(\hat{\nu}_e e) = \frac{G_F^2 m_e \omega}{2\pi} \left[\frac{1}{3}(2\sin^2\theta_W + 1)^2 + 4\sin^4\theta_W \right]. \quad (2.55)$$

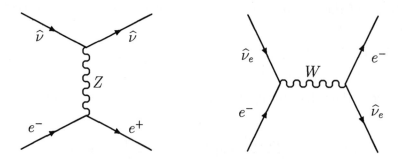

Figure 2.2: Tree-level diagrams that contribute to the $\widehat{\nu}_e e$ scattering in the standard model.

Putting in numbers, we get

$$\sigma(\widehat{\nu}_e e) = 0.378 \times 10^{-43} \left(\frac{\omega}{10\,\text{MeV}}\right)\,\text{cm}^2 \tag{2.56}$$

This cross section was measured [19].

Exercise 2.7 *Consider the scattering of two particles of mass m and μ, with 4-momentum given by p and k, to two particles with 4-momentum given by p' and k'. The general formula for the differential cross section for this process is given by*

$$d\sigma = \frac{(2\pi)^4 \delta^4(p + k - p' - k')}{4[p \cdot k - m^2\mu^2]^{\frac{1}{2}}} \frac{|\mathcal{M}|^2}{S} \frac{d^3p'}{(2\pi)^3 2p'_0} \frac{d^3k'}{(2\pi)^3 2k'_0}, \tag{2.57}$$

where $\mathcal{M}$ is the matrix element of the transition and S is the symmetry factor, which is n! if there are n identical particles in the final state. Use this to deduce the $\widehat{\nu}_e e$ cross section given above.

2.3.2 $\nu_\mu e$ and $\widehat{\nu}_\mu e$ scattering

For these scatterings, there is no W-contributions and only the neutral current contributes. The Feynman amplitude is now given by Eq. (2.36). Following the same arguments, here we get,

$$\frac{d\sigma_E(\nu_\mu e)}{dy} = \frac{G_F^2 m_e \omega}{2\pi} \left[(2\sin^2\theta_W - 1)^2 + 4(1 - y)^2 \sin^4\theta_W\right]$$

$$\frac{d\sigma_E(\widehat{\nu}_\mu e)}{dy} = \frac{G_F^2 m_e \omega}{2\pi} \left[(2\sin^2\theta_W - 1)^2(1 - y)^2 + 4\sin^4\theta_W\right]. \tag{2.58}$$

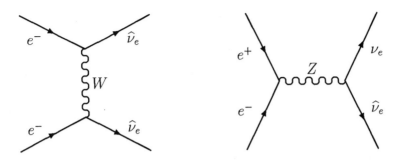

Figure 2.3: Tree-level diagrams contributing to the process $e^+e^- \to \nu\widehat{\nu}$.

They lead to

$$\sigma_E(\nu_\mu e) = \frac{G_F^2 m_e \omega}{2\pi}\left[(2\sin^2\theta_W - 1)^2 + \frac{4}{3}\sin^4\theta_W\right]$$

$$\sigma_E(\widehat{\nu}_\mu e) = \frac{G_F^2 m_e \omega}{2\pi}\left[\frac{1}{3}(2\sin^2\theta_W - 1)^2 + 4\sin^4\theta_W\right]. \qquad (2.59)$$

Numerically, we obtain,

$$\sigma_E(\nu_\mu e) = 0.15 \times 10^{-43}\left(\frac{\omega}{10\,\text{MeV}}\right)\text{cm}^2$$

$$\sigma_E(\widehat{\nu}_\mu e) = 0.14 \times 10^{-43}\left(\frac{\omega}{10\,\text{MeV}}\right)\text{cm}^2. \qquad (2.60)$$

Experimental studies [20] have verified these theoretical expectations with good accuracy.

2.3.3 Neutrino pair production

Before concluding this section, we would also like to present the cross sections for production of neutrinos in e^+e^- collision. Again, as in the case of $\nu_e e$ scattering, $e^+e^- \to \nu_e\widehat{\nu}_e$ receives contribution from both W and Z exchange graphs, as shown in Fig. 2.3. If we write a general coupling

$$\mathcal{H}_{\text{weak}} = \frac{4G_F}{\sqrt{2}}\overline{\nu}_\ell\gamma^\mu L\nu_\ell\,\overline{e}\gamma_\mu\frac{h_V - \gamma_5 h_A}{2}e\,, \qquad (2.61)$$

we get

$$\sigma(e^+e^- \to \nu\hat{\nu}) = \frac{G_F^2 s}{12\pi} \left(h_V^2 + h_A^2\right).$$

(2.62)

For the case of $e^+e^- \to \nu_e\hat{\nu}_e$, the parameters h_V and h_A are given by

$$h_V = \frac{1}{2} + 2\sin^2\theta_W, \quad h_A = \frac{1}{2},$$

(2.63)

leading to

$$\sigma(e^+e^- \to \nu_e\hat{\nu}_e) = \frac{G_F^2 s}{12\pi} \left(\frac{1}{2} + 2\sin^2\theta_W + 4\sin^4\theta_W\right)$$

(2.64)

For the case of muon neutrinos, since only the neutral current contributes, h_V and h_A are equal to $g_V^{(e)}$ and $g_A^{(e)}$ respectively. We then get

$$\sigma(e^+e^- \to \nu_\mu\hat{\nu}_\mu) = \frac{G_F^2 s}{12\pi} \left(\frac{1}{2} - 2\sin^2\theta_W + 4\sin^4\theta_W\right).$$

(2.65)

A typical estimate for these cross sections is $5 \times 10^{-43}\,\mathrm{cm}^2$ at the electron energy of $10\,\mathrm{MeV}$.

2.4 Neutrino-nucleon scattering in the standard model

2.4.1 Quasi-elastic $\nu_e N$ and $\hat{\nu}_e N$ scattering

In the low energy approximation ($\omega \ll m_N$ where m_N is the mass of the nucleon), the following interactions are possible:

$$\begin{aligned}
\nu_\ell + n &\to \ell^- + p \\
\hat{\nu}_\ell + p &\to \ell^+ + n,
\end{aligned}$$

(2.66)

where ℓ can be either e or μ. The matrix element for the first process, for example, can be written as

$$\begin{aligned}
\mathcal{M} &= \langle \ell^-(k'), p(p') | \mathcal{H}_{\mathrm{weak}} | \nu_\ell(k), n(p) \rangle \\
&= \frac{4G_F \cos\theta_c}{\sqrt{2}} \left[\overline{u}_{(\ell)}\gamma^\mu L u_{(\nu_\ell)}\right] \\
&\qquad \times \left[\overline{u}_{(p)}\gamma_\mu \left(\frac{1}{2}\left\{F_V(q^2) + F_A(q^2)\gamma_5\right\}\right) u_{(n)}\right]
\end{aligned}$$

(2.67)

Here, $\cos\theta_c$ is the element V_{11} of the mixing matrix V introduced in Eq. (1.4). In the nucleon matrix element, F_V and F_A are form factors which appear because the proton and the neutron are not elementary particles. In general, other form factors can appear in the matrix element, coupled to tensors of the form $\sigma_{\mu\nu}q^\nu$, $\sigma_{\mu\nu}\gamma_5 q^\nu$ etc, but their contribution to the final cross section is small in the regime of our interest. Although all these form factors depend on q^2, where q^μ is the momentum carried by the internal W-boson, we use their values at $q^2 = 0$ in view of the fact that we are examining the regime of small q^2. The cross sections therefore involve the vector and the axial weak charges of the neutron and the proton $g_V = F_V(0)$ and $g_A = F_A(0)$ respectively. The conserved vector current hypothesis says that $g_V = 1$, whereas g_A has been measured to be 1.26. For low energy neutrinos ($\omega \ll m_N$), the $\nu_e n$ cross section is given by

$$
\begin{aligned}
\sigma(\nu_e n) &= \frac{G_F^2 \omega^2}{\pi}\left[g_V^2 + 3g_A^2\right] \\
&\simeq 9.75 \times 10^{-42}\left(\frac{\omega}{10\,\mathrm{MeV}}\right)^2 \mathrm{cm}^2.
\end{aligned}
\tag{2.68}
$$

The same expression also applies to $\hat\nu_e p$ scattering, which is of more practical importance in astrophysical discussions. It is worth pointing out that these cross sections are of vital importance in designing neutrino detectors in various cases. For instance in the detection of neutrinos from supernovas or from the sun, one is interested in detecting neutrinos with energy in the range of 1 to 50 MeV. To get a better signal, one needs to rely on ν_e-nucleon scattering rather than ν_e-electron scattering. Since free nucleons cannot be used (unless in the case of water detectors), one has to study ν_e-nucleus scattering, which is in general more suppressed compared to free nucleon case. In the case of water detectors, a massive apparatus is required since the mean free path in water at an energy of 10 MeV is 1.5×10^{18} cm for $\hat\nu_e$ and 3.2×10^{19} cm for ν_e. Furthermore, nuclei have different threshold for neutrino absorption which is also important in the choice of detector material. Presently used targets are ^{37}Cl, Water, ^{71}Ga, ^{97}Mo. Use of several others such as ^{115}In, ^{7}Li etc are in the planning stage.

2.4.2 Deep inelastic scattering of neutrinos off nucleons

In §2.4.1, we considered the kinematical region of $q^2 \ll m_N^2$. Here, in contrast, we consider the region where $|q^2| \gg m_N^2$, but still $|q^2| \ll M_W^2$. The processes will now be of the generic form

$$\nu_\ell(k) + N(p) \to \ell^-(k') + X(p'), \qquad (2.69)$$

where the momenta of different particles are shown in parentheses, and X is a hadronic state for which there can be a large number of possibilities, depending on the value of q^2. We will consider the inclusive cross section, i.e., the cross section summed over all possible final hadronic states. The matrix element will now have a leptonic part which is the same as that occurring in Eq. (2.67), but the hadronic part will depend on the final state and hence must be parameterized differently. For this, we write the spin-summed square of the matrix element as

$$|\mathcal{M}|^2 = \left(\frac{4G_F}{\sqrt{2}}\right)^2 L_{\mu\nu} W^{\mu\nu}, \qquad (2.70)$$

where $L_{\mu\nu}$ and $W^{\mu\nu}$ come from the leptonic and the hadronic parts of the matrix element respectively. The leptonic part of the matrix element is known, from which we can write

$$
\begin{aligned}
L_{\mu\nu} &= \operatorname{tr}\left[\not{k}\gamma_\mu \mathsf{L} \not{k}'\gamma_\nu \mathsf{L}\right] \\
&= 2\left[k_\mu k'_\nu + k_\nu k'_\mu - g_{\mu\nu} k\cdot k' - i\varepsilon_{\mu\nu\lambda\rho} k^\lambda k'^\rho\right], \qquad (2.71)
\end{aligned}
$$

neglecting the charged lepton mass. For $W^{\mu\nu}$, on the other hand, we must use a parameterization since the matrix element itself is not known. Thus, we must write the most general tensor which depends on two vectors p and p', or alternatively on p and $q = p - p'$. This is given by

$$
\begin{aligned}
W^{\mu\nu} &= -g^{\mu\nu} W_1 + \frac{p^\mu p^\nu}{m_N^2} W_2 - \frac{i\varepsilon^{\mu\nu\lambda\rho} p_\lambda q_\rho}{2m_N^2} W_3 + \frac{q^\mu q^\nu}{m_N^2} W_4 \\
&\quad + \frac{(p^\mu q^\nu + p^\nu q^\mu)}{2m_N^2} W_5 + \frac{i(p^\mu q^\nu - p^\nu q^\mu)}{2m_N^2} W_6. \qquad (2.72)
\end{aligned}
$$

Here m_N is the nucleon mass, and W_1 to W_6 are form factors. All these form factors, being Lorentz scalars, can depend only on q^2 and $p\cdot q$, since these are the only kinematical scalar variables (note that $p^2 = m_N^2$, which is not a variable). Customarily, one chooses the independent variables to be

$$Q^2 \equiv -q^2, \qquad \nu \equiv \frac{p \cdot q}{m_N}. \qquad (2.73)$$

Thus, Q^2 is positive, and in the rest frame of the initial nucleon target, ν is the energy carried by the intermediate W-boson. More useful are the dimensionless *Bjorken scaling variables* x and y, defined by

$$x \equiv \frac{-q^2}{2p \cdot q} = \frac{Q^2}{2m_N \nu}, \qquad y \equiv \frac{p \cdot q}{p \cdot k}. \qquad (2.74)$$

Putting the expression of Eqs. (2.71) and (2.72) into Eq. (2.70), it is easy to see that the terms involving W_4, W_5 and W_6 are proportional to the charged lepton mass, and hence we neglect them. Then we are left with

$$L_{\mu\nu} W^{\mu\nu} = 4k \cdot k' W_1 + \frac{2}{m_N^2} \left[2p \cdot k\, p \cdot k' - m_N^2 k \cdot k' \right] W_2$$
$$+ \frac{2}{m_N^2} \left[p \cdot k\, q \cdot k' - q \cdot k\, p \cdot k' \right] W_3. \qquad (2.75)$$

In the lab frame, using $p^\mu = (m_N, 0, 0, 0)$ and $k^\mu = (\omega, 0, 0, \omega)$, we obtain

$$L_{\mu\nu} W^{\mu\nu} = 4m_N \omega xy\, W_1 + 2\omega[2(\omega - \nu) - m_N xy] W_2$$
$$+ 2\omega(2\omega - \nu) xy W_3. \qquad (2.76)$$

The differential cross section can now be calculated. It will be a function of W_1, W_2 and W_3. One obtains

$$\frac{d^2\sigma}{dx\, dy} = \frac{G_F^2 m_N \omega}{\pi(1 + Q^2/M_W^2)}$$
$$\times \left[xy^2 F_1 + \left(1 - y - \frac{m_N xy}{2\omega} \right) F_2 + y(1 - \frac{y}{2}) x F_3 \right], (2.77)$$

where

$$F_1 = m_N W_1, \qquad F_2 = \nu W_2, \qquad F_3 = \nu W_3. \qquad (2.78)$$

For $\hat{\nu} N$ interactions, the right-chirality projection operator R will appear instead of L in the expression for $L_{\mu\nu}$, so that the sign of the term containing the antisymmetric tensor in Eq. (2.71) will be reversed. In the expression for the cross section, this will reverse the sign of the F_3 term.

Experimental determination of the cross section gives information about the form factors, or *structure functions* as they are called. These are then analyzed to obtain the structure of the nucleon, e.g., the distribution of quarks in the nucleon. A detailed discussion of this procedure [21] is related to strong interaction physics and has little connection with neutrino physics which is the topic of discussion of this book.

2.5 Neutrino mass in the standard model

So far we have discussed the dynamical properties of the neutrino. Let us now turn to a discussion of its mass, which is the basis for a great deal of the static properties that the neutrino could acquire, such as mixings and oscillations, magnetic moments, decays etc. In the standard model only one helicity state of the neutrino per generation is present. Therefore, it could not have a Dirac mass, which requires both the helicity states. An alternative type of mass terms, called Majorana mass terms, could still be possible, which requires just one helicity state of a particle and uses the opposite helicity state of the antiparticle (see Ch. 4 for detailed discussion). However, this type of mass breaks lepton numbers by two units. But the standard model conserves lepton number separately for each generation, or more precisely the quantum number $B-L$. Therefore neither of these mass terms can arise at any level in perturbation theory or in the presence of non-perturbative effects.

To see that it is $B - L$, rather than L which is relevant for the discussion, we point out the triangle graphs of the type shown in Fig. 1.2 give rise to anomalies in the leptonic current. To show this, we refer to Eq. (1.48). Take, in particular, T^a to be the lepton number and both T^b and T^c to be the gauge generator I_3. Since the quarks have $L = 0$, the contribution comes only from leptons. The right handed fermions do not couple to the $SU(2)_L$ gauge bosons, so the only contribution to the left hand side of Eq. (1.48) is from the left handed lepton doublets. This contribution is $1/2$ for each generation, which shows that lepton number is anomalous. Even though it is conserved at the tree-level, non-perturbative quantum corrections break it. However, an identical anomaly is present in the baryon number current from three colors of quarks in the triangle diagram. Therefore, when we consider the current of $B - L$, we see that it is free of anomalies and is conserved. Note that in processes (or Lagrangian terms) that do not contain baryons, the above conserved current effectively becomes the lepton number.

Thus, we conclude that neutrino mass vanishes in the standard model. Because of this, its mixings, its magnetic moment etc also vanish. There is, however, one interesting property of neutrino induced by higher order weak interactions in the standard model, viz the charge radius of the neutrino. Even though the neutrino has zero charge and therefore has no interactions with the photon to lowest order, it can interact with the photon through higher order effects induced at the loop level. This interaction must however respect electromagnetic gauge invariance. If Γ_μ is the effective interaction vertex, then for neutral particles like the neutrino, Ward identity implies $\Gamma \cdot q = 0$ where q is the momentum of the photon at the vertex. Moreover, Γ cannot involve any right handed neutrino, since the latter are absent in the Standard model. Therefore, the interaction in momentum space will have the form [see Ch. 11 for details]:

$$\bar{\nu}\gamma_\mu \mathsf{L}\nu(q^2 g^{\mu\lambda} - q^\mu q^\lambda)\epsilon_\lambda^* . \tag{2.79}$$

Assuming, for the moment, that the form factor associated with this interaction is momentum independent, we get the following interaction in co-ordinate space:

$$\bar{\nu}\gamma_\mu \mathsf{L}\nu\partial_\lambda F^{\mu\lambda} . \tag{2.80}$$

Since this is non-vanishing only for off-shell photons, this kind of electromagnetic effect will only appear in processes such as $\nu_e e$ or $\nu_e q$ elastic scatterings. This kind of virtual coupling is called the charge radius of the neutrino since the strength of this coupling measures the $\langle r^2 \rangle$ of the neutrino. Since the coupling involves off-shell photon, there is no unique gauge invariant definition of neutrino charge radius [22].

Chapter 3

Massive neutrinos

The standard model predicts massless neutrinos, as we saw in the last chapter. We also discussed that the standard model has so far been very successful in explaining the various low-energy weak processes involving the charged- and neutral current interactions of the neutrinos.

One might then wonder, what is the motivation for considering neutrino masses? Is the question of the existence of neutrino masses an isolated one, or is it connected to other outstanding questions of particle physics? What are the general issues to be addressed? And finally, what sort of tests can be performed to know whether the neutrinos have mass? In this chapter we give an overview of such issues in a model independent manner [1].

3.1 Motivations for neutrino mass

Experimentally, we know only the following upper limits on neutrino masses:

$$
\begin{aligned}
m_{\nu_e} &< 12\,\mathrm{eV} \\
m_{\nu_\mu} &< 170\,\mathrm{keV} \\
m_{\nu_\tau} &< 24\,\mathrm{MeV}\,.
\end{aligned}
\tag{3.1}
$$

One group also claimed to have observed a lower limit of m_{ν_e}. If their result is confirmed, that alone shows that we must extend the standard model to accommodate for neutrino masses. However, the result is not confirmed by any other group, so at present zero neutrino masses are consistent with terrestrial experiments.

Nevertheless, the upper bounds are not tight enough to give us a feeling of satisfaction in assuming that all neutrinos are massless. The

ν_μ, for example, can be about half as heavy as the electron, and the ν_τ can be about 50 times heavier than the electron. Certainly it takes more than blind faith to accept that these particles must be massless. One must then pursue attempts to pin down the mass values with better accuracy.

This by itself can be a good enough motivation to open up a discussion on massive neutrinos. In any case, looking for neutrino mass is one way of looking for physics beyond the standard model. However, it turns out that there are other motivations too, which we will now describe.

3.1.1 Theoretical motivations in Particle physics

Even a cursory look at the standard model reveals that the masslessness of the neutrinos in this model is somewhat concocted. Why doesn't one, e.g., introduce a right handed neutrino field (ν_R) in the fundamental fermion content, as one does for all other fermions? If one did, the ν_R could have paired with the ν_L through the Higgs mechanism to produce a mass term for the neutrinos. [For details, see §5.1.1].

In fact, there is no fundamental reason why one does not do that. In other words, the ν_R is not introduced just because we want to predict massless neutrinos in the standard model. Things are arranged so as to produce that result. If one wants otherwise, one can just as well arrange things in some other way.

Contrast this issue with that of the masslessness of the photon. The latter is massless because of a conserved gauge symmetry which, in turn, governs the dynamics of the electromagnetic interaction. For the masslessness of the neutrinos, we see no such symmetry principle in the standard model. Hence the masslessness is unsatisfactory from a theoretical point of view.

It is also worth pointing out here that the standard model without right handed neutrinos has got the desirable property that anomaly cancellations imply charge quantization [2] if there is only one family of fermions. This property is lost in presence of right handed neutrinos, as well as of more than one fermion generation. However, it has been pointed out that if one includes right handed neutrinos and assumes that the neutrinos are Majorana particles, then anomaly cancellation implies charge quantization regardless of the number of generations. One could perhaps use this as an argument for neutrinos being both massive as well as Majorana type.

Secondly, one of the major aims of Particle Physics is unification of the fundamental interactions. This provides extra motivations for massive neutrinos since, as we will see in Ch. 7, many of the interesting unification models predict neutrino mass at some level or other.

The theoretical question is not only how one can extend the standard model to find models with massive neutrinos, but also how one can understand the smallness of the neutrino masses compared to the masses of the charged fermions. Of course, even the charged fermion masses vary widely. The top-quark mass is roughly five orders of magnitude larger than the electron mass, and the standard model does not even attempt to explain that. Why should then we be seeking a reason for the ν_e being at least five orders of magnitude lighter than the electron?

The point here is that the two examples of mass ratios quoted in the previous paragraph pose two different questions. The ratio m_t/m_e can be properly understood only through a proper understanding of the separation of the different fermion families. But the smallness of the neutrino mass remains a question even within a single family. Consider, for example, the first family. The electron mass is 0.511 MeV, and the current masses for u and d quarks are in the 5–10 MeV range. Thus they all conform within about an order of magnitude, whereas ν_e is at least 5 orders lighter. Similarly, in the second generation, we get $m_\mu = 105$ MeV, $m_s \simeq 150$ MeV, $m_c \simeq 1500$ MeV, again conforming to one another within about an order of magnitude, but ν_μ is at least 3 orders lighter than m_μ. A good theory of neutrino mass should also throw some light on the reason for this smallness.

3.1.2   Motivations from Astrophysics and Cosmology

Historically, a lot of motivation for serious consideration of massive neutrinos came from outside particle physics – notably from astrophysics and cosmology.

The astrophysical importance of neutrinos was pointed out by Bethe as early as 1939. He listed the chain of reactions that are responsible for burning hydrogen into helium in stellar cores. In these reactions, some electron neutrinos are produced. Since neutrinos interact only weakly, they have very low cross section for interaction with matter. So, once they are produced, they come out of the stars almost without any hindrance. These neutrinos thus bring information about the stellar cores.

Detailed calculations were performed to find out the flux of ν_e that one expects from the Sun. However, experiments detect only about one-third as much. It was speculated that this *solar neutrino puzzle* can be resolved if the ν_e produced in weak processes is not a mass eigenstate but is a superposition of several mass eigenstates. On the passage from the Sun to the earth, the ν_e can oscillate partly to some other flavor of neutrino. This phenomenon, called neutrino oscillations, was suggested by Pontecorvo in 1957 in analogy with neutral kaon oscillations. Clearly, a superposition of mass eigenstates can have a different physical property than the eigenstates themselves only if the eigenvalues are not all degenerate. Hence, if this suggestion has to work, all the neutrinos cannot be massless. Recently, it has been realized that the material medium of the solar interior can enhance the oscillation effects dramatically. However, to obtain a depletion in the observed rate, one still needs the oscillation in the vacuum, which calls for theories with massive neutrinos.

Observations of atmospheric neutrino fluxes also indicate some discrepancy with the theoretical calculations. Neutrino oscillation again may serve the key role to the resolution of this paradox.

Coming back to the solar neutrino puzzle, another way to explain it would be the conjecture of a magnetic moment of the neutrino. If this moment is large enough, the left-handed neutrinos produced in nuclear reactions can flip their helicity in the solar magnetic field and turn into right handed ones. Since the latter have much smaller cross section of interaction with matter, they would be hard to detect. This can explain the depletion of the observed flux. However, a massless neutrino cannot have a magnetic moment, as was pointed out in Ch. 2 and will be discussed in greater detail in Ch. 11. So, for this suggestion to work, one needs neutrino mass.

From stellar astrophysics, we now turn our attention to larger scales of gravitationally bound systems of stars, like galaxies, clusters, superclusters etc. It was known that the mass to light ratio for these different systems increases as one goes to larger and larger systems. This problem of missing light, more commonly called the dark matter problem, can also be resolved if neutrinos have masses of the order of a few eV. They can then be gravitationally bound to these systems and provide them with a non-luminous halo. The larger the system, the larger this halo, the larger the mass to light ratio.

However, if neutrinos have this kind of mass, they can be extremely important for cosmology as well. Standard big bang cosmology predicts the existence of relic background neutrinos all over the universe. The number density of such background neutrinos is about 8 orders of magnitude larger than the average number density of baryons in the universe. Hence, even if the neutrinos have a mass of about 10 eV, they can contribute a huge amount of energy density to the universe and thus affect the evolution of the universe as a whole.

Indeed, assuming sources of mass other than baryons is encouraged by various considerations. With a baryon-dominated universe, it is hard to understand how the initially uniform distribution of matter clumps into galaxies of masses roughly of the order of $10^{11} M_\odot$, where $M_\odot$ is the mass of the sun. Also, inflationary models which solve some of the outstanding problems in the standard big bang cosmology, predict the energy density of the universe as a function of its expansion rate. The expansion rate is not very well-known. But, if it is anywhere within the presently accepted range, it predicts an energy density which is about 10 to 100 times the energy density of baryons. This calls for new sources of energy, and neutrino mass can exactly be such a source. There are other suggestions for possible sources as well, but in a sense they are more exotic ones, because they involve hitherto unknown particles or structures. Hence, from this point of view, the possibility of massive neutrinos has to be taken quite seriously.

3.2 Questions related to neutrino mass

In view of the motivations discussed above, we will study the consequences of neutrino mass in this book. However, the question of neutrino mass cannot be discussed in isolation. It is intimately related to some other issues.

The most important of these issues is that of neutrino mixing. The electron neutrino, for example, is defined to be the object that couples to the electron via the charged weak current. Similarly, ν_μ connects to the muon and ν_τ to the tau. If neutrinos get mass, there is no fundamental reason why these objects should describe physical particles with a specified mass eigenvalue. Rather, in general these *gauge eigenstates* ν_e, ν_μ, ν_τ would be superpositions of the *mass eigenstates*. In other words, the mass eigenstates are mixtures of gauge eigenstates so that a given physical neutrino can couple to more than one charged lepton via the charged current. This is analogous to the mixing of quarks and called *neutrino mixing*.

If mixing occurs, the generational lepton numbers (i.e., the e-number, the μ-number and the τ-number) cannot remain as valid global symmetries. Unlike the predictions from the standard model, one should see processes that violate these numbers, even in processes where neutrinos do not appear in the initial and final states. Some such issues have been discussed in Ch. 13.

Another possible consequence of neutrino mixing is that, like the quark sector mixing matrix V defined in Eq. (1.4), the mixing matrix in the leptonic sector can also be complex in general. This would imply CP-violating phenomena in the leptonic sector, which again have been discussed in Ch. 13.

If the neutrinos are massive, a big question is whether they are Dirac particles or Majorana particles. They can of course be Dirac particles, like all charged fermions have to be. However, there is also the possibility that the neutrinos are their own antiparticles since they do not have any electric charge. This question is discussed in Ch. 4.

If the neutrinos are Majorana particles, then they must not have any additive quantum number, local or global. Thus, total lepton number symmetry must be broken. To be more precise, we should say that the symmetry that has to be broken is $B - L$, which is the only true global symmetry of the standard model in presence of neutrino mixing. One should then see processes violating $B - L$. One such process, called the neutrinoless double beta decay, has been discussed in detail in Ch. 12.

One can then ask whether $B - L$ is violated spontaneously. If it is, then because of the Goldstone theorem we would find a massless pseudo-scalar particle. Since the breaking of $B - L$ is related to Majorana masses for the neutrinos, the related Goldstone boson is called Majoron. The physical implications of the existence of such a particle is a rich and interesting topic by itself. In Ch. 5, we show how Majorons can arise in realistic models.

Another related issue is that of neutrino stability. If neutrinos are massless, they cannot decay. But if they have mass, all but the lightest neutrino should be unstable since in general, there would be no symmetry to prevent their decay. Only the lightest one, being the lightest fermion, would be stable. Some specific decay mechanisms for unstable neutrinos will be discussed in Ch. 5 and Ch. 11. The decays of the heavier neutrinos in turn can profoundly affect the cosmological scenarios. This question has been dealt with in Ch. 6 and Ch. 17 in some detail.

3.3 Tests of neutrino mass

Broadly speaking, there are two kinds of experiments in which neutrino masses can be detected. We discuss them one by one.

3.3.1 Kinematic tests

These are tests on processes which are allowed even in the standard model with $m_\nu = 0$. Take any such known process involving neutrinos in the final state. Calculate the rate as a function of neutrino mass. Try to see whether the observed rate differs significantly from the calculated rate with $m_\nu = 0$.

Obviously, sensitivity of such experiments increase with m_ν/Q, where $Q \equiv$ (initial energy) − (mass of all particles in the final state except the neutrinos). Some examples follow. Details will be discussed in Ch. 9.

- Nuclear β-decay: One can look at the Kurie plot. The shape of the curve can be calculated assuming $m_{\nu_e} = 0$. If, however the mass is not zero, the observed count will fall short of the calculated one. The lowest known Q value (18.6 keV) for this process occurs for ^{3}H β-decay. Extensive experimental studies of tritium decay have been conducted at various laboratories around the world to look for deviations from the Kurie plot, which will signal a non-vanishing neutrino mass. Lubimov et al claimed to have observed a nonzero mass from such experiments. The results were not confirmed by subsequent authors.

- Pion decay: One can look for the muon energy in $\pi^+ \to \mu^+\nu_\mu$ (or its charge conjugate decay). Obviously this energy depends on the ν_μ mass. Recent measurements give a nonzero ν_μ mass at one standard deviation level, but is consistent with a zero mass at the 90% confidence level.

- Tau decay: There are various decay modes of the tau. Since all the observed ones are consistent with the standard model, they all conserve the tau number L_τ. Thus, all these modes have a ν_τ in the final state. One can use the kinematics of the final state to find the mass of the ν_τ.

These and other processes have been described in Ch. 9.

3.3.2 Exclusive tests

This constitutes looking for processes which are forbidden if the neutrinos are massless. Observation of even one event of any such process would then be an unmistakable signature for neutrino mass. But often the specific tests involve extra assumptions about the nature of neutrino mass.

The most common of such assumptions is that of neutrino mixing. Although it is by no means a logical necessity, it is so hard to make gauge models of neutrino mass without having neutrino mixing that it is natural to accept the presence of mixing. Various processes might show the effect of this. Some examples follow.

- Neutrino oscillation: Suppose at $t = 0$, a ν_e is emitted in a particle interaction. If neutrino mixing is present, this ν_e is a superposition of various mass eigenstates with different masses. As time goes on, these different components evolve differently so that, after some time, the original beam becomes a different combination of the mass eigenstates. As a result, it might look somewhat like ν_μ or ν_τ, for example. It can then have charged current interactions with μ or τ. Such a phenomenon is called neutrino oscillation and will be discussed in detail in Ch. 10.

- Neutrino decays: The possibility was mentioned in §3.2. Specific processes which might have interesting consequences include the radiative decay of a heavier neutrino (ν_α) to a lighter one ($\nu_{\alpha'}$):

$$\nu_\alpha \to \nu_{\alpha'} + \gamma, \tag{3.2}$$

which has been discussed in Ch. 11. Another interesting possibility is to have three neutrinos in the final state, which may or may not be of the same kind:

$$\nu_\alpha \to \nu_{\alpha_1} \nu_{\alpha_2} \nu_{\alpha_3} . \tag{3.3}$$

This process receives some importance in the context of Ch. 6.

In models where the global $B - L$ symmetry is broken spontaneously, one obtains a Goldstone boson J which is usually called the Majoron. In such models, the mode

$$\nu_\alpha \to \nu_{\alpha'} + J \tag{3.4}$$

is also interesting. This has been discussed in some detail in Ch. 5.

Besides the above processes, one can look at others which do not crucially depend on neutrino mixing. The most important of such processes are the following:

- Electromagnetic properties: In Ch. 2, it has been mentioned that a massless neutrino has only one electromagnetic form factor, its charge radius. Observation of any other electromagnetic property, like a magnetic moment, would imply a massive neutrino. This question has been discussed in Ch. 11.

- Neutrinoless double β-decay: This process, abbreviated as $\beta\beta_{0\nu}$, is essentially two neutrons in a nucleus converting into two protons by emitting two electrons; or, in the quark picture,

$$d + d \to u + u + e + e . \tag{3.5}$$

Unlike the previous processes, it does not depend on neutrino mixing. However, it involves a violation of lepton number by 2 units. If the neutrinos are all Dirac particles, total lepton number is conserved and this process is forbidden. Only with Majorana neutrinos is such a process possible. A detailed discussion of this process, including the experimental efforts to measure its rate, is given in Ch. 12.

Chapter 4

Dirac vs. Majorana masses

The charged leptons and the quarks — i.e., all known fundamental fermions other than the neutrinos — are Dirac particles as a consequence of electric charge conservation. In other words, they obey the Dirac equation and are described by four component complex spinors. If the neutrinos are massless, it is well known that they can be described by two component complex spinors, alternatively called Weyl spinors. Our concern in this book, however, is to discuss the case of massive neutrinos, so we will not entertain this possibility.

If the neutrinos are massive, it is tempting to think that they are like any other fermion, and should therefore be Dirac spinors. However, there is an important difference between neutrinos and other fundamental fermions, viz, that the neutrinos do not carry any electric charge. This brings in a new theoretical possibility that the neutrinos might also be Majorana spinors [1, 2]. In this chapter we discuss what a Majorana spinor is, spelling out its differences with a Dirac spinor.

4.1 Two-component spinor field

To start with, consider a Dirac field, e.g., the electron. Its states can be described by four basic spinors. Two of these can be taken as the left and the right helicity states of the electron, e_L and e_R. As for the other two, we can use $\widehat{e}_L$ and $\widehat{e}_R$, the two helicity states of the antiparticle positron.

Now suppose that in our frame of reference, we find an electron moving in the z-direction. The z-component of its spin is $-\frac{1}{2}$. Thus, spin and momentum are anti-parallel, so we have a left-handed electron e_L. How does this particle appear to another observer who is running

49

faster than the electron in the z-direction? To him, the electron is moving in the negative z-direction, which is also the direction of its spin. Therefore, he sees a right-handed object. But we have two right-handed objects, e_R and $\hat{e}_R$, in our repertory. One can then ask which one of these two states is this other observer seeing.

The answer is easy here. The state $\hat{e}_R$ has opposite electric charge compared to the state e_L. As we know, electric charge is a Lorentz invariant quantity. By boosting to a different Lorentz frame, one thus cannot see a different charge on a particle. What is e_L in one frame cannot be $\hat{e}_R$ in another. The boosted observer, therefore, sees an e_R.

We now reopen the question in the context of a neutrino. We have a ν_L in our frame. If the neutrino is massive, its speed must be less than that of light, so we can imagine a boosted observer who runs faster. As before, this observer sees some right handed object. What is this object?

We must recall at this point that all experiments to date are consistent with the neutrino being left handed and the antineutrino being right handed. Thus, we have the states ν_L and $\hat{\nu}_R$ in our repertory. If we want to mimic the situation of the electron, we must postulate two more states, ν_R and $\hat{\nu}_L$, in this collection. The boosted observer will see a ν_R when we see a ν_L, just as in the electron case. In this case, the neutrino will be a Dirac particle with four complex degrees of freedom, as is apparent from the presence of the four basis spinor states.

But can we not do without postulating these two new spinor states? After all, the boosted observer sees a right-handed object, and we do have a right-handed object, viz, $\hat{\nu}_R$. Can't the boosted observer see the state $\hat{\nu}_R$? Unlike the case of electrons, ν_L and $\hat{\nu}_R$ have the same electric charge (viz, zero), so nothing goes wrong there. The only quantum number defined in the context of the standard model and which is different for ν_L and $\hat{\nu}_R$ is the lepton number (L). But the lepton number symmetry is a global symmetry. Unlike the electromagnetic $U(1)$ symmetry, it does not govern the dynamics. Rather, it is a consequence of the dynamics and the field contents of the standard model. So there is nothing sacred about the lepton number symmetry. If this is broken, there is thus no reason why ν_L and $\hat{\nu}_R$ cannot be the boosted counterparts of one another. These two spinors can thus constitute the left and right handed projections of the same fermionic field. Since just two basis spinors are needed here, we are talking about a two-component massive fermion field. This was originally proposed by Majorana in 1937. Therefore, such a field is called a Majorana field.

One might ask, how can one boost a certain particle and obtain its antiparticle? The point here is that the words *particle* and *antiparticle* are defined with respect to some conserved quantum number. If there are no conserved quantum numbers that distinguish the *particle* and the *antiparticle* states, the particle is identical to its own antiparticle. A Majorana neutrino *is* its own antineutrino.

The difference between a Majorana particle and a Weyl particle must be clearly understood. Both are two component spinors, but for two completely different reasons. A Weyl particle is massless. Thus, the ν_L moves at the speed of light. No observer can overtake it and view it as a right-handed object. So a right handed counterpart of ν_L is not necessary to obtain a Lorentz covariant picture. Similarly, a $\hat{\nu}_R$ does not require its left handed counterpart. But they can have different lepton numbers (or other quantum numbers) to distinguish themselves from each other.

A Majorana neutrino, on the other hand, has mass. But the ν is the same as $\hat{\nu}$. So the right handed counterpart of ν_L can be equivalently called ν_R or $\hat{\nu}_R$. Similarly, ν_L is the same as $\hat{\nu}_L$. That is why only ν_L and $\hat{\nu}_R$ suffice. They can be Lorentz transformed to each other and thus the neutrino cannot have any additive quantum number. This self-conjugacy is the reason why a Majorana particle has half as many degrees of freedom as a Dirac particle.

An analogy might clarify the case further. Consider the field of a charged pion. It is a complex scalar field since π^+ and π^- are obviously different. But the neutral pion π^0 is its own antiparticle, and thus is described by a real scalar field. A complex field amounts to two real fields, viz, the real and the imaginary parts of it. Thus, the neutral pion has one real degree of freedom whereas the charged pion field has two. Similarly, a Majorana particle has half as many degrees of freedom as a Dirac particle.

4.2 Mathematical definition of a Majorana field

First consider a free Dirac field. The field operator $\psi(x)$ can be written as

$$\psi(x) = \int \frac{d^3 p}{\sqrt{(2\pi)^3 2E_p}} \sum_{s=\pm\frac{1}{2}} \left(f_s(\boldsymbol{p}) u_s(\boldsymbol{p}) e^{-ip\cdot x} + \widehat{f}_s^\dagger(\boldsymbol{p}) v_s(\boldsymbol{p}) e^{ip\cdot x} \right) \quad (4.1)$$

in terms of the operator $f_s(\boldsymbol{p})$ that annihilates a single particle state of momentum $\boldsymbol{p}$ and spin component s in the direction of the momentum. Similarly $\widehat{f}_s(\boldsymbol{p})$ annihilates, i.e., $\widehat{f}_s^\dagger(\boldsymbol{p})$ creates an antiparticle state. The spinors $u_s(\boldsymbol{p})$ and $v_s(\boldsymbol{p})$ are the plane wave solutions of positive and negative energies respectively. They satisfy the equations

$$\begin{aligned}
(\gamma^\mu p_\mu - m)\, u_s(\boldsymbol{p}) &= 0 \\
(\gamma^\mu p_\mu + m)\, v_s(\boldsymbol{p}) &= 0 \,,
\end{aligned} \qquad (4.2)$$

where the γ^μ are a set of four matrices satisfying the relations

$$\{\gamma^\mu, \gamma^\nu\} = 2g^{\mu\nu} \qquad (4.3)$$

and

$$\gamma_\mu^\dagger = \gamma_0 \gamma_\mu \gamma_0 \,. \qquad (4.4)$$

If we write down $\psi^*(x)$, it will involve $f_s^\dagger(\boldsymbol{p})$ and $\widehat{f}_s(\boldsymbol{p})$, which will create a particle or annihilate an antiparticle.

Since the particle and the antiparticle are identical in the case of a Majorana neutrino, we feel that $\psi(x)$ must in some sense be related to $\psi^*(x)$. However, if we try to impose

$$\psi(x) = \psi^*(x) \,, \qquad (4.5)$$

that would be wrong in general, for the following reason.

A physically meaningful equation must be Lorentz covariant. It is easy to see that the above equation is not. Under a Lorentz transformation that changes the co-ordinates as

$$x'^\mu = x^\mu + \omega^\mu{}_\nu x^\nu \,, \qquad (4.6)$$

the spinor field changes as follows:

$$\psi'(x') = \exp\left(-\frac{i}{4}\sigma_{\mu\nu}\omega^{\mu\nu}\right)\psi(x) \tag{4.7}$$

where

$$\sigma_{\mu\nu} = \frac{i}{2}[\gamma_\mu, \gamma_\nu] \tag{4.8}$$

involves the commutator of the Dirac γ-matrices. Eq. (4.7) gives

$$\psi'^*(x') = \exp\left(\frac{i}{4}\sigma^*_{\mu\nu}\omega^{\mu\nu}\right)\psi^*(x), \tag{4.9}$$

which, in general, is not the same transformation law as Eq. (4.7) because the $\sigma_{\mu\nu}$ matrices are not purely imaginary. In that case, even if we impose Eq. (4.5) in one frame of reference, it will not be valid in another frame and will therefore lose its validity as a physical condition.

We therefore define a *conjugate* field

$$\hat{\psi}(x) \equiv \gamma_0 C \psi^*(x) \tag{4.10}$$

in terms of an yet undefined matrix C such that $\hat{\psi}(x)$ transforms the same way under Lorentz transformation as $\psi(x)$, i.e.,

$$\begin{aligned}
\hat{\psi}'(x') &= \exp\left(-\frac{i}{4}\sigma_{\mu\nu}\omega^{\mu\nu}\right)\hat{\psi}(x) \\
&= \exp\left(-\frac{i}{4}\sigma_{\mu\nu}\omega^{\mu\nu}\right)\gamma_0 C \psi^*(x).
\end{aligned} \tag{4.11}$$

However, using Eq. (4.7) directly on Eq. (4.10), we obtain

$$\begin{aligned}
\hat{\psi}'(x') &= \gamma_0 C \psi'^*(x') \\
&= \gamma_0 C \exp\left(\frac{i}{4}\sigma^*_{\mu\nu}\omega^{\mu\nu}\right)\psi^*(x).
\end{aligned} \tag{4.12}$$

Equating the above two expressions, we demand that the matrix C satisfies the relation

$$\gamma_0 C \exp\left(\frac{i}{4}\sigma^*_{\mu\nu}\omega^{\mu\nu}\right) = \exp\left(-\frac{i}{4}\sigma_{\mu\nu}\omega^{\mu\nu}\right)\gamma_0 C \tag{4.13}$$

for any arbitrary $\omega^{\mu\nu}$, which implies

$$\gamma_0 C \sigma^*_{\mu\nu} = -\sigma_{\mu\nu}\gamma_0 C. \tag{4.14}$$

The specific form of the matrix C depends on the representation of the γ-matrices and will be discussed in §4.3. At this point, it is enough to say that such a matrix C, and therefore $\widehat{\psi}$, can be defined. So we can change Eq. (4.5) by replacing the ψ^* on the right hand side by $\widehat{\psi}$. That will surely be a Lorentz-covariant relation to define a Majorana field. Since $\widehat{\psi}$ involves ψ^* and therefore $f_s^\dagger(\boldsymbol{p})$ and $\widehat{f}_s(\boldsymbol{p})$, its identification with ψ serves the purpose of identifying particles with antiparticles.

The true definition of a Majorana field is a little more general. We demand

$$\psi(x) = e^{i\theta}\widehat{\psi}(x) \tag{4.15}$$

since a phase can always be absorbed in the definition of the fermion field $\psi(x)$. We can always choose $\theta = 0$ by suitably defining the fermion field, but the freedom of the phase is sometimes convenient.

It can now be easily shown that the plane wave expansion of a Majorana field operator is

$$\psi(x) = \int \frac{d^3p}{\sqrt{(2\pi)^3 2E_p}} \sum_{s=\pm\frac{1}{2}} \left(f_s(\boldsymbol{p})u_s(\boldsymbol{p})e^{-ip\cdot x} + \lambda f_s^\dagger(\boldsymbol{p})v_s(\boldsymbol{p})e^{ip\cdot x} \right). \tag{4.16}$$

Notice that $\lambda f_s^\dagger(\boldsymbol{p})$ appears in the place of $\widehat{f}_s^\dagger(\boldsymbol{p})$ of the Dirac case. This signifies that the antiparticle is the same as the particle except for a phase λ. Thus λ is related to the angle θ of Eq. (4.15).

To find this relation and also to verify that Eq. (4.15) is indeed satisfied by Eq. (4.16), we start from the latter and evaluate $\widehat{\psi}(x)$:

$$\begin{aligned}
\widehat{\psi}(x) &= \int \frac{d^3p}{\sqrt{(2\pi)^3 2E_p}} \sum_{s=\pm} \gamma_0 C \left(f_s^\dagger(\boldsymbol{p})u_s^*(\boldsymbol{p})e^{ip\cdot x} + \lambda^* f_s(\boldsymbol{p})v_s^*(\boldsymbol{p})e^{-ip\cdot x} \right) \\
&= \lambda^* \int \frac{d^3p}{\sqrt{(2\pi)^3 2E_p}} \sum_{s=\pm} \left(\lambda f_s^\dagger(\boldsymbol{p})v_s(\boldsymbol{p})e^{ip\cdot x} + f_s(\boldsymbol{p})u_s(\boldsymbol{p})e^{-ip\cdot x} \right) \\
&= \lambda^* \psi(x).
\end{aligned} \tag{4.17}$$

We have used the fact that $|\lambda|^2 = 1$, along with the spinor relations

$$\begin{aligned}
\gamma_0 C u_s^*(\boldsymbol{p}) &= v_s(\boldsymbol{p}), \\
\gamma_0 C v_s^*(\boldsymbol{p}) &= u_s(\boldsymbol{p}),
\end{aligned} \tag{4.18}$$

which would be proven after we define C explicitly. We thus see that

$$\lambda = e^{i\theta} . \tag{4.19}$$

In the literature, λ is often called the creation phase factor since it goes with the creation operator in Eq. (4.16). However, we must emphasize that it is only a convention to put the phase in the creation part. It can just as well be put in the annihilation part.

Exercise 4.1 *The Lagrangian for a free Majorana field is*

$$\mathcal{L} = \frac{1}{2} \left(\overline{\psi} i \gamma^\mu \partial_\mu \psi - m \overline{\psi} \psi \right) . \tag{4.20}$$

Show that the propagator of a Majorana field is given by the same expression as that of a Dirac field. [Hint: the propagator of a scalar field is the same irrespective of whether the field is real or complex.]

4.3 Different representations of Dirac matrices

The Dirac matrices γ^μ are defined through the Hamiltonian

$$H = \gamma^0 \left(\boldsymbol{\gamma} \cdot \boldsymbol{p} + m \right) , \tag{4.21}$$

and they satisfy the anticommutation relations of Eq. (4.3) in order that Eq. (4.21) reproduces the relativistic energy momentum relation $E^2 = \boldsymbol{p}^2 + m^2$. Since the Hamiltonian is a hermitian operator, it also follows that γ^0 and $\gamma^0 \gamma^i$ must be hermitian, so that γ^i is anti-hermitian, which has already been succinctly expressed as Eq. (4.4). Apart from these constraints, the γ-matrices are arbitrary and there are different equivalent conventions to choose them, as we will shortly show with some examples.

As far as the matrix C is concerned, first of all we notice that in Eq. (4.10), if we demand that $\hat{\psi}$ is properly normalized when ψ is, then $\gamma_0 C$ must be a unitary matrix. This implies that C is unitary,

$$C^\dagger = C^{-1} , \tag{4.22}$$

using the properties of γ_0. Then, we note that Eq. (4.14) implies

$$C \sigma_{\mu\nu}^* C^{-1} = -\gamma_0 \sigma_{\mu\nu} \gamma_0 . \tag{4.23}$$

Now, it is easy to see that the above equation can be satisfied if we take

$$C^{-1}\gamma_\mu C = -\gamma_\mu^\mathsf{T} \tag{4.24}$$

as the defining relation for C. Finally, we demand that

$$\widehat{\widehat{\psi}} = \psi. \tag{4.25}$$

Using Eq. (4.24), this can be shown to imply $CC^* = -1$, which can be transformed into the form

$$C^\mathsf{T} = -C, \tag{4.26}$$

using Eq. (4.22).

With these properties, we can go back to the definition of conjugate fields in Eq. (4.10) and rewrite it as

$$\overline{\psi} \equiv \widehat{\psi}^\dagger \gamma_0 = \psi^\mathsf{T} C^\dagger \gamma_0^\dagger \gamma_0 = \psi^\mathsf{T} C^{-1}, \tag{4.27}$$

using Eq. (4.4), as well as $(\gamma_0)^2 = 1$ which follows from Eq. (4.3). This is an alternative form in which the conjugate field is often defined.

Exercise 4.2 *Using the definition in Eq. (4.10) for the conjugate fermion field and the properties of the matrix C, show that*

$$\begin{aligned} \overline{\psi}_\mathsf{L} \psi_\mathsf{R} &= \overline{\widehat{\psi}}_\mathsf{L} \widehat{\psi}_\mathsf{R} \\ \overline{\psi}_\mathsf{L} \gamma_\mu \psi_\mathsf{L} &= -\overline{\widehat{\psi}}_\mathsf{R} \gamma_\mu \widehat{\psi}_\mathsf{R}. \end{aligned} \tag{4.28}$$

Exercise 4.3 *Starting from the definition of a u-spinor in Eq. (4.2) and using the properties of the matrix C given above, show that $\gamma_0 C u^*$ satisfies the equation of a v-spinor.*

4.3.1   Dirac representation

For the γ-matrices, Dirac himself chose

$$\gamma^0 = \begin{pmatrix} I & 0 \\ 0 & -I \end{pmatrix}, \qquad \gamma^i = \begin{pmatrix} 0 & \sigma^i \\ -\sigma^i & 0 \end{pmatrix}, \tag{4.29}$$

where I is the 2×2 unit matrix and the σ^i's are the Pauli matrices. In this representation, the plane wave solutions are given by

$$u_s(\boldsymbol{p})e^{-ip\cdot x} \quad \text{or} \quad v_s(\boldsymbol{p})e^{ip\cdot x}, \tag{4.30}$$

which will be normalized by the relations

$$u_s^\dagger(\boldsymbol{p})u_{s'}(\boldsymbol{p}) = v_s^\dagger(\boldsymbol{p})v_{s'}(\boldsymbol{p}) = 2E_p\delta_{ss'} \,. \tag{4.31}$$

Using the explicit forms of the γ-matrices in the equations for the u-type and v-type spinors given in Eq. (4.2), we can find the solution in these representation. These are:

$$
\begin{aligned}
u_s(\boldsymbol{p}) &= \sqrt{E+m}\left(\begin{array}{c} \chi_s \\ \dfrac{\boldsymbol{\sigma}\cdot\boldsymbol{p}}{E+m}\chi_s \end{array}\right), \\
v_s(\boldsymbol{p}) &= \sqrt{E+m}\left(\begin{array}{c} \dfrac{\boldsymbol{\sigma}\cdot\boldsymbol{p}}{E+m}\chi_s' \\ \chi_s' \end{array}\right),
\end{aligned}
\tag{4.32}
$$

in the normalization defined in Eq. (4.31), where

$$\chi_{+\frac{1}{2}} = -\chi'_{-\frac{1}{2}}\left(\begin{array}{c}1\\0\end{array}\right), \qquad \chi_{-\frac{1}{2}} = \chi'_{+\frac{1}{2}} = \left(\begin{array}{c}0\\1\end{array}\right). \tag{4.33}$$

We now try to identify the matrix C in this representation. Notice that the defining relation in Eq. (4.24), or the associated properties given in Eqs. (4.22) and (4.26), still leaves the arbitrariness of an overall phase in the matrix C. We have to fix it with the relations between the spinors given in Eq. (4.18). With the choice of the spinors given above, we obtain

$$C = i\gamma_2\gamma_0 = \left(\begin{array}{cc} 0 & i\sigma^2 \\ i\sigma^2 & 0 \end{array}\right) \tag{4.34}$$

in this representation.

Exercise 4.4 *Verify that the definition of the matrix C and the spinors u and v satisfy Eq. (4.18). Use the fact that*

$$\sigma_i^* = -\sigma_2\sigma_i\sigma_2 \tag{4.35}$$

for the Pauli matrices.

4.3.2 Majorana representation

There is another representation which is particularly illuminating when we are talking of Majorana particles. Recall that we said in the previous section that we cannot take $\psi = \psi^*$ because the $\sigma_{\mu\nu}$ matrices are not imaginary in general. If they are so, however, in a particular representation of the Dirac matrices, then Eq. (4.5) becomes an acceptable condition. In such a representation, a Majorana particle is just a spinor whose components are all real. The analogy of Dirac and Majorana spinors with complex and real scalars become more apparent in such a representation.

Such a representation exists. It is given by

$$\gamma^0 = \begin{pmatrix} 0 & \sigma^2 \\ \sigma^2 & 0 \end{pmatrix} \quad , \quad \gamma^1 = \begin{pmatrix} i\sigma^3 & 0 \\ 0 & i\sigma^3 \end{pmatrix}$$

$$\gamma^2 = \begin{pmatrix} 0 & -\sigma^2 \\ \sigma^2 & 0 \end{pmatrix} \quad , \quad \gamma^3 = \begin{pmatrix} -i\sigma^1 & 0 \\ 0 & -i\sigma^1 \end{pmatrix} . \quad (4.36)$$

Notice that all the γ-matrices are imaginary. By Eq. (4.8), it assures that the $\sigma_{\mu\nu}$'s are also purely imaginary. The matrix C can be defined in this representation as

$$C = -\gamma^0 . \quad (4.37)$$

Since $(\gamma^0)^2 = 1$, it is easily seen that Eq. (4.14) is satisfied.

Exercise 4.5 *Check that the definitions of the matrix C given in Eqs. (4.34) and (4.37) satisfy the definition in Eq. (4.24).*

4.3.3 Other representations

In general, knowing one representation of the gamma matrices allows one to write down other ones, denoted by a prime, in the form

$$\gamma'_\mu = U\gamma_\mu U^\dagger , \quad (4.38)$$

where U is a unitary matrix. It is easy to see that the primed γ-matrices also satisfy the anticommutation relations of Eq. (4.3). In fact, it can be shown that if two sets of γ-matrices satisfy Eq. (4.3), they can always be related by a similarity transformation as in Eq. (4.38). The proof of this last statement is rather involved, and we are not even trying to indicate how the proof goes.

Starting from the Dirac representation, one can use

$$U = \frac{1}{\sqrt{2}} \begin{pmatrix} I & \sigma_2 \\ \sigma_2 & -I \end{pmatrix} \quad (4.39)$$

to define the Majorana representation. The plane wave solutions in the primed representation can be defined as

$$u'_s(\boldsymbol{p}) = U u_s(\boldsymbol{p}), \qquad v'_s(\boldsymbol{p}) = U v_s(\boldsymbol{p}), \qquad (4.40)$$

which can easily be shown to satisfy Eq. (4.2) with the primed γ-matrices. Similarly, in the primed representation, we can take the conjugation matrix to be

$$C' = UCU^{\mathsf{T}}, \qquad (4.41)$$

which satisfies Eq. (4.24). Since C is antisymmetric in the Dirac representation, this shows that it is antisymmetric in other representations as well.

Using Eqs. (4.40) and (4.41), it is trivial to check that if the relations between the spinors given in Eq. (4.18) are true in one representation, they are valid in the primed representation as well. Thus, these relations are valid in general, a fact we used before in deriving Eq. (4.17).

> **Exercise 4.6** *Show that the definition of the conjugate field given in Eq. (4.10) is independent of any representation of the γ-matrices.*

4.4 Majorana neutrinos and discrete symmetries of space-time

A Majorana neutrino is its own antiparticle. Therefore, it is expected that it possesses special properties under discrete symmetries like $\mathcal{C}, \mathcal{CP}$ and $\mathcal{CPT}$. We examine these properties one by one [3].

4.4.1 Properties under $\mathcal{C}$

Under the charge conjugation operation $\mathcal{C}$, a free fermion field transforms as

$$\mathcal{C}\psi(\boldsymbol{x}, t)\mathcal{C}^{-1} = \eta_C^* \gamma_0 C \psi^*(\boldsymbol{x}, t) = \eta_C^* \widehat{\psi}(\boldsymbol{x}, t), \qquad (4.42)$$

where η_C is a phase and the conjugate field $\widehat{\psi}$ has been defined in Eq. (4.10). While the above equation is true for any fermion field, for a Majorana field we can write

$$\mathcal{C}\psi(\boldsymbol{x}, t)\mathcal{C}^{-1} = (\eta_C \lambda)^* \psi(\boldsymbol{x}, t) \qquad (4.43)$$

using Eq. (4.17). Plugging in the plane wave expansion of a Majorana field given in Eq. (4.16), we obtain

$$
\begin{aligned}
\mathcal{C} f_s(\boldsymbol{p}) \mathcal{C}^{-1} &= (\eta_C \lambda)^* f_s(\boldsymbol{p}) \\
\mathcal{C} f_s^\dagger(\boldsymbol{p}) \mathcal{C}^{-1} &= (\eta_C \lambda)^* f_s^\dagger(\boldsymbol{p}) \,.
\end{aligned}
\tag{4.44}
$$

Using the unitarity of the operator $\mathcal{C}$, we observe that the consistency of these two equations demands that

$$
(\eta_C \lambda)^* = \eta_C \lambda \,.
\tag{4.45}
$$

If the vacuum is $\mathcal{C}$-symmetric, we can apply the second one of Eq. (4.44) on the vacuum to obtain

$$
\mathcal{C} \, |\boldsymbol{p}, s\rangle = \eta_C \lambda \, |\boldsymbol{p}, s\rangle \,,
\tag{4.46}
$$

where

$$
|\boldsymbol{p}, s\rangle = f_s^\dagger(\boldsymbol{p}) \, |0\rangle
\tag{4.47}
$$

is a one-particle state with momentum $\boldsymbol{p}$ and spin component s. Calling the $\mathcal{C}$ eigenvalue of the one-particle state as $\tilde{\eta}_C$, we thus obtain from Eq. (4.46) the relation

$$
\tilde{\eta}_C = \eta_C \lambda \,,
\tag{4.48}
$$

which is real due to Eq. (4.45).

A free Majorana particle is thus an eigenstate of the charge conjugation operator. The same cannot be said about a physical Majorana neutrino, because the interactions of a neutrino necessarily violate $\mathcal{C}$ by large amounts, so that it does not make any sense to talk about an eigenstate of $\mathcal{C}$. However, if instead of neutrinos one talks about some other Majorana fermions whose interactions conserve $\mathcal{C}$ (at least to a good extent), the particle would be an eigenstate of $\mathcal{C}$ (to the same good extent).

> **Exercise 4.7** *If the vacuum is C-symmetric, show that the operator $\mathcal{C}$ acting on the state $|\boldsymbol{p}, s\rangle$ of a Dirac particle produces an antiparticle with the same momentum and same spin. Thus, the C operation does not change the helicity of the particle.*

4.4.2   Properties under $\mathcal{CP}$

To discuss how a Majorana neutrino behaves under $\mathcal{CP}$, let us introduce the shorthand

$$\Xi \equiv \mathcal{CP} \tag{4.49}$$

for convenience. The transformation of a free fermion field under parity can be given as

$$\mathcal{P}\psi(\boldsymbol{x}, t)\mathcal{P}^{-1} = \eta_P \gamma_0 \psi(-\boldsymbol{x}, t), \tag{4.50}$$

η_P being the parity phase of the field. Combining with Eq. (4.42), we get

$$\Xi\psi(\boldsymbol{x}, t)\Xi^{-1} = \eta_\Xi^* \gamma_0 \widehat{\psi}(-\boldsymbol{x}, t) \tag{4.51}$$

where η_Ξ is a phase factor. For a Majorana field, we thus obtain

$$\Xi\psi(\boldsymbol{x}, t)\Xi^{-1} = (\eta_\Xi \lambda)^* \gamma_0 \psi(-\boldsymbol{x}, t). \tag{4.52}$$

Using the plane wave expansion of the Majorana field of Eq. (4.16) and changing the dummy variable $\boldsymbol{p}$ to $-\boldsymbol{p}$, we can write

$$\gamma_0\psi(-\boldsymbol{x}, t) = \int \frac{d^3p}{\sqrt{(2\pi)^3 2E_p}} \sum_{s=\pm\frac{1}{2}} \Big(f_s(-\boldsymbol{p})\gamma_0 u_s(-\boldsymbol{p})e^{-ip\cdot x} $$
$$+ \lambda f_s^\dagger(-\boldsymbol{p})\,\gamma_0 v_s(-\boldsymbol{p})e^{ip\cdot x} \Big). \tag{4.53}$$

We can now use the spinor relations

$$\gamma_0 u_s(-\boldsymbol{p}) = u_s(\boldsymbol{p}), \quad \gamma_0 v_s(-\boldsymbol{p}) = -v_s(\boldsymbol{p}), \tag{4.54}$$

whose validity can be explicitly checked in the Dirac representation and then generalized to other representations by the use of Eqs. (4.38) and (4.40). Comparing with Eq. (4.52), we then obtain

$$\begin{aligned} \Xi f_s(\boldsymbol{p})\Xi^{-1} &= (\eta_\Xi \lambda)^* f_s(-\boldsymbol{p}) \\ \Xi f_s^\dagger(\boldsymbol{p})\Xi^{-1} &= -(\eta_\Xi \lambda)^* f_s^\dagger(-\boldsymbol{p}). \end{aligned} \tag{4.55}$$

The consistency of these equations demands the condition

$$\eta_\Xi \lambda = -(\eta_\Xi \lambda)^*. \tag{4.56}$$

The free one-particle states then satisfy

$$\Xi \left|\boldsymbol{p}, s\right\rangle = \tilde{\eta}_\Xi \left|-\boldsymbol{p}, s\right\rangle ,\tag{4.57}$$

where

$$\tilde{\eta}_\Xi = \eta_\Xi \lambda ,\tag{4.58}$$

which are necessarily imaginary because of Eq. (4.56).

Once again, the above analysis strictly makes sense for physical particles only if $\mathcal{CP}$ is conserved, and can be used as a good approximation if $\mathcal{CP}$ is only slightly violated. In the quark sector, $\mathcal{CP}$ is indeed violated by a small amount compared to the amount of $\mathcal{C}$ or $\mathcal{P}$ violation. If the same is true in the leptonic sector, we can talk about Majorana neutrinos as $\mathcal{CP}$ eigenstates, as we will do for the most part in this book. However, it has to be borne in mind that this is at best an approximation.

> **Exercise 4.8** *Consider the decay $Z \to \nu\hat{\nu}$. For Majorana neutrinos, the final state contains two identical fermions. Show that they must be in the $^3P_1$ state (i.e., $L = 1$, $S = 1$ state). Since at the tree level, this decay is CP-conserving and since the CP eigenvalue of Z is $+1$, derive from this that the CP eigenvalues of Majorana neutrinos must be $\pm i$.*

4.4.3 Properties under $\mathcal{CPT}$

For the sake of convenience, we now introduce the notation

$$\Theta \equiv \mathcal{CPT} .\tag{4.59}$$

Then, for any fermion field,

$$\Theta \psi(x) \Theta^{-1} = -\eta_\Theta^* \gamma_5^\mathsf{T} \psi^*(-x) .\tag{4.60}$$

where η_Θ can be called the $\mathcal{CPT}$ phase of the fermion field. For a Majorana neutrino, we can use Eqs. (4.10) and (4.17) to write

$$\Theta \psi(x) \Theta^{-1} = -(\eta_\Theta \lambda)^* \gamma_5^\mathsf{T} C^{-1} \gamma_0 \psi(-x) .\tag{4.61}$$

We can now use the plane wave expansion of a Majorana field to obtain the Θ-transformation properties of the creation and annihilation operators. In order to simplify the right side of Eq. (4.61), we can use the spinor relations

$$\begin{aligned}
\gamma_5^\mathsf{T} C^{-1} \gamma_0 u_s(\boldsymbol{p}) &= (-1)^{s-\frac{1}{2}} u_{-s}^*(\boldsymbol{p}) , \\
\gamma_5^\mathsf{T} C^{-1} \gamma_0 v_s(\boldsymbol{p}) &= (-1)^{s+\frac{1}{2}} v_{-s}^*(\boldsymbol{p}) ,
\end{aligned}\tag{4.62}$$

which can be proved directly in the Dirac representation and can be shown to be valid in any other representation. This gives

$$\gamma_5^{\mathsf{T}} C^{-1} \gamma_0 \psi(-x) = \int \frac{d^3p}{\sqrt{(2\pi)^3 2E_p}} \sum_{s=\pm\frac{1}{2}} (-1)^{s-\frac{1}{2}} \left(f_s(\boldsymbol{p}) u_{-s}^*(\boldsymbol{p}) e^{ip\cdot x} \right.$$
$$\left. - \lambda f_s^\dagger(\boldsymbol{p}) v_{-s}^*(\boldsymbol{p}) e^{-ip\cdot x} \right)$$
$$= \int \frac{d^3p}{\sqrt{(2\pi)^3 2E_p}} \sum_{s=\pm\frac{1}{2}} (-1)^{s+\frac{1}{2}} \left(f_{-s}(\boldsymbol{p}) u_s^*(\boldsymbol{p}) e^{ip\cdot x} \right.$$
$$\left. - \lambda f_{-s}^\dagger(\boldsymbol{p}) v_s^*(\boldsymbol{p}) e^{-ip\cdot x} \right), \quad (4.63)$$

where in the last step, we have changed the dummy variable s to $-s$. To deal with the left side of Eq. (4.61), we need to remember that Θ is an anti-linear operator, i.e., for a complex number a and an operator A,

$$\Theta a A \Theta^{-1} = a^* \Theta A \Theta^{-1} = -a^* \Theta A \Theta^\dagger. \quad (4.64)$$

Using this, we get

$$\Theta f_s(\boldsymbol{p})\Theta^{-1} = \eta_\Theta^* \lambda^*(-1)^{s-\frac{1}{2}} f_{-s}(\boldsymbol{p})$$
$$\Theta f_s^\dagger(\boldsymbol{p})\Theta^{-1} = \eta_\Theta^* \lambda(-1)^{s+\frac{1}{2}} f_{-s}^\dagger(\boldsymbol{p}), \quad (4.65)$$

Taking now the hermitian conjugate of the second equation in Eq. (4.65) and using Eq. (4.64), we obtain

$$\Theta f_s(\boldsymbol{p})\Theta^{-1} = \eta_\Theta \lambda^*(-1)^{s+\frac{1}{2}} f_{-s}(\boldsymbol{p}). \quad (4.66)$$

Comparing this with the first equation in Eq. (4.65), we get

$$\eta_\Theta^* = -\eta_\Theta, \quad (4.67)$$

i.e., the phase η_Θ is $\pm i$. Applying Eq. (4.65) on the CPT-symmetric vacuum, we then get

$$\Theta |\boldsymbol{p}, s\rangle = \tilde{\eta}_\Theta^s |\boldsymbol{p}, -s\rangle, \quad (4.68)$$

where

$$\tilde{\eta}_\Theta^s = \eta_\Theta \lambda(-1)^{s-\frac{1}{2}}. \quad (4.69)$$

There is one sense in which this discussion about $\mathcal{CPT}$ is more fundamental than the previous ones concerning $\mathcal{C}$ and $\mathcal{CP}$. The point is that while $\mathcal{C}$ and $\mathcal{CP}$ are violated by some interactions, there are strong reasons to believe that all particle interactions conserve $\mathcal{CPT}$. So, whatever be the $\mathcal{CPT}$ properties of a free Majorana field, we can say that the same properties are present in the physical particle.

Thus, Eq. (4.61) can be taken as the definition of a Majorana particle with no sacrifice of rigor. We can thus say that a Majorana neutrino is a $\mathcal{CPT}$ eigenstate. In a world where neither $\mathcal{C}$ nor $\mathcal{CP}$ is conserved, this is the only way of saying that it is its own antiparticle [4].

4.5   The Majorana basis of mass terms

As far as $\mathcal{CPT}$ is concerned, a Majorana spinor is a self-contained item. It has a left-handed part and a right-handed one which are $\mathcal{CPT}$ conjugates of each other. It has half as many components as a Dirac spinor. Therefore, we can think of using this minimal representation as the basis of any mass term.

We first show that a Dirac spinor can be thought of to be made up of two such 2-component spinors. For this purpose, let us take a field ψ_L and its $\mathcal{CPT}$ conjugate $\widehat{\psi}_R$. They can form one Majorana spinor. Similarly, there is χ_L and $\widehat{\chi}_R$, which can do the same. However, suppose for some reason mass terms such as

$$\overline{\psi}_L \widehat{\psi}_R \quad \text{and} \quad \overline{\chi}_L \widehat{\chi}_R \tag{4.70}$$

are absent. On the other hand, there are cross terms like

$$\frac{1}{2} m \overline{\psi}_L \widehat{\chi}_R + \frac{1}{2} m' \overline{\chi}_L \widehat{\psi}_R + \text{h.c.}. \tag{4.71}$$

We can summarize these statements by putting the mass terms $\mathcal{L}_{\text{mass}}$ in a matrix form:

$$-\mathcal{L}_{\text{mass}} = \frac{1}{2} \begin{pmatrix} \overline{\psi}_L & \overline{\chi}_L \end{pmatrix} \begin{pmatrix} 0 & m \\ m' & 0 \end{pmatrix} \begin{pmatrix} \widehat{\psi}_R \\ \widehat{\chi}_R \end{pmatrix} + \text{h.c.} \tag{4.72}$$

In general, for a number of 2-component spinors ψ_{aL}, $\widehat{\psi}_{aR}$ we can write

$$-\mathcal{L}_{\text{mass}} = \frac{1}{2} \sum_{a,b} \overline{\psi}_{aL} M_{ab} \widehat{\psi}_{bR} + \text{h.c.} \tag{4.73}$$

and then M_{ab} is called the mass matrix in the Majorana basis. This basis is very useful for considering neutrino masses, as we will see in Part II of the book.

It is straightforward to show that written in this form, M_{ab} must be a symmetric matrix. For that, we use the definition of $\widehat{\psi}$ from Eq. (4.10) or Eq. (4.27) and the definition of the matrix C from Eq. (4.24) to write

$$
\begin{aligned}
-\mathcal{L}_{\text{mass}} &= \frac{1}{2} \sum_{a,b} \overline{\psi}_a M_{ab} \mathsf{R} \widehat{\psi}_b + \text{h.c.} \\
&= \frac{1}{2} \sum_{a,b} \widehat{\psi}_a{}^{\mathsf{T}} C^{-1} M_{ab} \mathsf{R} \gamma_0 C \psi_b^* + \text{h.c.} \\
&= -\frac{1}{2} \sum_{a,b} \widehat{\psi}_a{}^{\mathsf{T}} C^{-1} M_{ab} \mathsf{R} C \gamma_0^{\mathsf{T}} \psi_b^* + \text{h.c.} \\
&= \frac{1}{2} \sum_{a,b} \psi_b^\dagger \gamma^0 \left(C^{-1} \mathsf{R} C \right)^{\mathsf{T}} M_{ab} \widehat{\psi}_a + \text{h.c.} .
\end{aligned}
\tag{4.74}
$$

In going to the last step, we have used the following property of two spinors ψ, ψ' and a 4×4 matrix Γ:

$$
\begin{aligned}
\psi^{\mathsf{T}} \Gamma \psi' &= \sum_{\alpha,\beta} \psi_\alpha \Gamma_{\alpha\beta} \psi'_\beta \\
&= -\sum_{\alpha,\beta} \psi'_\beta \Gamma_{\alpha\beta} \psi_\alpha = -\psi'^{\mathsf{T}} \Gamma^{\mathsf{T}} \psi .
\end{aligned}
\tag{4.75}
$$

The spinor indices have been put in explicitly in the intermediate steps for the sake of clarity. The negative sign appears in the middle of this operation because of the interchange of the orders of two anti-commuting fermion fields. Now, since Eq. (4.24) implies

$$
C^{-1} \gamma_5 C = \gamma_5^{\mathsf{T}} ,
\tag{4.76}
$$

$\left(C^{-1} \mathsf{R} C \right)^{\mathsf{T}} = \mathsf{R}$, so that we can write Eq. (4.74) as

$$
\begin{aligned}
-\mathcal{L}_{\text{mass}} &= \frac{1}{2} \sum_{a,b} \overline{\psi}_{b\mathsf{L}} M_{ab} \widehat{\psi}_{a\mathsf{R}} + \text{h.c.} \\
&= \frac{1}{2} \sum_{a,b} \overline{\psi}_{a\mathsf{L}} M_{ba} \widehat{\psi}_{b\mathsf{R}} + \text{h.c.}
\end{aligned}
\tag{4.77}
$$

changing the dummy indices a, b in the last step. Comparing this with Eq. (4.73), we conclude that

$$M_{ab} = M_{ba} \,. \tag{4.78}$$

Looking back at Eq. (4.72), we realize that one must have $m' = m$ there. In that case, we can identify

$$\psi_L \to \Psi_L \quad , \quad \widehat{\chi}_R \to \Psi_R \quad , \quad \widehat{\psi}_R \to \widehat{\Psi}_R \quad , \quad \chi_L \to \widehat{\Psi}_L \tag{4.79}$$

and then Ψ is a 4-component Dirac spinor, because its mass term from Eq. (4.72) reads

$$\frac{1}{2}m \left(\overline{\Psi}_L \Psi_R + \overline{\widehat{\Psi}}_L \widehat{\Psi}_R \right) + \text{h.c.} = m\overline{\Psi}_L \Psi_R + \text{h.c.} \,, \tag{4.80}$$

by using the definition of $\widehat{\Psi}$. The last equation is the expected form of the mass term of a Dirac fermion.

> **Exercise 4.9** *The kinetic energy term of a Majorana field was given in Eq. (4.20). Show that, in the construction of a Dirac spinor from two Majorana spinors, the kinetic energy terms of the basis Majorana spinors add up to give the proper kinetic term for a Dirac spinor.*

From the discussion above, it is clear that the Dirac spinor is a very special form of two 2-component spinors. In general the mass matrix M_{ab} can be more complicated. Only when it is of the specific form of Eq. (4.72) with $m' = m$, a Dirac fermion results.

This can be stated in various ways. Diagonalization of the matrix of Eq. (4.72) yields two eigenvalues m and $-m$. Thus, we can say that for a general mass matrix M_{ab} in the 2-component basis, we will obtain a Dirac spinor for each pair of equal and opposite eigenvalues. Alternatively, one can say that if we find some symmetry that forbids the mass terms as in Eq. (4.70) for two different 2-component fields, we obtain a Dirac spinor in the spectrum. For charged fermions, electromagnetic gauge symmetry ensures this. For neutrinos, it has to be some global symmetry like the lepton number.

In the above example, if ψ_L is the left handed electron neutrino ν_{eL}, we can take $\widehat{\chi}_R$ to be a right handed neutrino state, absent in the standard model. In this case, the Dirac mass term conserves lepton number, which is equivalent to the electron number L_e for first generation fermions. However, there is also the possibility that $\widehat{\chi}_R$, instead of being a separate fermion absent in the standard model, could be the antiparticle of $\nu_{\mu L}$. In this case, the conserved global quantum number turns out to be $L_e - L_\mu$. This possibility was suggested by Zeldovich and independently by Konopinsky and Mahmoud [5]. Interest in this possibility has recently been revived in connection with the solar neutrino puzzle.

If the terms in Eq. (4.70) are not forbidden, the diagonalization would yield two different mass eigenvalues. The physical particles would be two Majorana spinors. We will encounter this situation quite often in our discussion of the models of neutrino mass.

When the number of basis Majorana spinors are not 2 but $2\mathcal{N}$, say, we can still use a mass matrix as shown in Eq. (4.72), where ψ_L, χ_L etc denote a collection of $\mathcal{N}$ spinors, and the element m, for example, is replaced by an $\mathcal{N} \times \mathcal{N}$ matrix M. In order to maintain the symmetric nature of the mass matrix, m' of Eq. (4.72) should now equal M^T. If the other elements are zero, then the diagonalization would yield $\mathcal{N}$ Dirac neutrinos. In this simple case, it is enough to diagonalize just the matrix M and not the entire $2\mathcal{N} \times 2\mathcal{N}$ mass matrix in the Majorana basis.

In general, the elements other than M are non zero, and diagonalization yields Majorana neutrinos. However, in view of the previous discussion, the matrix M is often called the Dirac mass matrix. The terms going in the diagonal elements of the matrix in Eq. (4.72) are called the Majorana mass terms. However, it is important to realize that these names do not imply anything about whether the mass eigenstates are Majorana or Dirac.

4.6 The two-component basis in a different notation

The chiral nature of weak interactions and the near masslessness of the neutrino allows a different two-component notation sometimes used in the literature. For the sake of completeness, we summarize this notation in this section. The basic mathematics is almost identical to that of §4.5.

It is useful to start with the SL(2,C) group, which is the set of complex linear transformations S with unit determinant. If we take

$z = (z_1, z_2)'$ as the two dimensional complex basis, then

$$z' = Sz. \tag{4.81}$$

The unit determinant constraint implies that

$$S_{11}S_{22} - S_{12}S_{21} = 1. \tag{4.82}$$

As is well-known, this condition enables us to identify S with the Lorentz transformations as follows. Define $\sigma_\mu = (1, \boldsymbol{\sigma})$ as a Lorentz 4-vector. Then

$$\sigma_\mu P^\mu = P^0 - \boldsymbol{\sigma} \cdot \boldsymbol{P}. \tag{4.83}$$

Note that

$$\det(\sigma_\mu P^\mu) = P^\mu P_\mu. \tag{4.84}$$

If we then require that $\sigma_\mu P^\mu$ transforms under SL(2,C) as

$$(\sigma_\mu P^\mu)' = (S^\dagger)^{-1}(\sigma_\mu P^\mu)S^{-1} \tag{4.85}$$

so that products of the form $z^\dagger(\sigma_\mu P^\mu)z$ remain invariant, we obtain

$$\det(\sigma_\mu P^\mu)' = \det(\sigma_\mu P^\mu). \tag{4.86}$$

By Eq. (4.84), this implies the invariance of $P^\mu P_\mu$, which is the essence of Lorentz transformations. Thus, one can identify the SL(2,C) transformations with the Lorentz transformations. A further check of this can be made from Eq. (4.82) which implies that the matrix S has six independent parameters (three complex ones), as in the case of Lorentz transformations.

The fundamental representations of SL(2,C) are two dimensional and are called two-component spinors, like z in Eq. (4.81). We can identify them with the two-component chiral fermions χ, i.e.,

$$\chi \equiv \begin{pmatrix} \chi_1 \\ \chi_2 \end{pmatrix}. \tag{4.87}$$

Under Lorentz transformation, $\chi \to \chi' = S\chi$. Note that χ^* is inequivalent to χ and transforms as $\chi^* \to \chi^{*\prime} = S^*\chi^*$. We will call χ a (2,1) representation and χ^* a $(1,2)$ representation. Then $\sigma_\mu P^\mu$ transforms as a $(2,2)$ representation under $SL(2,C)$. Components of χ will be denoted by χ_a with a lower index, and those of χ^* by $\overline{\chi}^a$ with an upper index.

We can now proceed to construct Lorentz invariant bilinears, which will enable us to define Majorana and Dirac neutrinos. The unit determinant property of S implies that, if χ and ψ are two 2-component spinors, then $\epsilon^{\alpha\beta}\chi_\alpha\psi_\beta$ is invariant under $SL(2,C)$, where ϵ is the antisymmetric tensor defined by $\epsilon^{11} = \epsilon^{22} = 0$, $\epsilon^{12} = -\epsilon^{21} = 1$. Note that the anti-commuting property of spinor fields allows one to write $\epsilon^{\alpha\beta}\chi_\alpha\chi_\beta$ also as a non-vanishing Lorentz-invariant bilinear. These invariants can be written as $i\chi^T\sigma_2\psi$ or $i\chi^T\sigma_2\chi$ in the matrix notation. Similar invariants exist for $\overline{\chi}$ and $\overline{\psi}$.

A 4-component Dirac spinor can in general be written as

$$\Psi = \begin{pmatrix} \chi \\ -i\sigma_2\psi^* \end{pmatrix}. \tag{4.88}$$

Thus, if we choose $\gamma_0 = \begin{pmatrix} 0 & 1 \\ 1 & 0 \end{pmatrix}$, a conventional Dirac mass term $\overline{\Psi}\Psi$ becomes

$$\overline{\Psi}\Psi = i\psi^T\sigma_2\chi - i\chi^T\sigma_2\psi^*. \tag{4.89}$$

Note that such a mass term is invariant under the $U(1)$ transformation under which

$$\psi \to e^{i\theta}\psi, \qquad \chi \to e^{-i\theta}\chi. \tag{4.90}$$

This $U(1)$ can be identified with the lepton number and if χ represents a lepton, ψ represents an antilepton.

A general mass term involving many two component spinors can be written as

$$-\mathcal{L}_{\text{mass}} = \frac{i}{2}\sum_{a,b} M_{ab}\psi_a^T\sigma_2\chi_b + \text{h.c.}, \tag{4.91}$$

where now the indices a and b label different spinors. It is then easy to check that $M = M^T$. Furthermore, it is also easy to see how the creation phase factor introduced in Eq. (4.16) emerges in this framework. To see this, let us assume that a free 2-component neutrino is described by the following Lagrangian:

$$\mathcal{L} = \nu^\dagger \sigma^\mu i \partial_\mu \nu - \frac{im}{2} e^{i\delta} \nu^T \sigma_2 \nu + \frac{im}{2} e^{-i\delta} \nu^\dagger \sigma_2 \nu^* . \tag{4.92}$$

This leads to the following field equation:

$$i\sigma^\mu \partial_\mu \nu + ime^{-i\delta} \sigma_2 \nu^* = 0 . \tag{4.93}$$

If we want to rewrite the Lagrangian of Eq. (4.92) in the 4-component notation such that the Lagrangian looks like in Eq. (4.20), then we need to write

$$\psi = \begin{pmatrix} \nu \\ -i\sigma_2 \nu^* e^{-i\delta} \end{pmatrix} , \tag{4.94}$$

and

$$\gamma_\mu = \begin{pmatrix} 0 & \overline{\sigma}_\mu \\ \sigma_\mu & 0 \end{pmatrix} , \tag{4.95}$$

where $\overline{\sigma}_\mu = (1, -\boldsymbol{\sigma})$. This representation is called the Chiral representation. The conjugation matrix in this representation is given by

$$C = i\gamma_2\gamma_0 = \begin{pmatrix} i\sigma^2 & 0 \\ 0 & -i\sigma^2 \end{pmatrix} . \tag{4.96}$$

Thus, using Eq. (4.94), we obtain

$$\widehat{\psi} = \gamma_0 C \psi^* = e^{i\delta} \psi . \tag{4.97}$$

This is the same as in Eq. (4.15). Thus, the 2-component formalism makes it clear that the phase arbitrariness in the conjugation operation arises when the mass matrices are complex. This can happen in gauge theories if either the Yukawa couplings or the vacuum expectation values of the Higgs fields are complex. This will also provide the origin of CP-violation in the leptonic sector, when all phases present cannot be removed by redefinition of complex 2-component neutrino fields.

Exercise 4.10 *Show that the choice of γ_μ of Eq. (4.95) can be obtained from the Dirac representation through Eq. (4.38) through a suitable choice of the matrix U.*

4.7 Diagonalization of fermion mass matrices

To find the eigenvalues of the Hamiltonian operator in quantum mechanics, we diagonalize it. For this, we use the theorem of matrix algebra that if a matrix A is normal (i.e., it commutes with its hermitian conjugate) then there exists a unitary matrix U such that $UAU^\dagger$ is diagonal. Hamiltonian operator for any system is hermitian and hence normal.

There are two reasons why this method of unitary transformation is insufficient for diagonalization of fermion mass matrices. First, the mass matrices in gauge theory are not necessarily normal. Second, even if they turn out to be normal in a specific situation, a unitary transformation does not guarantee that the diagonal elements would be non-negative. The non-negativity is essential in order to interpret the diagonal elements as masses of physical particles.

To this end, we prove another theorem of matrix algebra. Obviously, $AA^\dagger$ is hermitian for any matrix A. Denote

$$AA^\dagger = H^2.$$
(4.98)

It can be diagonalized by a unitary transformation:

$$UAA^\dagger U^\dagger = D^2.$$
(4.99)

Writing the matrix elements explicitly, we see that $(D^2)_{aa} = \sum_b |(UA)_{ab}|^2$, which are real and non-negative. Define a diagonal matrix D whose elements are the positive square roots of those of D^2. Then $H \equiv U^\dagger DU$ is a hermitian matrix satisfying Eq. (4.98). Using Eq. (4.98), it is now trivial to show that $H^{-1}A$ is a unitary matrix which we will call U'. Then $A = HU' = U^\dagger DV$, where $V = UU'$ is a unitary matrix. Thus, $UAV^\dagger = D$, which shows that:

for any matrix A, one can find two unitary matrices U and V such that $UAV^\dagger$ is diagonal with non-negative elements.

In the above proof, we assumed that H is invertible. For a square matrix A, this implies by Eq. (4.98) $\det A \neq 0$, i.e., A is invertible. However, the result is valid even for singular matrices. Succinctly, we can say that any matrix can be non-negatively diagonalized by a *biunitary transformation*.

Exercise 4.11 *Show that, for any $n \times n$ matrix A which has n' number of zero eigenvalues, one can define two matrices Q_L and Q_R such that*

$$Q_L^\mathsf{T} A Q_R = \begin{pmatrix} \mathbf{0}_{n' \times n'} & \mathbf{0}_{n' \times n-n'} \\ \mathbf{0}_{n-n' \times n'} & \mathbf{A}'_{n-n' \times n-n'} \end{pmatrix} . \tag{4.100}$$

From this, find a proof that A can be diagonalized by a bi-unitary transformation even if it is singular.

In physical terms, a matrix makes sense when it connects two vectors. A biunitary transformation then means that these two vectors must be *rotated* by different amounts in order to obtain the eigenvalues. It makes sense when the two vectors sandwiching the matrix are different. But this is always the case for a fermion mass matrix, which connects left chiral fermions to right chiral ones. Thus, a biunitary transformation can always be applied. Of course, it is important to realize that U and V mentioned above are not unique. If $A = U^\dagger D V$, then A is also equal to $\widetilde{U}^\dagger D \widetilde{V}$, where $\widetilde{U} = KU$ and $\widetilde{V} = KV$ for some diagonal unitary matrix K. But this arbitrariness relates only to the overall phases of the physical fermions states and not to their mass eigenvalues.

One special case of our interest is that of symmetric matrices. From $A = U^\dagger D V$, we get $A^\mathsf{T} = V^\mathsf{T} D U^*$. If $A = A^\mathsf{T}$, we thus see that we can choose $V = U^*$. In other words, for any symmetric matrix A, there exists a unitary matrix U such that $U A U^\mathsf{T}$ is diagonal with non-negative elements.

The last statement must be used with some caution. Consider a case when a Majorana mass matrix is real for some reason like CP-conservation. A symmetric real matrix is hermitian. It is therefore tempting to say that U is just an orthogonal matrix, since real hermitian matrices can be diagonalized by orthogonal transformation.

The point to remember is that, if A is real hermitian, $O A O^\mathsf{T}$ is diagonal for some orthogonal matrix O, but the diagonal elements are not necessarily non-negative. However, suppose it has a negative element at the nth place, then we can define $U = KO$, where K is a diagonal matrix whose n^th diagonal entry is i and the others are 1. Then $U A U^\mathsf{T}$ is non-negative diagonal, but this U is not an orthogonal matrix. In the CP-conserving case, it can be shown that the entries in K^2 are proportional to the CP-eigenvalues of the mass eigenstates.

Part II

Models of neutrino mass

Chapter 5

Neutrino mass in $SU(2)_L \times U(1)_Y$ models

The standard model itself is based on the gauge group $SU(2)_L \times U(1)_Y$. But this fixes only the gauge bosons of the model. The fermions and Higgs contents have to be chosen somewhat arbitrarily. In the standard model, these choices are made in such a way that the neutrinos are massless, as discussed in Ch. 2. However, even with the same gauge group as the standard model, one can conjecture extra fermions or Higgs bosons in the model so that the model predicts massive neutrinos. In this chapter, we will discuss some such simple modifications of the standard model.

5.1 Models with enlarged fermion sector

One of the peculiarities of the standard model, as commented before, is that it contains left and right chiral projections of all fermions except the neutrinos. This looks almost contrived.

To remedy this, let us add right-handed neutral fields $N_{\ell R}$ corresponding to each charged lepton ℓ. Like the other right-handed fields, they are assumed to be $SU(2)_L$ singlets. The definition of electric charge in Eq. (2.8) then implies that they also have $Y = 0$. We can summarize these statements as:

$$N_{\ell R} \quad : \quad (1,0) \,. \tag{5.1}$$

Thus, the $N_{\ell R}$ fields are singlets of the entire gauge group. In other words, they have no interaction with the gauge bosons. However, they affect the model non-trivially because of their other properties.

5.1.1 A simple model with Dirac neutrinos

The first non-trivial thing to notice is that the presence of these right-handed fields imply new gauge-invariant interactions in the Yukawa sector:

$$-\mathcal{L}'_Y = \sum_{\ell,\ell'} f_{\ell\ell'} \overline{\psi}_{\ell L} \hat{\phi} N_{\ell'R} + \text{h.c.}, \qquad (5.2)$$

where $f_{\ell\ell'}$ are new coupling constants, and $\psi_{\ell L}$ is the lepton doublet of Eq. (2.10). The Higgs multiplet, ϕ, is the same as that appearing in the standard model. With a vev $v/\sqrt{2}$ as in Eq. (2.3), this gives rise to the following mass terms:

$$-\mathcal{L}_{\text{mass}} = \sum_{\ell,\ell'} f_{\ell\ell'} \frac{v}{\sqrt{2}} \overline{\nu}_{\ell L} N_{\ell'R} + \text{h.c.}. \qquad (5.3)$$

This is a mass term for neutrinos if we identify the fields $N_{\ell R}$ as the right-handed components of the neutrinos, i.e., $N_{\ell R} = \nu_{\ell R}$. In the flavor space, it corresponds to a matrix with elements

$$M_{\ell\ell'} = \frac{v}{\sqrt{2}} f_{\ell\ell'}. \qquad (5.4)$$

Such a matrix is called a *mass matrix*. In general, this matrix is not diagonal so that the fields $\nu_{\ell L}$, $N_{\ell R}$ do not correspond to the chiral projections of physical fermion fields.

To obtain the physical fields, one has to find the eigenvectors of the matrix M. In general, this can be done by diagonalizing M with a biunitary transformation:

$$U^\dagger M V = m, \qquad (5.5)$$

where m is diagonal matrix. Thus, defining new states by the relations

$$\begin{aligned} \nu_{\ell L} &\equiv \sum_\alpha U_{\ell\alpha} \nu_{\alpha L}, \\ N_{\ell R} &\equiv \sum_\alpha V_{\ell\alpha} \nu_{\alpha R}, \end{aligned} \qquad (5.6)$$

the terms in Eq. (5.3) can be rewritten as

$$-\mathcal{L}_{\text{mass}} = \sum_\alpha \overline{\nu}_{\alpha L} m_\alpha \nu_{\alpha R} + \text{h.c.}, \qquad (5.7)$$

where m_α is the αth diagonal element of the matrix m. This equation shows that the fields ν_α are fields with definite masses m_α and are therefore physical particles.

5.1.2   Neutrino mixing

The simple model above shows one of the major consequences of generic
neutrino mass terms. The charged current interaction between the W
bosons and the leptons is given by

$$\frac{g}{\sqrt{2}} \sum_{\ell} \bar{\ell}_{\mathsf{L}} \gamma^{\mu} \nu_{\ell\mathsf{L}} W_{\mu}^{-} + \text{h.c.}, \qquad (5.8)$$

where $\nu_{\ell\mathsf{L}}$ is the field which appears in the same $SU(2)_L$ doublet as the
charged lepton ℓ.

Using Eq. (5.6), we can rewrite this in terms of the physical fields ν_α
as

$$\frac{g}{\sqrt{2}} \sum_{\ell} \sum_{\alpha} \bar{\ell}_{\mathsf{L}} \gamma^{\mu} U_{\ell\alpha} \nu_{\alpha\mathsf{L}} W_{\mu}^{-} + \text{h.c.}. \qquad (5.9)$$

This shows that in general all neutrinos can have charged current interaction with a given lepton ℓ (unless, of course, some particular element
of U vanishes). This is in parallel to mixing in quark sector where mixing arises due to the matrix of Eq. (1.4). In the present case, this is
appropriately called *neutrino mixing*.

An immediate consequence of neutrino mixing is that the generational lepton numbers like L_e, L_μ and L_τ are no more good global symmetries even at the classical level. This is because any mass eigenstate
is a mixture of ν_e, ν_μ, ν_τ etc. At the classical level, the only global
symmetry remaining in the leptonic sector is the total lepton number
$L_e + L_\mu + L_\tau$. This feature would give rise to various flavor violating
processes, e.g., $\mu \to e\gamma$, $\mu + e \to \tau + e$ and so on. These comments are
true not just specifically for the present model, but for any model with
neutrino mixing.

5.1.3   Shortcomings of the model

The model described above is very simple in the sense that it treats
neutrino mass on exactly the same footing as the masses of other
known fermions. The neutrinos are Dirac particles just like all other
known fermions, which makes life easier because calculations with Dirac
fermions are quite familiar to us. However, the model does not answer
some questions.

First, the matrix $f_{\ell\ell'}$, defined in Eq. (5.2), is completely arbitrary. Without any knowledge of its elements, we have neither any idea of its eigenvalues, i.e., the neutrino masses, nor about the magnitudes of neutrino mixings. In the context of this model, these are then completely arbitrary parameters which should be determined by experiments, to which the model does not provide any guideline whatsoever.

Secondly, it also provides no answer to the question about the lightness of neutrinos. Of course, if the coupling constants $f_{\ell\ell'}$ are very small compared to the corresponding coupling constants which generate lepton or quark masses, the neutrinos will turn out to be lighter, in agreement with experimental bounds. But there is no good reason why $f_{\ell\ell'}$ must be small in this model.

In addition, the model is incomplete in a sense, which is discussed next. As we will see, this brings in new issues, absent in the simple model.

5.1.4 The complete model with right-handed neutrinos

So far, we have considered only one type of term involving the right-handed neutrinos, viz., the Yukawa interaction of Eq. (5.2). But there can be more. Since the $N_{\ell R}$ fields are invariant under $SU(2)_L \times U(1)_Y$, so must be their conjugate fields $\widehat{N}_{\ell L}$. Thus, one can form gauge invariant bare mass terms

$$-\mathcal{L}_{\text{bare}} = \frac{1}{2} \sum_{\ell,\ell'} B_{\ell\ell'} \overline{\widehat{N}}_{\ell L} N_{\ell' R} + \text{h.c.} . \qquad (5.10)$$

These terms must be present if we write down the most general gauge-invariant Lagrangian involving the particles in the model. Of course, such terms violate $B - L$, but $B - L$ is a global symmetry that appears automatically with the particle content of the standard model and we are not obliged to impose it in the extended models for any fundamental reason. If one chooses to impose $B - L$ anyway, these terms are untenable and we go back to the model with Dirac neutrinos discussed earlier.

In the general case, the bare mass terms in Eq. (5.10) are allowed and add to the contributions of Eq. (5.3). Using the identity

$$\overline{\nu}_{\ell L} N_{\ell' R} = \overline{\widehat{N}}_{\ell' L} \widehat{\nu}_{\ell R} , \qquad (5.11)$$

we can write all the mass terms in the Majorana basis introduced in §4.5. We obtain

$$-\mathcal{L}_{\text{mass}} = \frac{1}{2} \begin{pmatrix} \overline{\nu}_{\text{L}} & \overline{\widehat{N}_{\text{L}}} \end{pmatrix} \begin{pmatrix} 0 & M \\ M^{\textsf{T}} & B \end{pmatrix} \begin{pmatrix} \widehat{\nu}_{\text{R}} \\ N_{\text{R}} \end{pmatrix} + \text{h.c.}, \qquad (5.12)$$

where M and B are $\mathcal{N} \times \mathcal{N}$ matrices for $\mathcal{N}$ generations of fermions, and N_{R}, ν_{L} etc are $\mathcal{N}$ element column-vectors containing fields from $\mathcal{N}$ generations. Notice that the mass matrix is symmetric, as it should be according to the discussion of §4.5.

Upon diagonalization of the mass terms, one now obtains $2\mathcal{N}$ Majorana neutrinos in general. To see this, consider the simple case of a single generation, i.e., $\mathcal{N} = 1$. Then the mass matrix in Eq. (5.12) can be written as

$$\mathcal{M} = \begin{pmatrix} 0 & M \\ M & B \end{pmatrix} \qquad (5.13)$$

where now M and B are simply numbers. Moreover, for the sake of simplicity, let us assume that both M and B are real and $B > 0$. Now choose an orthogonal matrix

$$O = \begin{pmatrix} \cos\theta & -\sin\theta \\ \sin\theta & \cos\theta \end{pmatrix}, \qquad (5.14)$$

with

$$\tan 2\theta = 2M/B. \qquad (5.15)$$

Then

$$O\mathcal{M}O^{\textsf{T}} = \begin{pmatrix} -m_1 & 0 \\ 0 & m_2 \end{pmatrix}, \qquad (5.16)$$

a diagonal matrix, where

$$m_{1,2} = \frac{1}{2}\left(\sqrt{B^2 + 4M^2} \mp B\right). \qquad (5.17)$$

Since $m_{1,2} \geq 0$, we still have a little problem, viz, the elements of the diagonal matrix in Eq. (5.16) are not all non-negative and therefore cannot be interpreted as the masses of physical fields. So let us write

$$\begin{pmatrix} -m_1 & 0 \\ 0 & m_2 \end{pmatrix} = \begin{pmatrix} m_1 & 0 \\ 0 & m_2 \end{pmatrix} \begin{pmatrix} -1 & 0 \\ 0 & 1 \end{pmatrix} \equiv mK^2, \qquad (5.18)$$

where m contains the positive mass eigenvalues and K^2 is a diagonal matrix of positive and negative signs. Then

$$\mathcal{M} = O^{\mathsf{T}} m K^2 O. \tag{5.19}$$

Now, if we define the column vectors

$$\begin{pmatrix} n_{1L} \\ n_{2L} \end{pmatrix} \equiv O \begin{pmatrix} \nu_L \\ \widehat{N}_L \end{pmatrix} = \begin{pmatrix} \cos\theta & -\sin\theta \\ \sin\theta & \cos\theta \end{pmatrix} \begin{pmatrix} \nu_L \\ \widehat{N}_L \end{pmatrix} \tag{5.20}$$

and

$$\begin{pmatrix} n_{1R} \\ n_{2R} \end{pmatrix} \equiv K^2 O \begin{pmatrix} \widehat{\nu}_R \\ N_R \end{pmatrix} = \begin{pmatrix} -\cos\theta & \sin\theta \\ \sin\theta & \cos\theta \end{pmatrix} \begin{pmatrix} \widehat{\nu}_R \\ N_R \end{pmatrix}, \tag{5.21}$$

the mass terms reduce to

$$-\mathcal{L}_{\text{mass}} = m_1 \bar{n}_{1L} n_{1R} + m_2 \bar{n}_{2L} n_{2R} + \text{h.c..} \tag{5.22}$$

Thus, for a single generation, we obtain two eigenstates. Furthermore, from Eqs. (5.20) and (5.21), we see that

$$\begin{aligned} n_1 = n_{1L} + n_{1R} &= \cos\theta(\nu_L - \widehat{\nu}_R) - \sin\theta(\widehat{N}_L - N_R) \\ n_2 = n_{2L} + n_{2R} &= \sin\theta(\nu_L + \widehat{\nu}_R) + \cos\theta(\widehat{N}_L + N_R). \end{aligned} \tag{5.23}$$

Since $\widehat{\nu}_R$ is the conjugate of ν_L and $\widehat{N}_L$ of N_R, this immediately proves

$$n_1 = -\widehat{n}_1 \quad, \quad n_2 = \widehat{n}_2 \tag{5.24}$$

so that both n_1 and n_2 are Majorana particles. For $\mathcal{N}$ generations, one obtains $2\mathcal{N}$ Majorana particles in general.

Alternatively, one can also diagonalize by using the unitary matrix $U = KO$, with $K = \text{diag}\,(i \quad 1)$. In this case, we can write $\mathcal{M} = U^{\mathsf{T}} m U$ instead of Eq. (5.19). The definition of mass eigenstates in Eq. (5.20) and Eq. (5.21) change in this case, but the physical implications are the same.

Exercise 5.1 *Will Eq. (5.24) be valid if we take $U = KO$ as indicated above? If not, show that the CP eigenvalues of the neutrino states come out to be the same in either formulation.*

One redeeming aspect of this model is that it can provide a reason for the smallness of neutrino masses. To see this, refer to the one-generation model once again. In the mass matrix, the quantity M arises from ϕ-coupling and therefore it is natural to assume that it is of the same order as the masses of other fermions in the same generation. Let us suppose $B \gg M$. Then the eigenvalues in Eq. (5.17) reduce to

$$m_1 \simeq \frac{M^2}{B} \quad , \quad m_2 \simeq B \, . \tag{5.25}$$

Since $B \gg M$, it follows that $m_1 \ll M$, which means that the neutrino is much lighter than the charged fermions. Of course, there is another neutrino whose mass, B, is much larger than the charged fermion masses. This mechanism of making one particle light at the expense of making another one heavy is called the *see-saw mechanism*.

But one point must be noticed here. Cosmological arguments restrict the mass of any stable neutrino to be less than about 30 eV [see Ch. 17 for details]. Thus, if even ν_τ has to be lighter than 30 eV, Eq. (5.25) demands, with $M \sim m_\tau$,

$$B \gtrsim 10^8 \, \text{GeV} \, . \tag{5.26}$$

Thus, successful implementation of this model requires a mass scale much larger than the weak scale, which is of order 10^2 GeV. In general, models with such widely disparate scales suffer from the theoretical problem of maintaining the mass scales. The point is that even if the classical solution of the vacuum has two widely different scales, the quantum corrections tend to merge them. This problem, called the *hierarchy problem*, plagues grand unified theories necessarily, but there one achieves unification at the cost of hierarchy problem. In the present context, the hierarchy problem makes the model unattractive unless we embed it into a grand unified theory.

5.2 Models with expanded Higgs sector

If we do not add any new fermion to the particle content of the standard model, we have, in each generation, only two degrees of freedom corresponding to uncharged fermions, viz, $\nu_{\ell L}$ and $\widehat{\nu}_{\ell R}$. Thus, in whatever way they pick up mass, the mass must always be Majorana type, i.e., the mass terms must violate $B - L$. Thus, we are motivated to introduce new Higgs bosons which can violate $B - L$ symmetry in their interactions. Also, the neutrino masses must somehow be induced by the Yukawa couplings. With these two considerations in mind, we examine what are the fermionic bilinears that have a net $B - L$ quantum number.

For this, we can combine some lepton field with some antilepton field. The leptonic fields are given in Eq. (2.10). Their antiparticles are given by

$$
\begin{aligned}
\widehat{\psi}_{\mathsf{R}} &= \gamma_0 C \epsilon \psi_{\mathsf{L}}^* &:& \quad (2,1) \\
\widehat{\ell}_{\mathsf{L}} &= \gamma_0 C \ell_{\mathsf{R}}^* &:& \quad (1,2) \,,
\end{aligned}
\tag{5.27}
$$

where $\epsilon = i\tau_2$ is the 2×2 antisymmetric matrix. Note that in the first equation here, we have to take the conjugation with respect to two sets of indices. For the spinor index, this is done as in Eq. (4.10), and for the $SU(2)_L$ index as in Eq. (2.15). It is now easy to see that one can obtain two fermion bilinears which have a net $B - L$ quantum number:

$$
\begin{aligned}
\overline{\widehat{\psi}_{\mathsf{L}}} \widehat{\psi}_{\mathsf{R}} &\sim (2,1) \times (2,1) = (1,2) + (3,2) \,, \\
\overline{\widehat{\ell}_{\mathsf{L}}} \ell_{\mathsf{R}} &\sim (1,-2) \times (1,-2) = (1,-4) \,,
\end{aligned}
\tag{5.28}
$$

where in parentheses we have put the transformations under $SU(2)_L \times U(1)_Y$. The Higgs multiplets that can directly couple to these bilinears to form gauge invariant Yukawa couplings are as follows [1]:

- a triplet $\boldsymbol{\Delta}$: $(3,-2)$

- a singly charged singlet : $(1,-2)$

- a doubly charged singlet : $(1,4)$.

We now see how the introduction of these scalars, one at a time or in combinations, can give rise to neutrino masses.

5.2.1 Adding a triplet $\boldsymbol{\Delta}$

With the value of Y equal to -2, the electric charges of the components of the triplet $\boldsymbol{\Delta}$ are as follows

$$
\Delta = \begin{pmatrix} \Delta_0 \\ \Delta_- \\ \Delta_{--} \end{pmatrix} \,.
\tag{5.29}
$$

and we have an additional Yukawa coupling outside the ones in the standard model:

$$-\mathcal{L}'_Y = \sum_{\ell,\ell'} f_{\ell\ell'} \overline{\psi}_{\ell L} \frac{1}{\sqrt{2}} \tau \cdot \Delta \widehat{\psi}_{\ell' R} + \text{h.c.} \tag{5.30}$$

where τ denotes the Pauli matrices. Note that the fermion bilinear combination involving the Pauli matrices is just the symmetric (i.e., triplet) combination of the two doublets $\psi_{\ell L}$ and $\widehat{\psi}_{\ell' R}$, whose dot product with the triplet Δ gives an $SU(2)_L$ singlet interaction in Eq. (5.30).

The Higgs potential now involves both Δ and the usual doublet ϕ. Let us assume that the parameters in this potential are such that, at the minimum,

$$\langle \phi_0 \rangle = \frac{v_2}{\sqrt{2}} \quad , \quad \langle \Delta_0 \rangle = \frac{v_3}{\sqrt{2}}. \tag{5.31}$$

In that case, after symmetry breaking, Eq. (5.30) produces Majorana mass terms for neutrinos

$$-\mathcal{L}_{\text{mass}} = \sum_{\ell,\ell'} \overline{\nu}_{\ell L} M_{\ell\ell'} \widehat{\nu}_{\ell' R} + \text{h.c.} \tag{5.32}$$

where

$$M_{\ell\ell'} = \frac{v_3}{\sqrt{2}} f_{\ell\ell'}. \tag{5.33}$$

In Ch. 4, we proved that any mass matrix in the Majorana basis must be symmetric. To see that the statement is true here, we will have to show that $f_{\ell\ell'}$ is symmetric under the generation indices. To show this, we rewrite Eq. (5.30) in the following way, using the definition of the conjugate field from Eq. (4.27):

$$\sum_{\ell,\ell'} f_{\ell\ell'} \widehat{\psi}_{\ell R}^{\mathsf{T}} C^{-1} \epsilon \frac{1}{\sqrt{2}} \tau \cdot \Delta \widehat{\psi}_{\ell' R}. \tag{5.34}$$

Written this way, it is clear that it is a bilinear involving $\widehat{\psi}_R$ twice, and therefore it must obey Fermi statistics. The $SU(2)_L$ indices are contracted by the $\epsilon\tau$ matrix, which is symmetric. The spinor indices are contracted by the matrix C^{-1}, which is antisymmetric because C is antisymmetric, as shown in Eq. (4.26). The only other indices are the generation indices, and they must be symmetric in order that the entire combination is antisymmetric. Thus, $f_{\ell\ell'} = f_{\ell'\ell}$, which shows that the mass matrix is symmetric.

The diagonalization gives the mass eigenvalues and eigenstates. However, since the couplings $f_{\ell\ell'}$ are unknown, one cannot predict any pattern of neutrino mixing, just as in the model of §5.1.1. But, unlike that model, a different vev (viz, that of the triplet) is responsible for neutrino masses than for other fermion masses. Thus, one explanation for the lightness of the neutrinos could be forwarded here, viz, that the neutrinos could be light because $v_3 \ll v_2$.

Indeed, v_3 has to be somewhat smaller than v_2 from another consideration. Since Δ_0 couples to the W and the Z, its vev contributes to the masses of these gauge bosons. Instead of the standard model formulas, we now obtain

$$
\begin{aligned}
M_W^2 &= \frac{1}{4}g^2(v_2^2 + 2v_3^2) \\
M_Z^2 &= \frac{1}{4}(g^2 + g'^2)(v_2^2 + 4v_3^2)\,,
\end{aligned}
\tag{5.35}
$$

so that

$$
\rho \equiv \frac{M_W^2}{M_Z^2 \cos^2\theta_W} = \frac{1 + 2v_3^2/v_2^2}{1 + 4v_3^2/v_2^2}\,.
\tag{5.36}
$$

Using the experimental bounds on ρ from Eq. (2.34), we obtain

$$
\frac{v_3}{v_2} < 0.07\,.
\tag{5.37}
$$

To provide an explanation of why neutrino masses are at least three orders of magnitude smaller than charged fermion masses in the same generation, v_3 in fact must be quite a bit smaller than this upper limit.

Exercise 5.2 *Verify Eq. (5.35).*

Exercise 5.3 *In this model, identify the quadratic terms of the form $W_\mu^+ \partial^\mu H_-$ coming from the covariant derivative terms after symmetry breaking. This H_-, properly normalized, would then represent the unphysical Higgs boson eaten up by W_-. Show that*

$$
H_- = \frac{v_2\phi_- + \sqrt{2}v_3\Delta_-}{\sqrt{v_2^2 + 2v_3^2}}\,.
\tag{5.38}
$$

An important question that arises is that of the global quantum number $B - L$. It must be somehow broken in the model since Majorana neutrinos have resulted. The Yukawa coupling of Eq. (5.30) as such does not imply $B - L$ violation. Although the Higgs boson in this case couples to a fermion bilinear whose $B - L$ number equals 2, we can assign a $B - L$ value of -2 to Δ and then the Yukawa coupling is $B - L$ invariant. However, once this quantum number is carried by Δ, it is broken when Δ_0 acquires a vev. In fact, it breaks $B - L$ by 2 units, which is exactly what is necessary for the generation of Majorana mass terms.

5.2.2 Model with a singly charged singlet

This model was proposed by Zee [2]. He introduced an $SU(2)_L$ singlet particle h_- in the Higgs sector. The Yukawa coupling of this particle is

$$-\mathcal{L}'_Y = \sum_{\ell,\ell'} f_{\ell\ell'} \overline{\psi}_{\ell L} \widehat{\psi}_{\ell' R} h_- + \text{h.c.}. \tag{5.39}$$

The coupling constants $f_{\ell\ell'}$ must be antisymmetric because of Fermi statistics.

As in the previous model, we can assign a $B - L$ quantum number of -2 to the field h_-. Since h has electric charge, its vev must vanish in a physically acceptable ground state, because otherwise electromagnetic gauge symmetry would be spontaneously broken. So, in order to generate Majorana masses for the neutrinos, one must have some alternative sources of $B - L$ violation.

This can now only come from the Higgs potential, since kinetic energy terms and gauge interactions do not violate any $U(1)$ quantum number. In the Higgs potential, if apart from h_- we have just one doublet ϕ as in the standard model, it is impossible to write down any $B - L$ violating term. Consequently, neutrinos are massless just as in the standard model.

The situation changes if there are more than one doublets in the theory. For example, let us say we have a doublet ϕ' in addition to the standard model doublet ϕ. If ϕ' couples to quarks and leptons in the same way as ϕ does, it must have a vanishing $B - L$ quantum number. In that case, the trilinear coupling

$$\mu \phi^{\mathsf{T}} \epsilon \phi' h_- + \text{h.c.} \tag{5.40}$$

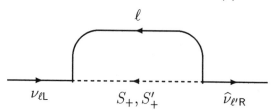

Figure 5.1: Self-energy diagrams giving neutrino masses in Zee's model.

violates $B - L$ by 2 units. Such a coupling cannot exist with just one doublet because $\phi^T \epsilon \phi$, being the antisymmetric combination, vanishes.

There are now three different singly charged scalars in the theory: ϕ_+, ϕ'_+ and h_+. At the symmetry breaking, some combination of the first two of these will be eaten up by the W^+ gauge boson. So the physical spectrum will contain two charged scalars S and S'. Each of these would be mixtures of h_+ and the ϕ_+, ϕ'_+. Therefore, these physical scalars are not eigenstates of $B - L$. Thus, neutrinos will obtain Majorana masses at the one-loop level from the self-energy diagrams, shown in Fig. 5.1, involving these physical scalars S and S'.

By a naive power counting, one might be tempted to say that the diagram in Fig. 5.1 is divergent, so that it gives an infinite contribution to neutrino masses. This is not true because of the renormalizability of the theory, which requires that all infinities arising in the calculation can be absorbed into the definitions of the parameters in the classical Lagrangian. In this case, the tree-level Lagrangian does not have any neutrino mass term. So, if Fig. 5.1 gives an infinite mass term, we cannot absorb it in any tree-level term. However, the renormalizability of the model is guaranteed by the general proof of 'tHooft. Therefore, Fig. 5.1 must give finite masses.

The mass terms are particularly simple if we assume that only one of the two Higgs doublets couples to leptons [3]. A straightforward calculation shows [see also §5.3]

$$M_{\ell\ell'} = A f_{\ell\ell'} (m_\ell^2 - m_{\ell'}^2) \tag{5.41}$$

where A is a constant. Since $f_{\ell\ell'} = -f_{\ell'\ell}$, as commented before, $M_{\ell\ell'}$ is symmetric like any mass matrix in the Majorana basis.

Unlike the models described before, this model gives a pattern of neutrino masses and mixings. To see that, let us define

$$\tan \alpha = \frac{f_{\mu\tau}}{f_{e\tau}}\left(1 - \frac{m_\mu^2}{m_\tau^2}\right)$$

$$\sigma = \frac{f_{e\mu}}{f_{e\tau}}\frac{m_\mu^2}{m_\tau^2}\cos \alpha. \tag{5.42}$$

Neglecting the electron mass which is tiny compared to the masses of the muon and the tau, the mass matrix of Eq. (5.41) can be rewritten as

$$M = m_0 \begin{pmatrix} 0 & \sigma & \cos \alpha \\ \sigma & 0 & \sin \alpha \\ \cos \alpha & \sin \alpha & 0 \end{pmatrix}. \tag{5.43}$$

where $m_0 = A m_\tau^2 f_{\tau e}/\cos \alpha$. We now assume that $\sigma \ll 1$, which seems plausible because it is proportional to m_μ^2/m_τ^2. Unless $f_{e\mu} \gtrsim 10^4 f_{e\tau}$, which there is no reason to assume, σ must be small. Then the diagonalization of the matrix M gives the eigenvalues

$$m_1 = -m_0 \sigma \sin 2\alpha$$
$$-m_2 = m_0\left(1 - \frac{1}{2}\sigma \sin 2\alpha\right)$$
$$m_3 = m_0\left(1 + \frac{1}{2}\sigma \sin 2\alpha\right) \tag{5.44}$$

correct upto first order terms in σ. To eliminate the negative eigenvalue $-m_2$ (and maybe also m_1, depending on the sign of $\sigma \sin 2\alpha$), we can again choose a matrix K described in §5.1.4 appropriately, so that m_2 is the mass of the physical eigenstate. Eq. (5.44) now shows that there must be two physical neutrinos whose masses, m_2 and m_3, are very close, whereas the other mass m_1 is much smaller.

5.2.3 Model with doubly charged singlet

The doubly charged singlet can have Yukawa couplings like

$$-\mathcal{L}_Y' = \sum_{\ell,\ell'} F_{\ell\ell'}\overline{\hat{\ell}}_L \ell_R' k_{++} + \text{h.c.} \tag{5.45}$$

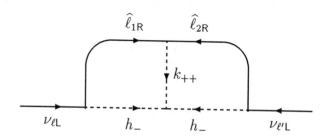

Figure 5.2: 2-loop diagram in Babu's model, giving rise to Majorana masses of the neutrinos.

This assigns a $B - L$ quantum number of 2 to the field k_{++}. In order to generate Majorana masses for the neutrinos, we need $B - L$ violation. This is not possible with any number of doublet Higgses, since there can be no trilinear coupling like the one involving the singly charged Higgs bosons, as shown in Eq. (5.40). Thus, neutrinos are massless in all orders of perturbation theory [1].

The situation changes, as was pointed out by Babu [4], if the model contains the singly charged scalar h_- of §5.2.2 alongwith the k_{++}. Due to their Yukawa couplings, both h_+ and k_{++} carry 2 units of $B - L$ quantum number. With just one doublet ϕ, the trilinear $\phi\phi h$ coupling vanishes, as discussed in the previous section. However, there is one trilinear coupling, viz,

$$\mu h_- h_- k_{++} + \text{h.c.} \tag{5.46}$$

which breaks $B - L$. Thus, neutrino masses can be generated from 2-loop diagrams as shown in Fig. 5.2. It gives a mass matrix of the form

$$m_{\ell\ell'} = 8\mu \sum_{\ell_1,\ell_2} m_{\ell_1} m_{\ell_2} f_{\ell\ell_1} F_{\ell_1\ell_2} f_{\ell'\ell_2} I_{\ell_1\ell_2} \tag{5.47}$$

where

$$I_{\ell_1\ell_2} = \int \frac{d^4p}{(2\pi)^4} \int \frac{d^4q}{(2\pi)^4} \frac{1}{p^2 - m_{\ell_1}^2} \frac{1}{q^2 - m_{\ell_2}^2}$$
$$\frac{1}{p^2 - m_h^2} \frac{1}{q^2 - m_h^2} \frac{1}{(p-q)^2 - m_k^2}. \tag{5.48}$$

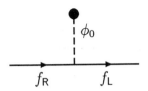

Figure 5.3: A graphic representation of how fermion masses arise.

Notice that the eigenvalues of m are naturally small here because of the suppression factors that appear in the 2-loop integration. This provides a rationale for the smallness of neutrino masses.

Another interesting prediction of this model is that, since the coupling matrix f is antisymmetric, the determinant of the matrix m vanishes if there are three generations. Thus, there must be one zero eigenvalue.

This does not exactly mean that there will be a massless neutrino, because the expression for $m_{\ell\ell'}$ has corrections coming from higher loops. Such corrections should be smaller than the 2-loop contribution in the perturbative regime. Nevertheless, they will modify the eigenvalues by small amounts. But in any case, we expect one eigenvalue to be much smaller than the other two.

5.3 The method of flavor diagrams

In calculations for obtaining physical quantities like the S-matrix elements, it is most convenient to use Feynman diagrams, where the external lines correspond to physical particles. However, in order to understand questions regarding mass and symmetry breaking, it is often beneficial to use another basis, which we call the *flavor basis*. The method and its advantages will be clear from the examples below.

In this method, we draw Feynman-like diagrams using the fields that appear in the gauge-invariant Lagrangian. For the fermions, it means that we talk of the left- and right-chiral fields separately. For the Higgs field ϕ, we use the original classical field which appears, say, in Eq. (2.2), and not the quantum field that is obtained by the fluctuation of this classical field from its vev. Since the charged fermions are mass-

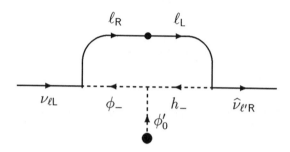

Figure 5.4: The flavor diagram for neutrino masses in Zee's model.

less before symmetry breaking, the generation of mass after symmetry breaking can be represented as in Fig. 5.3. The blob at the end of the ϕ_0 line represents its annihilation into the vacuum, which manifests the fact that ϕ_0 has a vev. The figure now is a graphical illustration of the fact that a charged fermion mass is given by its Yukawa coupling with ϕ, multiplied by the vev of ϕ_0, as expressed algebraically in Eq. (2.17).

If neutrinos are Dirac particles and obtain their mass through their coupling with ϕ as in the model of §5.1.1, then neutrino masses are also represented by Fig. 5.3, and we get no new insight. However, if the neutrinos get Majorana masses as in Zee's model described in §5.2.2, we can see the power of the method of flavor diagrams by trying to derive the neutrino mass matrix. In this case, we should draw, instead of Fig. 5.1, the neutrino mass diagram as in Fig. 5.4. Notice that here we use the neutrino states $\nu_{\ell L}$ and $\widehat{\nu}_{\ell R}$ which couple to one flavor of charged lepton via charged current interactions. The blob on the internal fermion line is a shorthand for Fig. 5.3 representing the mass of the charged lepton. The introduction of the mass is necessary on the internal fermion line to flip the helicity of $\widehat{\ell}'$, without which we cannot obtain a right-handed particle on the right end of the diagram.

For the sake of simplicity, let us assume that the Yukawa couplings of ϕ' are much smaller than those of ϕ, so that charged lepton masses come mainly through their coupling with ϕ. In this case, just a cursory look at Fig. 5.4 tells us that its contribution to $m_{\ell\ell'}$ must have the following features:

1. it has a factor of $f_{\ell\ell'}$ from the vertex with h_-;

2. it has a factor of m_ℓ from the mass insertion;

3. it has a factor of v' from the vev of ϕ_0';

4. it has a factor of the Yukawa coupling of ℓ with ϕ, which is m_ℓ/v.

Taking all these factors together, we obtain a contribution of $f_{\ell\ell'}m_\ell^2$, apart from factors which do not depend on fermion generations. Similarly, there would be one flavor diagram where the ϕ_- attaches to the right end of the fermion line, and the h_- line attaches to the left end. This gives a contribution proportional to $f_{\ell'\ell}m_{\ell'}^2$. Using the antisymmetry of f, we now obtain the form of the mass matrix given in Eq. (5.41), which shows the power of the method.

The flavor diagrams are also useful for examining the breaking of global quantum numbers. For example, if we compare Fig. 5.1 and Fig. 5.4, it is much easier to tell from the latter that $B − L$ is violated at the $h\phi\phi'$ vertex. Because of this, we use this method extensively in the next section and later in the book.

Exercise 5.4 *By drawing flavor diagrams, show that the neutrino mass matrix in the model of §5.2.3 contains two factors of charged lepton masses, as shown in Eq. (5.47). Verify it by evaluating the Feynman diagrams for neutrino masses.*

5.4   Models with spontaneous $B − L$ violation

In all but one models of neutrino mass discussed in this chapter, the mass is of Majorana type, arising out of $B − L$ violation. One might therefore wonder as to how exactly these models violate $B − L$. There are two ways to break global $B − L$ symmetry: i) by including explicit $B − L$ breaking terms in the Lagrangian; ii) by making the vacuum non-invariant under $B − L$ symmetry. The latter case, which was proposed by Chikashige, Mohapatra and Peccei (CMP) in 1980, implies the existence of a massless pseudo-scalar particle called the Majoron which we will denote by the symbol J. Even though a priori the existence of the Majoron would imply long range forces between particles to which it couples, it was pointed out [5] that only spin dependent forces between matter arise in the non-relativistic limit from Majoron exchange.

The laboratory limits on such forces are very weak, implying that the scale of $B - L$ breaking could be low. This is very interesting because, new physics associated with $B - L$ breaking could then be accessible to ongoing experiments. Low energy manifestations of the Majoron of course depends on its transformation property under the weak $SU(2)_L$ group. Three classes of Majoron models have been discussed in the literature. The original Majoron of CMP transforms like an isosinglet and is dubbed the singlet Majoron. Subsequently, a Majoron model was proposed by Gelmini and Roncadelli [6] using the triplet Higgs field. This is known as the triplet Majoron. Several models also have been constructed where the Majoron has doublet transformation property [7, 8]. Below we discuss the present status of these models.

5.4.1   Constraints on Majoron models

The phenomenological constraints on Majoron models derive from known upper bounds of flavor changing currents in the leptonic sector. For example, in a model with Majoron, one expects the muon decay mode

$$\mu \to e + J \qquad (5.49)$$

at some level. Experimental results on this mode is [9]

$$\Gamma(\mu \to e + J) < 3 \times 10^{-4} . \qquad (5.50)$$

There are similar bounds from τ-decays but they are in general much weaker. Another class of bounds come from the absence of Majoron bremsstrahlung from final states particles in physical processes. For example, analysis of leptonic decays of kaons and pions [10] imply that the Majoron coupling to neutrinos must be smaller than about 10^{-2} or so.

In practice, the strongest constraints on Majoron models come from astrophysical considerations. The point is that Majorons can be produced in stars through the process

$$\gamma + e \to e + J , \qquad (5.51)$$

and they can come out of the star, carrying some energy out. This thus contributes to the energy loss mechanism of stars and one should be able to put bounds on the process from known values of stellar luminosities.

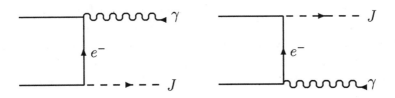

Figure 5.5: Photoproduction of Majoron.

Here we give a very simple derivation of the quantitative bound. The process in Eq. (5.51) occurs, at the tree level, through the diagrams in Fig. 5.5. For stellar core temperatures satisfying $T \ll m_e$, the cross-section of this process is easily seen to be

$$\sigma \simeq \frac{(eg_{eJ})^2}{12\pi} \frac{\omega^2}{m_e^4}, \tag{5.52}$$

where ω is the incident photon energy and g_{eJ} is the coupling of the electron with the Majoron J. The number of occurrences of the photoproduction reaction per unit volume per unit time is given by $n_e n_\gamma \sigma$, where n_e and n_γ are the number densities of the electron and the photon respectively. As a rough order-of-magnitude estimate, we can put $\omega \sim T$ and $n_\gamma \sim T^3$. Since at each occurrence, we obtain a Majoron of energy ω' of order T, the amount of energy going into Majorons per unit time per unit mass, L_{majo}, is given by

$$L_{\text{majo}} \simeq \frac{n_e n_\gamma \sigma \omega'}{\rho} \simeq \frac{\alpha g_{eJ}^2 n_e T^6}{3\rho m_e^4}, \tag{5.53}$$

where ρ is the mass per unit volume. For a rough estimate, we can neglect the contribution of neutrons to the mass and write the proton contribution as $\rho \simeq m_p n_p$. For $T \ll m_e$ which is the case of present interest, no positrons are present. Therefore, the proton number density n_p must equal n_e since the stellar core is electrically neutral. Thus we finally obtain

$$L_{\text{majo}} \simeq \frac{\alpha g_{eJ}^2 T^6}{3m_e^4 m_p}. \tag{5.54}$$

Let us assume that all this energy is coming out of the star, the Majoron being so weakly interacting that its mean free path is larger than the stellar radius. Then the stellar energy loss per unit time per unit mass because of Majoron emission is given by the last expression. For the sun, for example, the core temperature is $T \simeq 15.7 \times 10^6$ K. In the standard model of the sun, the energy loss rate per unit mass at the center is calculated [11] to be $17.5 \, \mathrm{erg \, s^{-1} \, g^{-1}}$. Since the standard model works quite well, we can quite conservatively demand that the energy loss due to any new mechanism like Majoron emission is smaller than the above amount. This gives the bound [12]

$$g_{eJ} \lesssim 10^{-10} . \tag{5.55}$$

For red-giant stars which have higher core temperature, one obtains a more stringent bound,

$$g_{eJ} \lesssim 10^{-12} , \tag{5.56}$$

although it is less certain than the solar bound since the dynamics of the red giants is not as well understood as that of the sun. Observation of neutrinos from the supernova SN1987A puts bounds on Majoron coupling to neutrinos as well [13]. These bounds severely constrain many Majoron models, some examples of which follow.

5.4.2 Majoron in the model with right-handed neutrinos

We start with the complete model with right-handed neutrinos, described in §5.1.4. In the model, the couplings of the doublet ϕ conserves $B - L$, but the bare mass terms of Eq. (5.10) break $B - L$ by two units and give rise to Majorana neutrinos. If $B - L$ breaks that way through the explicit occurrence of the bare mass term, we do not expect any Majoron. However, consider an alternative situation where the model contains a gauge singlet, spin-0 field S having the following Yukawa coupling:

$$-\mathcal{L}'_Y = \sum_{\ell,\ell'} b_{\ell\ell'} \overline{\widehat{N}_{\ell L}} N_{\ell' R} S + \mathrm{h.c.} . \tag{5.57}$$

If S develops vev, neutrinos obtain mass just as in §5.1.4, where

$$B_{\ell\ell'} = b_{\ell\ell'} \langle S \rangle . \tag{5.58}$$

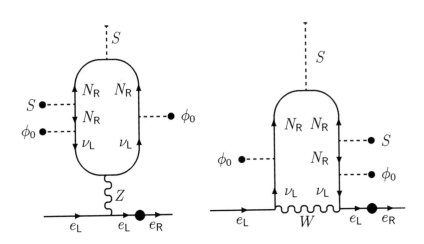

Figure 5.6: Majoron coupling with charged fermions in the singlet Majoron model.

The difference with the earlier model is that here all terms involving fermions conserve $B - L$ provided we assign 2 units of $B - L$ quantum number to S. In the scalar potential also, let us forbid any term that violates $B - L$. Then $B - L$ would be a global symmetry of the entire Lagrangian. But since the scalar S carries this quantum number, $B - L$ will be broken spontaneously if S develops a vev. As a result, we will obtain a Majoron J. In the simple case when all vevs are real, J would simply be the imaginary part of the complex field S. At one-loop level, couplings are induced through the diagrams of Fig. 5.6. From these flavor diagrams, we can identify the following characteristics of the effective coupling g_{eJ}:

- it must involve a factor of G_F from the propagator and the couplings of the W or the Z;

- it must be proportional to m_e which is necessary to get a helicity flip on the electron line;

- it should involve two powers of M from the neutrino coupling to ϕ and the associated ϕ vevs.

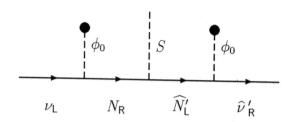

Figure 5.7: Flavor diagram for neutrino decay via a Majoron emission.

Since the heavy mass in the loop is of order B, it provides the damping scale, so that we can write

$$g_{eJ} \sim \frac{1}{16\pi^2} G_F m_e \frac{M^2}{B} \sim \frac{1}{16\pi^2} G_F m_e m_\nu, \qquad (5.59)$$

which easily satisfies the astrophysical bound of Eq. (5.56) for any acceptable value of neutrino masses. With such small effective couplings with charged fermions, rates for $\mu \to e + J$ etc are also very slow.

However, as we discussed before, the model is unattractive because of the presence of hierarchy of mass scales provided the neutrinos are stable. One can, however, question the stability of neutrinos in this model, since all but the lightest of the neutrinos can decay through a Majoron emission:

$$\nu \to \nu' + J. \qquad (5.60)$$

Naively, from the flavor diagram of Fig. 5.7 one might expect that the amplitude for this process is proportional to two powers of the mixing between the light ν and the heavy N states, i.e., goes as $(M/B)^2$. A detailed analysis shows that there is an accidental cancellation of the $(M/B)^2$ term [14]. This cancellation persists even if one considers loop effects [15]. The leading non-vanishing term in the amplitude [16, 15] goes like $(M/B)^4$. This makes the process so slow that the lifetime, for all allowed values of ν_μ and ν_τ masses, exceeds the age of the universe by many orders of magnitude. Thus, in the time-scale of the age of the universe, the light neutrinos are indeed stable in this model and the model has the hierarchy problem built in it.

5.4.3 Majorons in models with extended Higgs sector

If neutrinos obtain mass through their coupling with a triplet Higgs Δ, as discussed in §5.2.1, the triplet must carry 2 units of $B - L$ quantum number and develop a vev at the minimum of the scalar potential. In that case, $B - L$ is spontaneously broken if it were a symmetry of the classical Lagrangian.

The latter condition is not satisfied if we insist on writing down the full gauge-invariant potential involving ϕ and Δ, which contains a term

$$A\phi^{\mathsf{T}} \epsilon \tau \cdot \Delta \phi + \text{h.c.} \tag{5.61}$$

that breaks $B-L$ explicitly. However, if we forbid this term by imposing $B-L$, then v_3, the vev of Δ_0, produces a Majoron J. This variant of the triplet model, proposed by Gelmini and Roncadelli [6], has interesting consequences, since the Majoron can mediate ν-ν scattering at tree level and therefore gives a large cross section.

In the model with h_- of §5.2.2, the trilinear coupling $\phi^{\mathsf{T}} \epsilon \phi' h_-$ of Eq. (5.40) breaks $B - L$ only if we assume that ϕ' is neutral under $B - L$ just as ϕ is. However, one can also contemplate that, since h_- carries 2 units of $B - L$, ϕ' carries -2 units and the trilinear coupling conserves $B-L$. One then will have to forbid other possible $B-L$ violating terms like $(\phi'^{\dagger}\phi)(\phi'^{\dagger}\phi)$ in the scalar potential. If that is done, then the vev v' of ϕ'_0 breaks $B - L$ spontaneously and a Majoron results [8].

Both these models fall in a general class in the sense that in both models, the Majoron carries non-trivial quantum numbers of the gauge group $SU(2)_L \times U(1)_Y$. This fact has interesting consequences, some of which rule out such models with present experimental data.

To show that, let us specialize to the model of Gelmini and Roncadelli. In this model, the vev of the field Δ_0 breaks $B - L$ spontaneously. But the same vev also contributes to the mass of the Z, so the unphysical Higgs eaten up by Z is a linear superposition of Δ_0 and ϕ_0, the latter being the component of the usual doublet. In fact, the combination eaten up by Z is

$$\frac{v_2 \text{Im}\, \phi_0 + 2v_3 \text{Im}\, \Delta_0}{\sqrt{v_2^2 + 4v_3^2}}, \tag{5.62}$$

assuming v_2 and v_3 to be real. The Majoron must be an orthogonal combination, and therefore is given by

$$J = \frac{2v_3 \text{Im}\,\phi_0 - v_2 \text{Im}\,\Delta_0}{\sqrt{v_2^2 + 4v_3^2}} \simeq -\text{Im}\,\Delta_0 + \frac{2v_3}{v_2}\text{Im}\,\phi_0, \qquad (5.63)$$

where in the last step we have used the fact that $v_3/v_2 \ll 1$, as argued in Eq. (5.37). Thus J contains a small part of ϕ_0 and therefore couples directly to charged fermions. Since the ϕ-coupling to the electron is $\sqrt{2}m_e/v_2$, the coupling of J is

$$g_{eJ} = \frac{\sqrt{2}m_e}{v_2} \cdot \frac{2v_3}{v_2}. \qquad (5.64)$$

The astrophysical bound of Eq. (5.56) then gives

$$v_3/v_2 \lesssim 10^{-6}, \qquad (5.65)$$

which gives $v_3 \lesssim 20\,\text{keV}$.

With such a small vev, the mass of the field containing mostly $\text{Re}\,\Delta_0$ would be naturally of order v_3. In that case, Z can easily decay into the Majoron and $\text{Re}\,\Delta_0$. This decay mode contributes twice as much to the total width of Z as a pair of neutrino and antineutrino does [17]. For the doublet Majoron model, the same argument can be used to show that $v'/v \lesssim 10^{-6}$. Here, Z can decay to the Majoron and the $\text{Re}\,\phi_0'$, and the rate is half of that of a neutrino antineutrino pair. Measured widths of the Z-boson does not have any room for these particles. So, both these models, although interesting at the time they were proposed, have been ruled out by now. One can, of course, make more complicated viable Majoron models where the Majoron is a mixture of, say, a doublet and a singlet.

Exercise 5.5 *Calculate the decay width of the Z boson decaying to the Majoron and the real part of the same multiplet when the Majoron transforms like a n-dimensional representation of $SU(2)_L$. Compare this with the rate of $Z \to \nu\hat{\nu}$ derived in Exercise 2.4.*

In contrast, if we want to break $B - L$ spontaneously in the model involving h_- and k_{++}, we can introduce a gauge singlet Higgs field s whose important interaction is a quartic coupling

$$\lambda h_- h_- k_{++} s + \text{h.c.}, \qquad (5.66)$$

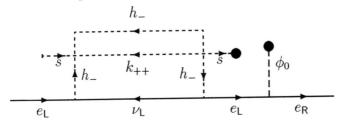

Figure 5.8: Majoron coupling with electron in no-hierarchy Majoron model.

which assigns a $B − L$ number 2 to s. The Im s appears as the Majoron when Re s develops a vev [18]. The implication for neutrino mass is the same as in Babu's model of §5.2.3 if we identify

$$\mu = \lambda \langle s \rangle , \tag{5.67}$$

where μ is the coupling constant defined in Eq. (5.46), associated with the trilinear scalar interaction of Babu's model. Thus, neutrino masses are small because they arise at 2-loop level. On the other hand, in absence of right-handed neutrinos in the model, s does not couple to any fermions at tree level. The coupling to electrons is generated at the 2-loop level through the diagram in Fig. 5.8 which can easily satisfy the astrophysical bound even if $\langle s \rangle$ is of the same order as the weak scale. Thus, this model does not suffer from any hierarchy problem.

Chapter 6

Neutrino mass in Left-Right symmetric models

An attractive extension of the standard model which may manifest itself in the multi-TeV (or perhaps higher) range of energies is the left-right symmetric model of weak interaction. As the name implies, in this model the left and right chiralities of fermions are assumed to play an identical role prior to symmetry breaking (or at high energies above all symmetry breaking scales). It therefore follows that in the symmetric phase of this model, weak interactions conserve parity [1], a property already shared by electromagnetic, strong and gravitational interactions. This is perhaps closer to the spirit of unified theories than the standard model.

Left-right symmetric treatment of weak interactions requires that all left handed fermions must have a right handed partner in the spectrum. Thus, an immediate consequence of these models is the existence of a new particle, the right handed neutrino, denoted by ν_R or sometimes by N_R. This model therefore obeys complete quark-lepton symmetry in its spectrum and automatically leads to massive neutrinos.

The smallest gauge group that implements the hypothesis of left-right symmetry of weak interactions is $SU(2)_L \times SU(2)_R \times U(1)_{B-L}$. In the original versions of the model [1], the $U(1)$ symmetry was not identified with $B - L$. This point was realized subsequently [2], when it was pointed out that the electric charge formula for these models acquires a rather attractive form:

$$Q = I_{3L} + I_{3R} + \frac{B - L}{2}, \tag{6.1}$$

which looks quite similar to the Gell-Mann Nishijima relation for electric charge in the case of strong interactions. A number of very important implications of parity symmetry breaking can be deduced from this equation [3]. For instance, at an energy scale where $SU(2)_L$ is unbroken but $SU(2)_R$ is broken, one finds from Eq. (6.1) that

$$\Delta I_{3R} = -\frac{1}{2}\Delta(B - L).$$ (6.2)

This equation, for purely baryon number conserving interactions (i.e., $\Delta B = 0$), implies that $|\Delta L| = 2|\Delta I_{3R}|$. If symmetry breaking is chosen so as to give $\Delta I_{3R} = 1$, we find $\Delta L = 2$, which implies Majorana neutrinos and neutrinoless double beta decays. Even if $\Delta I_{3R} = \frac{1}{2}$, Majorana neutrinos can arise in higher orders. Similarly, for purely lepton number conserving interactions, this implies violation of baryon number by two units, leading to processes such as neutron-antineutron oscillation [3, 4]. Our concern in this chapter will be neutrino masses in left-right symmetric models.

6.1 The gauge sector

6.1.1 Symmetry breaking

In the left-right symmetric models, the quarks and leptons are assigned to the following irreducible representations of the gauge group $SU(2)_L \times SU(2)_R \times U(1)_{B-L}$:

$$q_{\mathsf{L}} = \begin{pmatrix} u_{\mathsf{L}} \\ d_{\mathsf{L}} \end{pmatrix} : (2, 1, 1/3), \qquad q_{\mathsf{R}} = \begin{pmatrix} u_{\mathsf{R}} \\ d_{\mathsf{R}} \end{pmatrix} : (1, 2, 1/3)$$

$$\psi_{\mathsf{L}} = \begin{pmatrix} \nu_{e\mathsf{L}} \\ e_{\mathsf{L}} \end{pmatrix} : (2, 1, -1), \qquad \psi_{\mathsf{R}} = \begin{pmatrix} N_{e\mathsf{R}} \\ e_{\mathsf{R}} \end{pmatrix} : (1, 2, -1).$$ (6.3)

As in §5.1, we write the right handed neutrino with a different symbol than the left handed one to keep open the possibility that they might end up being parts of two different Majorana particles.

The gauge invariant Lagrangian for the quarks and leptons leads to the following gauge boson interactions with fermions:

$$
\begin{aligned}
\mathcal{L}_{\text{gauge}} &= g_L \left[\overline{q}_{\mathsf{L}} \gamma_\mu \frac{\boldsymbol{\tau}}{2} q_{\mathsf{L}} + \overline{\psi}_{\mathsf{L}} \gamma_\mu \frac{\boldsymbol{\tau}}{2} \psi_{\mathsf{L}} \right] \cdot \mathbf{W}_L^\mu \\
&\quad + g_R \left[\overline{q}_{\mathsf{R}} \gamma_\mu \frac{\boldsymbol{\tau}}{2} q_{\mathsf{R}} + \overline{\psi}_{\mathsf{R}} \gamma_\mu \frac{\boldsymbol{\tau}}{2} \psi_{\mathsf{R}} \right] \cdot \mathbf{W}_R^\mu \\
&\quad + g' \left[\frac{1}{6} \overline{q} \gamma_\mu q - \frac{1}{2} \overline{\psi} \gamma_\mu \psi \right] B^\mu
\end{aligned}
$$ (6.4)

where $\mathbf{W}_L^\mu$, $\mathbf{W}_R^\mu$ and B^μ are the gauge bosons corresponding to the groups $SU(2)_L$, $SU(2)_R$ and $U(1)_{B-L}$ respectively, whereas g_L, g_R and g' are the corresponding gauge coupling constants. We now require that the theory be invariant under parity operation $\mathcal{P}$ under which various fields transform as follows:

$$\psi_L \leftrightarrow \psi_R, \quad q_L \leftrightarrow q_R, \quad \mathbf{W}_L \leftrightarrow \mathbf{W}_R. \tag{6.5}$$

This requires $g_L = g_R = g$, reducing the number of arbitrary gauge coupling constants to two as in the standard model. As in the standard model, we can parameterize g and g' in terms of two parameters: the electric charge of the positron, e and the Weinberg angle, θ_W, defined by

$$g = e/\sin\theta_W, \tag{6.6}$$

which then leads to

$$g' = \frac{e}{\sqrt{\cos 2\theta_W}}. \tag{6.7}$$

Exercise 6.1 *For any gauge theory, if the electric charge operator is given by the combination $Q = \sum_n a_n T_n$, where T_n are generators and a_n are numerical coefficients, show that the electric charge is given by [5]*

$$\frac{1}{e^2} = \sum_n \frac{a_n^2}{g_n^2}, \tag{6.8}$$

where g_n is the coupling constant accompanying T_n. Use this equation to derive Eq. (6.7) from Eqs. (6.1) and (6.6).

Let us now consider the breaking of this gauge symmetry. In order to maintain left-right symmetry, we must choose Higgs multiplets which are left-right symmetric. The first candidate for this purpose seems to be

$$\Phi = \begin{pmatrix} \phi^0 & \phi'^+ \\ \phi^- & \phi'^0 \end{pmatrix} \quad : \quad (2, 2, 0), \tag{6.9}$$

which can couple to the fermion bilinears $\bar{q}_L q_R$ and $\bar{\psi}_L \psi_R$. Thus, after symmetry breaking, non-zero vevs of the electrically neutral components of Φ,

$$\langle \Phi \rangle = \begin{pmatrix} \kappa & 0 \\ 0 & \kappa' \end{pmatrix}, \tag{6.10}$$

can give masses to quarks and leptons. However, these vevs are not
sufficient to break the gauge symmetry. Since Φ is neutral under $B-L$,
$U(1)_{B-L}$ is not broken by the vevs of Φ. Moreover, from Eq. (6.1), we
see that the electrically neutral components of Φ have $I_{3L} + I_{3R} = 0$.
Thus, the gauge symmetry is broken to $U(1)_{I_{3L}+I_{3R}} \times U(1)_{B-L}$, and not
to $U(1)_Q$ as observed in the real world. One must then introduce more
Higgs multiplets to obtain the desired symmetry breaking pattern.

There are various ways to achieve this goal. In the early days of
the development of the left-right symmetry, the breaking of the gauge
symmetry [1, 6] was implemented by choosing the Higgs multiplets

$$\chi_L : (2,1,1) \quad , \quad \chi_R : (1,2,1) . \tag{6.11}$$

Later, it was shown that in order to understand the smallness of neutrino
masses, it is preferable to introduce the following set of fields [7]:

$$\Delta_L : (3,1,2) \quad , \quad \Delta_R : (1,3,2) . \tag{6.12}$$

The choice in Eq. (6.11) implies Dirac neutrinos whereas the choice in
Eq. (6.12) leads to Majorana neutrinos. We first discuss the Majorana
neutrino alternative and comment later on the Dirac neutrino case.

The gauge symmetry breaking proceeds in two stages. In the first
stage, the electrically neutral component of Δ_R, denoted by Δ_R^0, acquires
a vev v_R and breaks the gauge symmetry down to $SU(2)_L \times U(1)_Y$ where

$$\frac{Y}{2} = I_{3R} + \frac{B-L}{2} . \tag{6.13}$$

The parity symmetry breaks down at this stage. In the second stage,
the vevs of the electrically neutral components of Φ break the symmetry
down to $U(1)_Q$. Experimental constraints force the relation that $\kappa, \kappa' \ll$
v_R, as we will see later. Here we take κ and κ' real for simplicity. Making
them complex leads to an interesting model of CP-violation.

At the first stage, the charged right handed gauge bosons denoted by
$W_R^\pm$ and a neutral gauge boson called Z' acquire masses proportional to
v_R and become much heavier than the usual left handed $W_L^\pm$ and the Z
bosons which pick up masses proportional to κ and κ' only at the second
stage. In general the different gauge bosons mix and lead to a 2×2 mass
matrix describing the W_L-W_R system and a 3×3 mass matrix describing

the neutral gauge bosons W_{3L}, W_{3R} and B. The charged gauge boson mass matrix turns out to be

$$
\begin{pmatrix} \frac{1}{2}g^2(\kappa^2 + \kappa'^2 + 2v_L^2) & g^2\kappa\kappa' \\ g^2\kappa\kappa' & \frac{1}{2}g^2(\kappa^2 + \kappa'^2 + 2v_R^2) \end{pmatrix}. \tag{6.14}
$$

Here v_L denotes the vev of Δ_L^0 which is assumed to be much smaller than κ, κ'. The eigenstates of this matrix are

$$
\begin{aligned} W_1 &= W_L \cos\zeta + W_R \sin\zeta \\ W_2 &= -W_L \sin\zeta + W_R \cos\zeta, \end{aligned} \tag{6.15}
$$

where

$$
\tan 2\zeta = \frac{2\kappa\kappa'}{v_R^2 - v_L^2}. \tag{6.16}
$$

In what follows, we will assume that $\kappa' \ll \kappa$, although strictly speaking it is not required from phenomenological considerations. With this assumption, ζ is small, i.e., the physical charged gauge bosons W_1 and W_2 are the same as W_L and W_R to a good approximation. The masses of these gauge bosons can be obtained from the relations

$$
\begin{aligned} M_{W_L}^2 &\simeq M_{W_1}^2 \cos^2\zeta + M_{W_2}^2 \sin^2\zeta \\ M_{W_R}^2 &\simeq M_{W_1}^2 \sin^2\zeta + M_{W_2}^2 \cos^2\zeta, \end{aligned} \tag{6.17}
$$

where $M_{W_L}^2$ and $M_{W_R}^2$ are the diagonal elements of the mass matrix in Eq. (6.14). The charged current weak interactions can be written, suppressing generation indices for the sake of clarity, as

$$
\begin{aligned} \mathcal{L}_{cc} = \frac{g}{\sqrt{2}} [&(\bar{u}_L \gamma_\mu d_L + \bar{\nu}_L \gamma_\mu e_L) W_L^\mu \\ &+ (\bar{u}_R \gamma_\mu d_R + \overline{N}_R \gamma_\mu e_R) W_R^\mu] + \text{h.c.}. \end{aligned} \tag{6.18}
$$

It is clear that for $M_{W_L} \ll M_{W_R}$, the charged current weak interactions will appear nearly maximally parity violating at low energies. Any deviation from the pure left-handed (or $V - A$) structure of charged weak current will constitute evidence for the right-handed currents and therefore for a left-right symmetric structure of weak interactions.

For the discussion of neutral currents and gauge bosons, it is convenient to define the following basis:

$$
\begin{aligned}
A &= \sin\theta_W (W_{3L} + W_{3R}) + \sqrt{\cos 2\theta_W}\, B \\
Z &= \cos\theta_W W_{3L} - \sin\theta_W \tan\theta_W W_{3R} - \tan\theta_W \sqrt{\cos 2\theta_W}\, B \\
Z' &= \frac{\sqrt{\cos 2\theta_W}}{\cos\theta_W} W_{3R} - \tan\theta_W B .
\end{aligned}
\tag{6.19}
$$

Here, A is the photon, which remains massless after symmetry breaking.

Exercise 6.2 *Write the mass matrix for the neutral vector bosons when the symmetry is broken by the vevs of Φ, Δ_L and Δ_R. Show that the photon eigenstate given in the text is indeed a zero-mass eigenstate of this mass matrix.*

Exercise 6.3 *Show that when χ_L and χ_R are used instead of Δ_L and Δ_R for symmetry breaking, the photon is still given by the same combination.*

The two massive neutral gauge bosons can be called Z_1 and Z_2, which are mixtures of Z and Z':

$$
\begin{aligned}
Z_1 &= Z\cos\xi + Z'\sin\xi \\
Z_2 &= -Z\sin\xi + Z'\cos\xi .
\end{aligned}
\tag{6.20}
$$

so that the masses of these gauge bosons can be calculated from the relations

$$
\begin{aligned}
M_Z^2 &= M_{Z_1}^2 \cos^2\xi + M_{Z_2}^2 \sin^2\xi \\
M_{Z'}^2 &= M_{Z_1}^2 \sin^2\xi + M_{Z_2}^2 \cos^2\xi .
\end{aligned}
\tag{6.21}
$$

Here,

$$
\begin{aligned}
M_Z^2 &= \frac{g^2}{2\cos^2\theta_W}(\kappa^2 + \kappa'^2 + 4v_L^2) \\
M_{Z'}^2 &= \frac{g^2}{2\cos^2\theta_W \cos 2\theta_W}\big(4v_R^2 \cos^4\theta_W \\
&\qquad + (\kappa^2 + \kappa'^2)\cos^2 2\theta_W + 4v_L^2 \sin^4\theta_W\big),
\end{aligned}
\tag{6.22}
$$

and the mixing angle ξ, neglecting v_L and using $\kappa^2 + \kappa'^2 \ll v_R^2$, is given by

$$
\tan 2\xi \simeq \frac{(\cos 2\theta_W)^{3/2}}{2\cos^4\theta_W} \frac{\kappa^2 + \kappa'^2}{v_R^2} \simeq 2\sqrt{\cos 2\theta_W}\, M_Z^2/M_{Z'}^2 .
\tag{6.23}
$$

Obviously, as $v_R \to \infty$, $\xi \to 0$, so that $M_{Z_1}^2 \to M_Z^2$. In this limit, neglecting v_L, the masses of W_1 and Z_1 satisfy the standard model relation $M_{W_1} = M_{Z_1} \cos \theta_W$. The interaction of W_1 and Z_1 also reduce to those of the standard model in this limit. The $1/v_R^2$ corrections to the four-Fermi interactions can be obtained by considering the neutral current Lagrangian:

$$
\begin{aligned}
\mathcal{L}_{\mathrm{nc}} &= \frac{g}{\cos \theta_W} \left[K_L^\mu Z_\mu + \frac{1}{\sqrt{\cos 2\theta_W}} \left\{ \sin^2 \theta_W K_L^\mu + \cos^2 \theta_W K_R^\mu \right\} Z'_\mu \right] \\
&\simeq \frac{g}{\cos \theta_W} \left[\left\{ K_L^\mu - \frac{\xi}{\sqrt{\cos 2\theta_W}} \left(\sin^2 \theta_W K_L^\mu + \cos^2 \theta_W K_R^\mu \right) \right\} Z_{1\mu} \right. \\
&\qquad \left. + \frac{1}{\sqrt{\cos 2\theta_W}} \left\{ \sin^2 \theta_W K_L^\mu + \cos^2 \theta_W K_R^\mu \right\} Z_{2\mu} \right], \quad (6.24)
\end{aligned}
$$

where

$$
K_L^\mu = \sum_f \overline{f} \gamma^\mu \left[I_{3L} \mathsf{L} - Q \sin^2 \theta_W \right] f, \quad (6.25)
$$

and K_R^μ can be obtained by replacing I_{3L} by I_{3R} and changing the sign of γ_5.

Exercise 6.4 *In the four-Fermi limit, find the leading-order corrections to the neutral current parameters $g_V^{(e)}$ and $g_A^{(e)}$ defined in Eq. (2.28).*

6.1.2 Constraints on the masses of the gauge bosons

We are now ready to discuss the bounds on the masses of the gauge bosons W_2 and Z_2 as well as the left-right mixing parameter ζ. The most model-independent limit is on the mass of the Z_2 boson. This is obtained by analyzing neutrino neutral current data, where one searches for deviations from the predictions of the standard model. The present experimental accuracy in the neutral current data implies that [8, 9]

$$
M_{Z_2} \geq 389\,\mathrm{GeV}. \quad (6.26)
$$

A Z_2 with leptonic couplings given in Eq. (6.24) and mass in the hundreds of GeV range should also be detectable in $p\overline{p}$ collider experiments. Analysis of these data leads to a somewhat more stringent bound [10] of $445\,\mathrm{GeV}$ on M_{Z_2}. The implication for future e^+e^- machines have also been studied extensively [11].

To obtain limits on M_{W_2} and ζ, an obvious thing to do is to look for deviations from the predictions of the $V - A$ theory for muon decay. However, since right-handed leptonic charged currents involve the right-handed neutrino field, one needs the mass of the right-handed neutrino in order to carry out the analysis. Since, as discussed in Ch. 4, the neutrinos can be Majorana particles, the N_R can have a different mass than the ν_L. For right-handed neutrinos to contribute to muon decay, N_R mass must satisfy $m_N \ll m_\mu$. Assuming this to be true, the muon decay parameters were analyzed [12] using existing data to obtain bounds on M_{W_2} and ζ. Subsequently, more refined experiments have been carried out. The most stringent limits at this moment come from the measurement of the ξ-parameter in μ-decay [13] using 100% stopped polarized muons. The limits are

$$M_{W_2} \geq 432\,\text{GeV} \qquad \text{for arbitrary } \zeta$$
$$\zeta \leq 0.035 \qquad \text{for } M_{W_2} \to \infty. \qquad (6.27)$$

In general these two bounds are correlated and one gets elliptical regions in the M_{W_2}-ζ plane which give the allowed and forbidden values for the above parameters. In Fig. 6.1, we have summarized the various constraints on M_{W_2} and ζ.

For values of right-handed neutrino mass close to or bigger than the mass of the muon, the above analysis does not shed light on the strength of the right-handed interactions and one must look at weak processes involving only hadrons. The most stringent bound arises from consideration of K_L-K_S mass difference, where one finds [15]

$$\Delta m_{K_1^0 \to K_2^0} \simeq \left(\Delta m_{K_1^0 \to K_2^0}\right)_{\text{SM}} \left[1 - 430\left(\frac{M_{W_L}^2}{M_{W_R}^2}\right)\right], \qquad (6.28)$$

where the short distance contribution calculated in the standard model has been denoted by the symbol $\left(\Delta m_{K_1^0 \to K_2^0}\right)_{\text{SM}}$. The standard model contribution can account for the bulk of the observed mass difference and gives the right sign as well. Thus, Eq. (6.28) implies

$$\frac{M_{W_L}^2}{M_{W_R}^2} < \frac{1}{430}, \qquad (6.29)$$

or $M_{W_R} \geq 1.6\,\text{TeV}$, which is the most stringent bound on M_{W_R}. The original derivation of this result, though important, was incomplete in

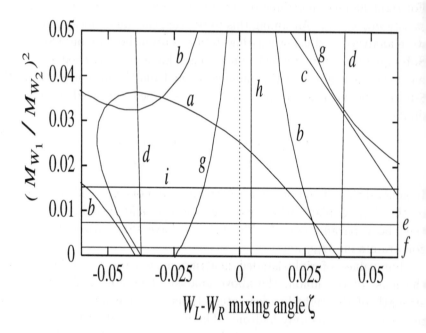

Figure 6.1: Various constraints between the heavy charged gauge boson mass M_{W_2} and the gauge boson mixing angle ζ. For each line, only the region towards the origin is allowed by a particular experiment. The constraints come from the following sources (see Ref. [14]): (a) endpoint spectrum in muon decay; (b) asymmetry parameter in the β-decay of ^{19}Ne; (c) electron polarization in Gamow-Teller type β decay; (d) Michel parameter ρ in muon decay; (e) neutrinoless double beta decay and vacuum stability; (f) K_L-K_S mass difference; (g) relative lepton polarization in Fermi and Gamow-Teller decays; (h) unitarity; (i) direct collider search.

three ways: (*a*) the effect the intermediate *t*-quark was not taken into account; (*b*) the Higgs boson effects, which are part of the complete left-right models, were not included; (*c*) the diagrams on which the analysis was based did not form a gauge invariant set. These issues were examined in subsequent papers [16, 17]. For the case of equal mixing angles in the left and right handed fermionic sectors, it is by now accepted that $M_{W_R} \geq 1.6\,\mathrm{TeV}$. These bounds have been further strengthened by nearly a factor of 2 by the inclusion of higher order QCD effects [18]. It has been noted that if the neutrinos are Dirac particles or if they are Majorana particles with $m_{N_R} \lesssim 10\,\mathrm{MeV}$, observation of the neutrino signal from the supernova SN1987A provides a lower limit [19] on M_{W_R} and ζ: $M_{W_R} \geq 22\,\mathrm{TeV}$ and $\zeta \leq 10^{-5}$.

6.2 Majorana neutrinos

6.2.1 The see-saw mechanism

It has been pointed out in §5.1.4 that a simple way to understand the smallness of neutrino mass is to assume it to be a Majorana particle and use the see-saw mechanism. In the left-right model, the see-saw mechanism appears as follows. We work with the Higgs multiplets [7] Φ along-with the triplets introduced in Eq. (6.12). The most general Yukawa couplings involving the leptons are given by

$$
\begin{aligned}
-\mathcal{L}_Y = \ & \sum_{a,b} h_{ab}^{(\ell)} \overline{\psi}_{aL} \Phi \psi_{bR} + \tilde{h}_{ab}^{(\ell)} \overline{\psi}_{aL} \tilde{\Phi} \psi_{bR} \\
& + f_{ab} \left[\psi_{aL}^{\mathsf{T}} C^{-1} \epsilon \boldsymbol{\tau} \cdot \boldsymbol{\Delta}_L \psi_{bL} + (L \to R) \right] + \text{h.c.}, \quad (6.30)
\end{aligned}
$$

where $\tilde{\Phi} = \tau_2 \Phi^* \tau_2$, and a, b now label different generations. It is then clear that at the first stage of symmetry breaking, we have $\langle \Delta_R^0 \rangle = v_R \neq 0$, leading to a heavy Majorana mass for the right-handed neutrinos, given by the matrix $f_{ab} v_R$. At the second stage, once the neutral components in Φ develop non-zero vevs as in Eq. (6.10), we obtain the following form for the mass matrix of the neutrinos:

$$
\begin{pmatrix} 0 & m_D \\ m_D^{\mathsf{T}} & f v_R \end{pmatrix}, \quad (6.31)
$$

where

$$m_D = h^{(\ell)}\kappa + \tilde{h}^{(\ell)}\kappa' . \tag{6.32}$$

For $\mathcal{N}$ generations of fermions, Eq. (6.31) gives a $2\mathcal{N} \times 2\mathcal{N}$ matrix, where all the elements shown are $\mathcal{N} \times \mathcal{N}$ blocks. One can block diagonalize this matrix by a similarity transformation using the orthogonal matrix [20]

$$\begin{pmatrix} 1 - \frac{1}{2}\rho\rho^{\mathsf{T}} & \rho \\ -\rho^{\mathsf{T}} & 1 - \frac{1}{2}\rho^{\mathsf{T}}\rho \end{pmatrix} . \tag{6.33}$$

where $\rho = m_D f^{-1}/v_R$. This diagonalization is correct up to terms smaller than of order ρ^2 and one obtains the mass matrix for the light neutrinos to be

$$m^{\text{light}} = \frac{1}{v_R} m_D f^{-1} m_D^{\mathsf{T}} . \tag{6.34}$$

Of course, this matrix needs further diagonalization to obtain the light neutrino eigenvalues and eigenstates. If we ignore mixing between generations in the first approximation and assume that $m_D \simeq m_\ell$, where m_ℓ is the mass of the charged lepton, we obtain from Eq. (6.34) the relation

$$m_{\nu_\ell} \simeq \frac{m_\ell^2}{M_{N_\ell}} \tag{6.35}$$

where M_{N_ℓ} is the mass of the heavy right-handed neutrino. If M_{N_ℓ} is assumed to be generation independent, one gets the following quadratic mass formula for neutrino masses:

$$m_{\nu_e} : m_{\nu_\mu} : m_{\nu_\tau} = m_e^2 : m_\mu^2 : m_\tau^2 . \tag{6.36}$$

Here and elsewhere, when we talk about the "mass" of ν_e, for example, we imply the mass eigenvalue of the physical state which contains mostly the ν_e component, assuming mixings are small. It is clear from Eq. (6.34) that as $v_R \to \infty$, $m_\nu \to 0$. Therefore the smallness of the neutrino mass is connected to the suppression of $V + A$ currents in this scenario [7, 21].

Furthermore, if v_R is in the TeV region, one expects an eV-keV-MeV type spectrum for neutrinos. In other words, if $m_{\nu_e} \approx 1\,\text{eV}$, one gets $m_{\nu_\mu} \approx 40\,\text{keV}$ and $m_{\nu_\tau} \approx 12\,\text{MeV}$. This spectrum is accessible to a variety of laboratory experiments such as the end point spectrum in tritium decay, neutrinoless double beta decay, spectrum of μ and τ decay, etc. There are, however, severe cosmological constraints on such a spectrum.

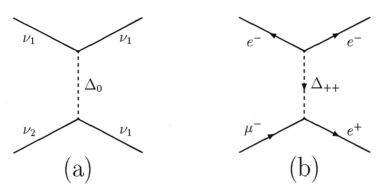

Figure 6.2: Tree-level diagram for (a) $\nu_2 \to 3\nu_1$ and (b) $\mu \to 3e$ in the left-right model.

6.2.2 Constraints on the eV-keV-MeV spectrum

It has long been known [22] that the sum of the masses of stable neutrinos must be bounded above by about $100\,\text{eV}$ in order to satisfy the cosmological mass density constraints [for details, see Ch. 17]. The ν_μ and the ν_τ in our model must therefore by unstable. For unstable neutrinos, again to prevent the decay products from exceeding the closure density, one must satisfy the following inequality [23]

$$m_\nu \left(\frac{\tau_\nu}{t_0}\right)^{1/2} \leq 100\,\text{eV} . \qquad (6.37)$$

where m_ν and τ_ν denote the mass and lifetime of any unstable neutrino and t_0 stands for the present age of the universe [for details, see Ch. 17]. To see whether the ν_μ and the ν_τ in the left-right model is consistent with these cosmological constraint with v_R in the range of a few TeV, we have to search for their decay modes.

Let us first discuss the decays of the ν_μ. In the minimal model, it has the following decay modes: $i)$ $\nu_\mu \to \nu_e + \gamma$, and $ii)$ $\nu_\mu \to 3\nu_e$. The radiative decay proceeds via loop graphs and have been calculated [24]. The details of this calculation will be given in Ch. 11. Here, we simply note that this decay leads to lifetimes of order $10^{21}\,\text{s}$ for $m_{\nu_\mu} \sim 100\,\text{keV}$. This does not satisfy the inequality of Eq. (6.37) which requires $\tau_\nu \lesssim 10^{11}\,\text{s}$ for $m_\nu \sim 100\,\text{keV}$.

In the left-right model an alternative decay mode for ν_μ, viz, $\nu_\mu \to 3\nu_e$ exists [25] via the exchange of the Δ_L^0 boson as shown in Fig. 6.2a.

The decay rate via this mode can be written as

$$\Gamma_{\nu_\mu \to 3\nu_e} = \left(\frac{f_{11} f_{12}}{M_{\Delta_L^0}^2}\right)^2 \frac{m_2^5}{192\pi^3} . \tag{6.38}$$

To satisfy the bound of Eq. (6.37), one thus needs

$$\frac{f_{11} f_{12}}{M_{\Delta_L^0}^2} \geq \left(\frac{m_2}{40\,\text{keV}}\right)^{-3/2} \cdot 6 \times 10^{-3} G_F . \tag{6.39}$$

We thus see that either Δ_L^0 must be light or the couplings f must not be too small. However, the same set of couplings f give rise to $\mu^- \to e^- e^- e^+$ (commonly called $\mu \to 3e$) decay [26] via the diagrams shown in Fig. 6.2b. The upper limit on this decay rate is very stringent:

$$B(\mu \to 3e) \leq 1.0 \times 10^{-12} . \tag{6.40}$$

The rate for $\mu \to 3e$ derived from Fig. 6.2b is

$$\Gamma_{\mu \to 3e} = \left(\frac{f_{ee} f_{e\mu}}{M_{\Delta^{++}}^2}\right)^2 \frac{m_\mu^5}{192\pi^3} , \tag{6.41}$$

where $M_{\Delta^{++}}$ is the mass of the lighter doubly charged particle. The bound of Eq. (6.40) now implies

$$\frac{f_{ee} f_{e\mu}}{M_{\Delta^{++}}^2} \leq 10^{-6} G_F . \tag{6.42}$$

Since the Z-decay modes do not reveal the existence of any neutral scalar triplet, Δ_L^0 must be heavier than 45 GeV. On the other hand, since Δ_L^0 and Δ_L^{++} belong to the same $SU(2)_L$ multiplet, their mass difference should be of order the $SU(2)_L$ breaking, i.e., a few hundred GeV at most. With these constraints, the masses of Δ_L^0 and Δ_L^{++} must be within a factor of about 10 of each other. The bounds in Eq. (6.39) and Eq. (6.42) are therefore not easy to satisfy simultaneously [26].

One point, however, must be noticed [27]. The two equations in question contain couplings f in two different basis. The relation between these two sets of numbers is given by

$$f_{\alpha\alpha'} = \sum_{\ell,\ell'} f_{\ell\ell'} \mathcal{U}_{\ell\alpha} \mathcal{U}_{\ell'\alpha'} , \tag{6.43}$$

$\mathcal{U}$ being the mixing matrix for the light neutrinos. Thus, for example, the $\mu \to 3e$ rate might be suppressed if, say, $f_{ee} < 10^{-6}$. The question is then, what happens to f_{11}? Looking at Eq. (6.43), we see that f_{11} can receive contribution from other terms not involving f_{ee}. However, these terms involve off-diagonal terms of the mixing matrix. In the eV-keV-MeV mass spectrum that we are considering here, experimental constraints [see §10.1.2 for details] imply that the off-diagonal elements of the mixing matrix cannot be larger than $O(10^{-2})$. Thus, even if the other elements of the form $f_{\ell\ell'}$ are of order unity, f_{11} cannot be larger than $O(10^{-2})$. Referring back to Eq. (6.39), we now need the Δ_L^0 particle to be reasonably light, $M_{\Delta_L^0} \lesssim 100 \, \text{GeV}$. This has the interesting implication that, since Δ_L^+ and Δ_L^{++} are members of the same isomultiplet as Δ_L^0, their masses must be in the 200 to 300 GeV range.

There are two interesting physical processes mediated by the exchange of Δ_L^+ and Δ_L^{++}. The Δ_L^+ exchange leads to an anomalous muon decay of the following type:

$$\mu^- \to e^- + \nu_e + \widehat{\nu}_\mu \,. \tag{6.44}$$

Note that the conventional muon decay is $\mu^- \to e^- + \widehat{\nu}_e + \nu_\mu$. The Δ_L^{++} exchange, on the other hand, leads to muonium to antimuonium transition:

$$\mu^+ e^- \to \mu^- e^+ \,. \tag{6.45}$$

The rates of these processes are given in terms of the following effective couplings:

$$G_{\mu \to e\nu_e \widehat{\nu}_\mu} \simeq \frac{f_{ee} f_{\mu\mu}}{M_{\Delta_L^+}^2}$$

$$G_{\mu^+ e^- \to \mu^- e^+} \simeq \frac{f_{ee} f_{\mu\mu}}{M_{\Delta_L^{++}}^2} \,. \tag{6.46}$$

The $\nu_\mu \to 3\nu_e$ constraints can be translated into lower bounds for the above processes:

$$G_{\mu \to e\nu_e \widehat{\nu}_\mu} \geq 10^{-2} G_F \quad , \quad G_{\mu^+ e^- \to \mu^- e^+} \geq 2.5 \times 10^{-3} G_F \,. \tag{6.47}$$

The present experimental upper limits for these processes are at the level of $10^{-2} G_F$ and $10^{-3} G_F$ respectively [28, 29].

Let us now turn to the decay of the ν_τ. If ν_τ is above an MeV, it can have the following decay modes:

$$\nu_\tau \; \to \; \nu e^+ e^- \;\; \text{or} \;\; \nu\gamma \;\; \text{or} \;\; 3\nu \,, \tag{6.48}$$

where in each case the neutrinos in the decay products can be either electron or muon neutrino. The first decay mode must be highly suppressed in order to be consistent with astrophysical constraints [30]. The second decay mode [see Ch. 11 for details] is much too long to be useful in satisfying Eq. (6.37). The $\nu_\tau \to 3\nu$ decay is mediated via the Δ_L^0 exchange. Therefore the rough order of magnitude of its lifetime can be obtained by just scaling Eq. (6.38). We get $\tau_{\nu_\tau} \simeq \tau_{\nu_\mu}(m_{\nu_\mu}/m_{\nu_\tau})^5$, which gives a lifetime of about 1 s for $m_{\nu_\tau} \simeq 10\,\text{MeV}$. It is interesting to note that this lifetime not only satisfies Eq. (6.37) but also the somewhat stronger constraints from nucleosynthesis [31].

6.3 Physics involving right-handed neutrinos

The left-right symmetric model with Majorana neutrinos has several interesting implications on the properties of neutrinos, which we discuss below. To understand these properties, it is useful to restrict our attention only to one generation. In this case the quantities m_D and $M \equiv f v_R$, appearing in the matrix of Eq. (6.31), become just numbers.

6.3.1 Flavor changing neutral currents

The diagonalization of the mass matrix was discussed in §5.1.4. The eigenvalues of the matrix are found to be

$$m_\nu \simeq m_D^2/M \quad \text{and} \quad m_N \simeq M \,, \tag{6.49}$$

where ν and N denote the light and heavy Majorana neutrinos corresponding to a given generation.

The coupling with the Z-boson can be written in the flavor basis as

$$\frac{g}{2\cos\theta_W} Z^\mu \overline{\nu}_{eL} \gamma_\mu \nu_{eL} \,, \tag{6.50}$$

since the N_e-field does not couple to the Z-boson. Recalling Eq. (5.20), we can write this coupling in terms of the mass eigenstates as follows:

$$\frac{g}{2\cos\theta_W}Z^\mu \left[\cos^2\theta\overline{\nu}\gamma_\mu L\nu + \sin^2\theta\overline{N}\gamma_\mu LN \right.$$
$$\left. + \cos\theta\sin\theta(\overline{N}\gamma_\mu L\nu + \overline{\nu}\gamma_\mu LN)\right] \qquad (6.51)$$

where θ is the mixing angle, given in Eq. (5.15). Since $m_N \gg m_D$, we can approximately write

$$\theta \simeq \frac{m_D}{M} \simeq \left(\frac{m_\nu}{m_N}\right)^{1/2}. \qquad (6.52)$$

For Z coupling to a single Majorana fermion, the vector part of the coupling vanishes and only the axial vector part contributes. However, Eq. (6.51) shows that Z can also couple to two different fermions. In other words, Z has flavor changing neutral currents (FCNC) so far as its coupling to neutral fermions are concerned. This is unlike the case in the standard model. For $m_N < M_Z$, this will lead to the following decay width for $Z \to \nu N$:

$$\Gamma(Z \to \nu N) \approx \theta^2 \cdot 165\,\text{MeV}. \qquad (6.53)$$

If we assume $m_D \simeq m_e$ as in the previous section, we get $m_{N_1} \geq 20\,\text{GeV}$ from the experimental constraint $m_{\nu_1} \leq 12\,\text{eV}$. From Eq. (6.52), this implies $\theta \leq 2 \times 10^{-5}$. Thus the ν-N mixing is small in the see-saw limit. If, however, the see-saw limit is relaxed, one could get bigger values for θ, which would then lead to an appreciable decay width of the Z through Eq. (6.53). This decay mode has a distinctive signature since, as we will see below, decay modes of N_1 can involve charged leptons or hadrons which can be easily detected.

6.3.2 Decay of the right-handed neutrinos

In the see-saw picture, the right-handed neutrino mass is in the tens of GeV range. If we give up the see-saw picture, its mass could be arbitrary. It is therefore important to know its decay properties so that one could restrict its mass and coupling from observations discussed in this section. If we consider only one generation, we find that N has the

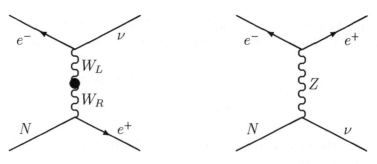

Figure 6.3: Diagram for the decay $N \to \nu + e^+ + e^-$.

following possible decay modes if we restrict the discussion to modes having not more than three particles in the final state:

$$N \quad \to \quad \begin{cases} \nu + e^+ + e^- \\ \nu + \gamma \\ \nu + \gamma + \gamma \\ 3\nu \\ e^{\pm} + \text{hadrons} \, . \end{cases} \qquad (6.54)$$

The hadronic decay becomes predominant above the pion threshold, i.e., for $m_N \gtrsim 140 \, \text{MeV}$. For masses below that threshold, the only decay modes are leptonic and radiative. Let us first discuss these non-hadronic modes.

• $N \to \nu + e^+ + e^-$

This decay proceeds via the W_L-W_R mixing graph shown in Fig. 6.3a and the FCNC violating neutral current graph in Fig. 6.3b. The partial decay width of N into this channel is given by

$$\Gamma(N \to \nu + e^+ + e^-) \simeq \frac{G_F^2 m_N^5}{192\pi^3}(\zeta^2 + \theta^2 \Delta) \, , \qquad (6.55)$$

where $\Delta = \frac{1}{4} - \sin^2\theta_W + 2\sin^4\theta_W$.

• $N \to \nu + \gamma$

This decay is absent at the tree level and is induced at the one loop level via the W_L-W_R mixing graph of Fig. 6.4. The decay amplitude for

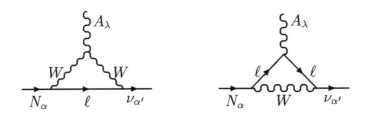

Figure 6.4: Diagram for the decay $N \to \nu + \gamma$.

this process can be written as

$$\bar{\nu}\sigma^{\mu\rho}(a + b\gamma_5)N q_\rho \epsilon_\mu^* \,, \tag{6.56}$$

where ϵ_μ is the polarization and q is the momentum of the photon. Consider the simple case of CP-conserving theories. The Majorana neutrinos have specific CP eigenvalues in this case. If, for example, the CP eigenvalue of N and ν are opposite, one obtains $a = 0$ in the above equation. Detailed calculation, described in §11.3.3, shows that

$$b = - \frac{eG_F}{\sqrt{2}\pi^2} \sum_\ell m_\ell U_{\ell\nu} V_{\ell N} \,, \tag{6.57}$$

where U and V are the mixing matrices in the left and right handed sectors respectively. Due to the presence of the factor m_ℓ in Eq. (6.57), decay into the ν_τ mode is likely to dominate and we find [32] the width to be

$$\Gamma(N \to \nu_\tau + \gamma) \simeq 8 \times 10^{-22} \zeta^2 \theta_{e\tau}^2 \left(\frac{m_N}{1\,\mathrm{MeV}}\right)^3 \mathrm{MeV} \,, \tag{6.58}$$

where $\theta_{e\tau}$ stands for the mixing.

• $N \to 3\nu$

This decay proceeds through the flavor changing Z-exchange graph similar to that in Fig. 6.3b and leads to the following decay width:

$$\Gamma(N \to 3\nu) \simeq \frac{G_F^2}{192\pi^3} \theta^2 m_N^5 \,. \tag{6.59}$$

Using Eq. (6.58), we find the relative branching ratios

$$\frac{\Gamma(N \to 3\nu)}{\Gamma(N \to \nu\gamma)} \simeq 2 \times 10^{-3} \left(\frac{\theta}{\zeta}\right)^2 \left(\frac{m_N}{1\,\mathrm{MeV}}\right)^2. \tag{6.60}$$

• $N \to e^{\pm} + hadrons$

This decay mode occurs for $m_N \geq 150\,\mathrm{MeV}$ or so [33]. Due to the Majorana nature of N, one obtains

$$\Gamma(N \to e^+ + \mathrm{hadrons}) = \Gamma(N \to e^- + \mathrm{hadrons}) \tag{6.61}$$

Depending on the value of m_N, one or two generation of quarks will contribute. If only the first generation contributes, we get

$$\Gamma(N \to e^+ + \mathrm{hadrons}) = \frac{G_F^2}{192\pi^3} \cdot 9m_N^5 \left(\frac{M_{W_L}}{M_{W_R}}\right)^4. \tag{6.62}$$

For $M_{W_R} \simeq 2\,\mathrm{TeV}$, if the charm threshold is open, we get

$$\Gamma(N \to e^+ + \mathrm{hadrons}) = 1.3 \times 10^{-15} \left(\frac{m_N}{5\,\mathrm{GeV}}\right)^5 \mathrm{GeV}. \tag{6.63}$$

6.4   Naturalness of the see-saw formula

The see-saw picture discussed in §6.2.1 has a problem of naturalness associated with it. The neutrino mass matrix written in Eq. (6.31) is derived from the Eq. (6.30) on the assumption that the Higgs field Δ_L has got a zero vev. However, an actual analysis of the Higgs potential does not bear out this result. To see this, let us write the full Higgs potential involving Φ, Δ_L and Δ_R. For this purpose, it is useful to introduce the matrices $\Delta_L \equiv \epsilon\tau \cdot \Delta_L$ and a similar matrix form of Δ_R. Along-with the matrix Φ defined in Eq. (6.9), we can write the potential now:

$$V(\Phi, \Delta_L, \Delta_R) = -\sum_{i,j} \mu_{ij}^2 \operatorname{Tr} \Phi_i^\dagger \Phi_j - \mu^2 (\operatorname{Tr} \Delta_L^\dagger \Delta_L + \operatorname{Tr} \Delta_R^\dagger \Delta_R)$$

$$+ \sum_{ijkl} \lambda_{ijkl} \operatorname{Tr} (\Phi_i^\dagger \Phi_j) \operatorname{Tr} (\Phi_k^\dagger \Phi_l) + \sum_{ijkl} \lambda'_{ijkl} \operatorname{Tr} (\Phi_i^\dagger \Phi_j \Phi_k^\dagger \Phi_l)$$

$$+ \rho_1 [(\operatorname{Tr} \Delta_L^\dagger \Delta_L)^2 + (\operatorname{Tr} \Delta_R^\dagger \Delta_R)^2]$$

$$+ \rho_2 [(\operatorname{Tr} \Delta_L^\dagger \Delta_L \Delta_L^\dagger \Delta_L) + (\operatorname{Tr} \Delta_R^\dagger \Delta_R \Delta_R^\dagger \Delta_R)]$$

$$+ \rho_3 (\operatorname{Tr} \Delta_L^\dagger \Delta_L)(\operatorname{Tr} \Delta_R^\dagger \Delta_R)$$

$$+ \rho_4 [(\operatorname{Tr} \Delta_L^\dagger \Delta_L^\dagger)(\operatorname{Tr} \Delta_L \Delta_L) + (\operatorname{Tr} \Delta_R^\dagger \Delta_R^\dagger)(\operatorname{Tr} \Delta_R \Delta_R)]$$

$$+ \sum_{ij} \alpha_{ij} (\operatorname{Tr} \Phi_i^\dagger \Phi_j)(\operatorname{Tr} \Delta_L^\dagger \Delta_L + \operatorname{Tr} \Delta_R^\dagger \Delta_R)$$

$$+ \sum_{ij} \beta_{ij} \operatorname{Tr} [\Phi_i^\dagger \Phi_j (\Delta_L^\dagger \Delta_L + \Delta_R^\dagger \Delta_R)]$$

$$+ \sum_{ij} [\gamma_{ij} \operatorname{Tr} \Delta_L^\dagger \Phi_i \Delta_R \Phi_j^\dagger + \text{h.c.}], \tag{6.64}$$

where the sums over i, j, k and l always runs from 1 to 2, with $\Phi_1 = \Phi$ and $\Phi_2 \equiv \epsilon \Phi^* \epsilon$. In the presence of the γ_{ij} terms, minimization of the Higgs potential leads to an equation [7]

$$v_L = \left(\frac{\gamma_{12}}{2(\rho_1 + \rho_2) - \rho_3} \right) \frac{\kappa^2}{v_R}, \tag{6.65}$$

where $v_L = \langle \Delta_L^0 \rangle$ and we have assumed $\kappa' \ll \kappa$ for simplicity. We thus see that $v_L \simeq \gamma \kappa^2 / v_R$, leading to non-vanishing entries in the upper left corner of the mass matrix of Eq. (6.31). To see the implications of this term, let us restrict our attention to the case of one generation. We then see that the light neutrino mass is given by

$$m_\nu \simeq \gamma \frac{f \kappa^2}{v_R} - \frac{m_D^2}{f v_R}. \tag{6.66}$$

First point to realize is that, as $v_R \to \infty$, the light neutrino mass goes to zero. Thus, the analytical connection between vanishing of neutrino mass and the absence of $V + A$ interactions is preserved. However, if we look at the magnitude of the neutrino masses, we see that, for $v_R \simeq 10\,\text{TeV}$ and $f \simeq 10^{-2}$ or so, we require $\gamma \lesssim 10^{-7}$ to get m_{ν_e} in the

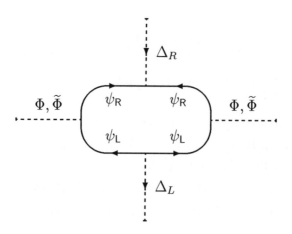

Figure 6.5: One-loop diagram giving rise to a non-trivial correction to the tree level value of γ_{ij}.

electron-volt range. This clearly requires a fine tuning of parameters and we ought to look for ways to avoid this unpleasant situation.

Let us ask ourselves what happens if we set $\gamma_{ij} = 0$ in the tree level potential. One finds that it is induced by the diagram of Fig. 6.5 which is logarithmically divergent and therefore one must introduce tree level counterterms to absorb this infinity. Thus, setting it to zero at the tree level does not solve our problem. It is however interesting that the diagram of Fig. 6.5 involves only Yukawa couplings and if we assume that the theory has a natural cutoff at $\Lambda \simeq M_{\text{Pl}}$, we estimate the magnitude of the one-loop corrections to be $\gamma_{1-\text{loop}} \simeq (f^2 h^2/16\pi^2) \ln(\Lambda/M_{W_R})$. Assuming the third generation to make the dominant contribution, we get $\gamma \simeq 10^{-7}$ for our choice of parameters, as required. Thus, choosing such a small value of γ is not as bad as it might appear.

There is however a more natural way to solve this problem [34] if we introduce a parity odd real scalar field σ into the theory [35]. In the presence of this particle, the Higgs potential acquires the following terms in addition to those already presented in Eq. (6.64):

$$V_\sigma = -\mu_\sigma^2 \sigma^2 + \lambda \sigma^4 + m_0 \sigma (\text{Tr}\, \Delta_L^\dagger \Delta_L - \text{Tr}\, \Delta_R^\dagger \Delta_R) + \dots \quad (6.67)$$

Minimization of the full potential $V + V_\sigma$ gives

$$\langle \sigma \rangle = M_p \neq 0 \,. \tag{6.68}$$

Since σ is parity odd, its vev breaks only the discrete parity symmetry but keeps the gauge symmetry intact. If $M_p \gg M_{W_R}$, then one finds

$$v_L \simeq \gamma \frac{\kappa^2 v_R}{M_p^2} \tag{6.69}$$

instead of Eq. (6.65). Thus, for example, if $M_p \geq 10^4 v_R$, this makes the $\nu_L \nu_L$ mass term in the see-saw matrix very small, helping to restore naturalness of the see-saw picture. In Ch. 7, we will see that this mechanism of decoupling parity and $SU(2)_R$ breaking scales has a smooth realization in the $SO(10)$ model of grand-unification. An important phenomenological consequence of this idea is that the Δ_L-triplet acquires masses of order M_p and therefore cannot play any role in neutrino decay. Thus, some new mechanism for neutrino decay must be introduced into the theory in order to avoid cosmological problems [36].

6.5 Dirac neutrinos

In the original version of the left-right symmetric models [1, 6], though the neutrinos were considered to be Dirac particles, the smallness of their mass was not understood. More recently, a version of these models have been developed [37, 38] wherein the neutrino mass vanishes at the tree level but arises at one or two loop level, thus making it naturally much smaller than the charged fermion masses. An interesting feature of these models is that the neutrino masses scale linearly with the charged fermion mass and one has a mass formula of the type

$$m_{\nu_\ell} \simeq 10^{-7} m_\ell \,, \tag{6.70}$$

ignoring generational mixings. These models are inherently different from the class of left-right models that lead to small Majorana masses for neutrinos. We provide a brief sketch of these models in this section.

These models use heavy, vectorlike singlet quark and leptons in their construction and use the see-saw mechanism for quarks and charged leptons instead of the neutrino. The idea of using partial see-saw mechanism for down-type quarks and charged leptons was already discussed (e.g. in [39]). But later it was proposed [40] that the entire fermion spectrum be generated via the see-saw mechanism by postulating the following heavy fermions in the left-right symmetric models in addition to the already existing light fermions of Eq. (6.3):

$$
\begin{aligned}
&\mathcal{U}_{\mathsf{L,R}}\left(1,1,\tfrac{4}{3}\right), &&\mathcal{D}_{\mathsf{L,R}}\left(1,1,-\tfrac{2}{3}\right), \\
&E_{\mathsf{L,R}}^{-}\left(1,1,-2\right), &&E_{\mathsf{L,R}}^{0}\left(1,1,0\right).
\end{aligned}
\tag{6.71}
$$

An advantage of these models is their simple Higgs structure. Only one pair of Higgs doublets $\chi_{L,R}$ and a parity odd real singlet scalar (needed to generate the left-right asymmetry) are sufficient:

$$
\chi_L\left(2,1,1\right), \quad \chi_R\left(1,2,1\right), \quad \sigma(1,1,0).
\tag{6.72}
$$

The most general gauge invariant coupling is given by

$$
\begin{aligned}
-\mathcal{L}_Y =\; & \left[h^d \bar{q}_{\mathsf{L}}\chi_L \mathcal{D}_{\mathsf{R}} + h^u \bar{q}_{\mathsf{L}}\tilde{\chi}_L \mathcal{U}_{\mathsf{R}} + h^\ell \overline{\psi}_{\mathsf{L}}\chi_L E_{\mathsf{R}}^- \right. \\
& \left. + h^\nu \overline{\psi}_{\mathsf{L}}\tilde{\chi}_L E_{\mathsf{R}}^0 + (L \leftrightarrow R)\right] + \text{h.c.} \\
& + \overline{\mathcal{U}}\left(f_{\mathcal{U}} + i\gamma_5\sigma f_{\mathcal{U}}'\right)\mathcal{U} + \overline{\mathcal{D}}\left(f_{\mathcal{D}} + i\gamma_5\sigma f_{\mathcal{D}}'\right)\mathcal{D} \\
& + \overline{E^-}\left(f_{E^-} + i\gamma_5\sigma f_{E^-}'\right)E^- + \overline{E^0}\left(f_{E^0} + i\gamma_5\sigma f_{E^0}'\right)E^0 \\
& + (E_{\mathsf{L}}^{0\mathsf{T}}C^{-1}\mathcal{M}^L E_{\mathsf{L}}^0 + E_{\mathsf{R}}^{0\mathsf{T}}C^{-1}\mathcal{M}^R E_{\mathsf{R}}^0 + \text{h.c.}).
\end{aligned}
\tag{6.73}
$$

where we have dropped the generation indices for the sake of clarity. At the minimum of the Higgs potential, one has $\langle \chi_{L,R}\rangle = v_{L,R}$, leading to the charged lepton and quark mass matrices of the form

$$
\begin{pmatrix} 0 & h v_L \\ h^{\mathsf{T}} v_R & M \end{pmatrix}.
\tag{6.74}
$$

It is clear from this that the light quark masses are given by

$$
m_{\text{light}} \simeq h^2 v_L v_R / M.
\tag{6.75}
$$

The interesting aspect of the above quark and charged lepton mass formula is that, if we chose $v_R/M \simeq 10^{-1}$ and $h \simeq 10^{-2}$, one can understand small masses of the first generation fermions. This ameliorates somewhat the severe finetuning problem of the standard model. A very exciting development is that within this scenario, a mechanism has been

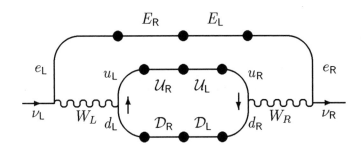

Figure 6.6: 2-loop diagrams giving rise to Dirac neutrino masses.

proposed [41] wherein the mass and mixing patterns among quarks and leptons receive a very plausible explanation.

Turning now to the neutrino sector, one can use [40] a generalized see-saw mechanism for neutrino masses leading to Majorana neutrinos. But to understand the smallness of neutrino masses, one needs the scale of new physics (i.e., the singlet quark masses, the right handed scale etc) to be in the range of 10^{10} GeV or so. However, if the E^0 is omitted from the heavy fermion sector, the neutrino mass vanishes at the tree level and arise only at the two loop level [37]. A typical relevant two loop graph is shown in Fig. 6.6. This graph leads to small and finite Dirac neutrino masses and leads to a linear mass formula given in Eq. (6.70) for $M_{W_R} \simeq 10$ TeV. In this picture, therefore, by increasing the W_R scale by one order of magnitude, one may, if one wishes, accommodate the matter oscillation solution of the solar neutrino problem [see Ch. 15 for details].

In this class of models, one does not expect any new charged Higgs bosons that couple to quarks although there may be charged Higgs bosons coupling to leptons in some versions [37]. Furthermore, in this model one expects tree level flavor changing neutral current signals such as $Z \to b\widehat{s}$ etc.

Chapter 7

Neutrino mass in Grand unified models

At the present time, there is no solid experimental evidence for physics beyond the standard model. It has, therefore, proved useful to explore new physics beyond the standard model using two basic strategies. Firstly, one can try to understand some of the puzzles of the standard model such as fermion masses and mixings or origin of weak symmetry breaking, solution to strong CP-problem. Secondly, one can imagine scenarios beyond the standard model by demanding aesthetic requirements such as the unification of all forces and matter at short distances. While these requirements might sound severe, they still allow many possible choices. These choices are narrowed down once we require the theory to provide a natural understanding of small neutrino masses. In this chapter, we focus only on three popular grand-unification groups: $SU(5)$, $SO(10)$ and E_6, and see how detailed consideration of neutrino masses restrict the choice of Higgs structure, possible existence of intermediate scale, $\sin^2\theta_W$ etc. For other details of the grand-unified theories, we refer the reader to existing books on the subject [1].

7.1 SU(5)

In the $SU(5)$ model [2], the fermions of each generation are assigned to $\{10\}$ and $\overline{\{5\}}$ representations, denoted respectively by T and F:

$$
F = \begin{pmatrix} \widehat{d}_1 \\ \widehat{d}_2 \\ \widehat{d}_3 \\ e \\ \nu \end{pmatrix}_L \quad , \quad T = \begin{pmatrix} 0 & \widehat{u}_3 & -\widehat{u}_2 & u_1 & d_1 \\ & 0 & \widehat{u}_1 & u_2 & d_2 \\ & & 0 & u_3 & d_3 \\ & & & 0 & e^+ \\ & & & & 0 \end{pmatrix}_L . \tag{7.1}
$$

Here $\hat{u}$ denotes the conjugate state so that $\hat{u}_L$ for instance is the CPT-conjugate state of the right-handed helicity states of the up quark. The indices 1,2,3 stand for three different colors. The matrix T is antisymmetric, so we did not write down the elements below the diagonal.

The breaking down of the gauge symmetry to $SU(3)_c \times U(1)_Q$ is achieved by a choice of only two Higgs multiplets:

$$H \equiv \overline{\{5\}} \quad , \quad \Phi \equiv \{24\} \,. \tag{7.2}$$

The vacuum expectation value of Φ is chosen such that it breaks $SU(5)$ down to $SU(3)_c \times SU(2)_L \times U(1)_Y$ as follows

$$\langle \Phi \rangle = \mathrm{diag}\left(V, V, V, -\frac{3}{2}V, -\frac{3}{2}V\right) \,. \tag{7.3}$$

Denoting the gauge bosons of the $SU(5)$ model by

$$\left(\begin{array}{ccc|cc}
\multicolumn{3}{c|}{\frac{1}{\sqrt{2}}\sum_{a=1}^{8}\lambda^a G'^a + \sqrt{\frac{2}{15}}B_{24}} & X_1^{4/3} & Y_1^{1/3} \\
 & & & X_2^{4/3} & Y_2^{1/3} \\
 & & & X_3^{4/3} & Y_3^{1/3} \\
\hline
X_1^{-4/3} & X_2^{-4/3} & X_3^{-4/3} & \multicolumn{2}{c}{\frac{1}{\sqrt{2}}T\cdot\mathbf{W} - \sqrt{\frac{3}{10}}B_{24}} \\
Y_1^{-1/3} & Y_2^{-1/3} & Y_3^{-1/3} & &
\end{array} \right) \tag{7.4}$$

we find that

$$M_X^2 = M_Y^2 = \frac{25}{8}g^2V^2 \,. \tag{7.5}$$

The final stage of symmetry breaking down to $U(1)_Q$ occurs via the non-zero vev of the $\{5\}$-dimensional Higgs field H:

$$\langle H \rangle = \begin{pmatrix} 0 \\ 0 \\ 0 \\ 0 \\ v/\sqrt{2} \end{pmatrix} \,. \tag{7.6}$$

This gives masses:

$$M_W = \frac{1}{2}gv \quad , \quad M_Z = \frac{gv}{2\cos\theta_W} \,, \tag{7.7}$$

where θ_W is the Weinberg angle. In this case, since there is only one gauge coupling constant, Weinberg angle is predicted at the unification scale to satisfy

$$\tan \theta_W = \sqrt{3/5}. \tag{7.8}$$

Exercise 7.1 *Using any one of the representations in Eq. (7.1), show that*

$$\text{Tr}\left[(Y/2)^2\right] = \frac{5}{3}\text{Tr}\left[(I_{3L})^2\right]. \tag{7.9}$$

From this, deduce Eq. (7.8).

The unification scale V is obtained from the constraint that the three different gauge couplings g_3, g_2 and g_1 at low energies unify to a single gauge coupling g_U above the unification scale. The evolution of the gauge couplings with the mass scale μ is given via the renormalization group equations for each gauge group which in the one-loop approximation looks as follows

$$\frac{dg_n}{d\ln \mu} = b_n \frac{g_n^3}{16\pi^2} \qquad \text{for } n = 1, 2, 3. \tag{7.10}$$

The coefficient b_n is given by $\frac{4}{3}\mathcal{N} - \frac{11}{3}n$ for $n = 2, 3$ if we neglect the Higgs contribution and assume that the only fermions are $\mathcal{N}$ generations of the fields shown in Eq. (7.1). Under the same assumptions, $b_1 = \frac{4}{3}\mathcal{N}$. The values of the three gauge couplings at the scale M_Z have been experimentally determined from the LEP and SLC experiments to be:

$$
\begin{aligned}
\alpha_1^{-1}(M_Z) &= 58.89 \pm 0.11; \\
\alpha_2^{-1}(M_Z) &= 29.75 \pm 0.11 \\
\alpha_3(M_Z) &= 0.121 \pm .004 \pm .001.
\end{aligned}
\tag{7.11}
$$

It is then easy to show that the three gauge couplings of the standard model do not unify as they are extrapolated to high scales. To see this, let us write down explicitly the solutions to the 1-loop evolution equations for the gauge couplings:

$$
\begin{aligned}
\alpha_3^{-1}(M_Z) &= \alpha_3^{-1}(M_U) - \frac{7}{2\pi}\ln\left(\frac{M_U}{M_Z}\right); \\
\alpha_2^{-1}(M_Z) &= \alpha_2^{-1}(M_U) - \frac{19}{12\pi}\ln\left(\frac{M_U}{M_Z}\right); \\
\alpha_1^{-1}(M_Z) &= \alpha_1^{-1}(M_U) + \frac{41}{20\pi}\ln\left(\frac{M_U}{M_Z}\right).
\end{aligned}
\tag{7.12}
$$

Unification condition is that $\alpha_i(M_U) = \alpha_U$ for all i. One can then eliminate the two unknowns α_U and $\ln(M_U/M_Z)$ to obtain the following constraint on $\alpha_i(M_Z)$:

$$1.9\alpha_3^{-1}(M_Z) - 2.9\alpha_2^{-1}(M_Z) + \alpha_1^{-1}(M_Z) = 0 \qquad (7.13)$$

which is not satisfied by the LEP and SLC data for the gauge couplings. Therefore, $SU(5)$ is not acceptable as a viable model of grand unification unless it is "polluted" with extra Higgs bosons, in which case, the beauty and predictivity of the model is lost. Regardless of this unfortunate lack of unification however, the $SU(5)$ is an excellent tool for studying many generic properties of grand unified theories. Furthermore, the lack of unification in the non-supersymmetric $SU(5)$ model does not mean that the idea of grand unification is not right. As we will discuss subsequently, the $SO(10)$ model as well as the supersymmetric version of the $SU(5)$ model do lead to unification of coupling constants in an interesting manner. We therefore turn to the discussion of fermion masses in this model and more specifically the neutrino masses.

Exercise 7.2 *Write down the evolution equations for $\alpha_n = g_n^2/4\pi$, beginning with Eq. (7.10). Solve these to obtain Eq. (7.12).*

7.2 Neutrino masses in $SU(5)$ model

The most general gauge invariant Yukawa couplings in the $SU(5)$ model are given by

$$\mathcal{L}_Y = h_1 T_{ij}^T C^{-1} F^i H^j + h_2 \, \epsilon^{ijklm} T_{ij}^T C^{-1} T_{kl} H_m + \text{h.c.}, \quad (7.14)$$

where the displayed indices are $SU(5)$ indices. We have suppressed generation indices for the sake of clarity. Also, complex conjugation is implied by lowering and raising of indices. Thus, for example, F^i denotes the complex conjugate of F_i.

Once the $\{5\}$-dimensional Higgs boson H acquires a vev, Eq. (7.14) leads to

$$\begin{aligned} M^{(d)} &= M^{(\ell)} = h_1 v/\sqrt{2} \\ M^{(u)} &= h_2 v/\sqrt{2}, \end{aligned} \qquad (7.15)$$

where, as in Ch. 2, $M^{(u)}$ and $M^{(d)}$ denote the mass matrices of the positively and negatively charged quarks respectively, whereas $M^{(\ell)}$ is the mass matrix for charged leptons. The neutrinos remain massless. To see that it never acquires a mass through radiative corrections, we note that $\mathcal{L}_Y$ is invariant under the global $U(1)_{\mathcal{F}}$ symmetry, if we assign $\mathcal{F}(H_i) = -2/3$, $\mathcal{F}(T_{ij}) = +1/3$ and $\mathcal{F}(F_i) = 1$. This combines with the $U(1)_Y$ to generate the $B - L$ symmetry:

$$B - L = \frac{3}{5}\mathcal{F} - \frac{2}{\sqrt{15}}\lambda_{24}, \qquad (7.16)$$

where λ_{24} is the generator which couples to the gauge boson B_{24} introduced in Eq. (7.4). It is easy to see that the Higgs fields developing vevs do not have any $B - L$ quantum number. Thus $B - L$ remains unbroken even after symmetry breaking. As a result, neutrinos cannot acquire any Majorana mass, which requires $B - L$ violation. It cannot acquire Dirac mass since there is no right-handed neutrino in the model. Thus, $m_\nu = 0$ in the simple $SU(5)$ model.

The simplest way to modify the $SU(5)$ model to generate a massive neutrino is to include a $\{15\}$-dimensional Higgs boson $S_{ij} = S_{ji}$ (symmetric in the $SU(5)$ indices). In the presence of this multiplet, there is a Yukawa coupling of the form:

$$\mathcal{L}'_Y = f F_i^T C^{-1} F_j S^{ij} + \text{h.c.} \qquad (7.17)$$

The multiplet S under $SU(3)_c \times SU(2)_L \times U(1)_Y$ contains an isotriplet Higgs boson Δ_L used in §5.2.1 for extending the standard model to generate neutrino mass. If we assign $\mathcal{F}(S) = +2$, then the model has exact $B - L$ number symmetry. By assigning a non-zero vev to S_{55}, i.e.

$$\langle S_{55} \rangle = \frac{u}{\sqrt{2}}, \qquad (7.18)$$

the global $B - L$ symmetry is spontaneously broken, leading to a triplet Majoron [3]. Measurements of the Z-width [4] have ruled out this model. Let us therefore consider a variation of this model where $B - L$ is explicitly broken in the Higgs potential by the following term:

$$V = \mu_S H_i H_j S^{ij} + \text{h.c.}. \qquad (7.19)$$

This breaks $U(1)_{\mathcal{F}}$ symmetry and therefore the $B - L$ symmetry explicitly. The vev of S in this case is determined to be

$$\langle S_{55} \rangle = \frac{u}{\sqrt{2}} \simeq \frac{\mu_S v^2}{V^2}. \qquad (7.20)$$

Even if μ_S is chosen to be of order V, we get $u \sim 10^{-12}v \simeq 10^{-1}\,\mathrm{eV}$. Thus, in this model, smallness of neutrino mass arises naturally, and the neutrino is a Majorana particle.

This model of course has the same problem as the minimal $SU(5)$ model as far as unification is concerned. It is, therefore, phenomenologically not viable. Yet we presented it to illustrate the techniques and ideas connected with generating neutrino mass in grand-unified theories.

7.3 SO(10)

The next higher symmetry useful for grand-unification of particle interactions is $SO(10)$. An interesting aspect of $SO(10)$ is that it contains the left-right symmetric gauge group $SU(2)_L \times SU(2)_R \times SU(4)_C$ and, therefore, it automatically contains the right-handed neutrino. Since it has quarks and leptons in the same irreducible representation, the mechanism responsible for generating quark lepton masses automatically makes the neutrino massive. Thus, unlike the $SU(5)$ model, in $SO(10)$ model massive neutrinos arise naturally. Below we will describe ways to understand their smallness and study the kind of constraints they impose on the details of $SO(10)$ grand-unification.

> **Exercise 7.3** *Show that the generators of an orthogonal group are antisymmetric matrices. For the $SO(2n)$ groups, the generators in the fundamental (i.e., 2n-dimensional) representation can be chosen to be the matrices T_{ab}, whose elements are given by*
>
> $$(T_{ab})_{pq} \equiv -i(\delta_{ap}\delta_{bq} - \delta_{bp}\delta_{aq})\,.$$
>
> *Deduce the algebra of $SO(2n)$ groups from this. Show that the mutually commuting generators are $T_{12}, T_{34}, T_{56}\ldots$.*

To begin with, the two maximal continuous subgroups of $SO(10)$ are:

- $G_{224} = SU(2)_L \times SU(2)_R \times SU(4)_C$;

- $G_5 = SU(5) \times U(1)$.

The spinor representation [5] of $SO(10)$ is $\{16\}$-dimensional and decomposes under G_{224} and G_5 as follows:

$$
\begin{aligned}
G_{224} \quad &: \{16\} \supset (2,1,4) + (1,2,\overline{4}) \\
G_5 \quad &: \{16\} \supset \{10\}_1 + \{\overline{5}\}_{-3} + \{1\}_5\,.
\end{aligned}
\tag{7.21}
$$

From the $SU(5)$ content of the spinor representation, it is clear that all the known fermions of a single generation can be fitted into the spinor. The $SU(5)$ singlet piece of the spinor can be identified with the right-handed neutrino.

Let us first discuss the symmetry breaking of $SO(10)$ down to the standard model. Since $SO(10)$ has rank 5, there are many possible chains of symmetry breaking including one in which $SU(5)$ appears at an intermediate scale. Since $SU(5)$ is ruled out as a grand-unification symmetry, we will focus on the more interesting chains which contain the left-right symmetric group $SU(4)_C \times SU(2)_L \times SU(2)_R$ as an intermediate symmetry. It is important to realize that the actual maximal subgroup of $SO(10)$ is [6]:

$$SO(10) \supset SU(4)_C \times SU(2)_L \times SU(2)_R \times D, \qquad (7.22)$$

where D is a discrete symmetry under which $f_L \to \hat{f}_L$, i.e., it interchanges the $(2,1,4)$ and the $(1,2,\overline{4})$ sub-multiplets of the $SO(10)$ spinor. It essentially plays the role of charge conjugation, at least on the fermion fields. This symmetry was called D-parity [6] and it plays an important role in understanding neutrino mass, as already alluded to in Ch. 6. Besides, it also has important bearings on the question of baryon asymmetry and domain wall problem in $SO(10)$ models [7]. Two of the important implications of D-parity are: a) it implies $g_L = g_R$ and b) it requires $\eta_B = \eta_{\overline{B}}$ where η_B is the number of baryons in the universe.

Exercise 7.4 *Show that the following identification of the generators of the subgroups of the $SO(10)$ grand-unification group satisfies the algebras of the subgroups:*

$$SU(2)_L \quad : \quad \frac{1}{2}(T_{89} + T_{70}), \frac{1}{2}(T_{97} + T_{80}), \frac{1}{2}(T_{78} + T_{90})$$

$$SU(2)_R \quad : \quad \frac{1}{2}(T_{89} - T_{70}), \frac{1}{2}(T_{97} - T_{80}), \frac{1}{2}(T_{78} - T_{90})$$

$$SU(3)_c \quad : \quad \frac{1}{2}(T_{23} - T_{14}), \frac{1}{2}(T_{31} - T_{24}), \frac{1}{2}(T_{12} - T_{34}),$$

$$\frac{1}{2}(T_{25} - T_{16}), \frac{1}{2}(T_{51} - T_{26}), \frac{1}{2}(T_{45} - T_{36}),$$

$$\frac{1}{2}(T_{53} - T_{46}), \frac{1}{2\sqrt{3}}(T_{12} + T_{34} - 2T_{56})$$

$$B - L \quad : \quad \propto T_{12} + T_{34} + T_{56}. \qquad (7.23)$$

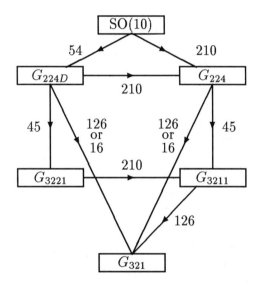

Figure 7.1: Some interesting chains of symmetry breaking from $SO(10)$ down to the standard model group which includes the left-right symmetric group as an intermediate stage. The representation whose vev can induce a certain breaking has also been shown.

If we demand the intermediate symmetry to contain an $SU(2)_L \times SU(2)_R$ group, then some of the possible symmetry breaking chains are listed in Fig. 7.1, where we have also listed the dimensionality of Higgs bosons responsible for each stage of symmetry breaking. To appreciate the effect of a particular kind of Higgs multiplet on symmetry breaking from G to G', we give the decomposition of the multiplets used in Fig. 7.1 under the group G'. For instance, for the case of $\{45\}$, $G = G_{224}$ or G_{224D} and $G' = G_{3221}$. So we give the decomposition of the $\{45\}$-dimensional representation of $SO(10)$ under $SU(3)_c \times SU(2)_L \times SU(2)_R \times U(1)_{B-L}$:

$$\{45\} \xrightarrow{G_{3221}} (1,3,1,0) + (1,1,3,0) + (8,1,1,0) + (1,1,1,0)$$
$$+ (3,2,2,\tfrac{2}{3}) + (\overline{3},2,2,-\tfrac{2}{3}) + (3,1,1,\tfrac{1}{3}) + (\overline{3},1,1,-\tfrac{1}{3}).$$
$$(7.24)$$

We see that $\{45\}$ contains a singlet of the group G_{3221}. If the multiplet develops a vev in that direction, then we obtain G_{3221} as the unbroken subgroup.

Let us now examine the multiplets which can break the $SO(10)$ symmetry down to G_{224} or G_{224D}. For this, we need the multiplets which contains a singlets under the broken symmetries. Note that

$$\{54\} \xrightarrow{G_{224}} (2,2,6) + (1,1,20) + (3,3,1) + (1,1,1) \qquad (7.25)$$
$$\{210\} \xrightarrow{G_{224}} (1,1,15) + (2,2,20) + (3,1,15) + (1,3,15)$$
$$+ (2,2,6) + (1,1,1) \qquad (7.26)$$
$$\{16\} \xrightarrow{G_{321}} (3,2,\tfrac{1}{3}) + (1,2,-1) + (\overline{3},1,-\tfrac{4}{3}) + (\overline{3},1,\tfrac{2}{3})$$
$$+ (1,1,2) + (1,1,0) . \qquad (7.27)$$

For the $\{126\}$-dimensional representation, we first give its reduction under G_{224}:

$$\{126\} \xrightarrow{G_{224}} (1,1,6) + (2,2,15) + (3,1,10) + (1,3,\overline{10}) . \quad (7.28)$$

The multiplet responsible for breakdown of G_{3221} or G_{224} down to the standard model is clearly contained in $(1,3,\overline{10})$.

Again, as in the case of $SU(5)$, the gauge bosons contained in the quotient group $SO(10)/SU(4)_C \times SU(2)_L \times SU(2)_R$ become massive at the first stage of symmetry breaking and contribute to baryon non-conserving processes such as proton decay. There also exist detailed renormalization group analysis including two loop contributions [7, 8], which gives the predictions for $\sin^2 \theta_W$ for different symmetry breaking chains. In contrast with the minimal non-supersymmetric $SU(5)$ model, in this model the coupling constants truly unify [8].

Two chains important for the discussion of neutrino masses are summarized here.

- The first chain is

$$SO(10) \xrightarrow[\{54\}]{M_U} G_{224D} \xrightarrow[\{210\}]{M_p} G_{224} \xrightarrow[\{210\}]{M_C=M_{W_R^+}} G_{2113} \xrightarrow[\{126\}]{M_{Z'}} G_{321} \, . \ (7.29)$$

In this case, the grand-unification scale $M_U \simeq 10^{16.6}\,\text{GeV}$, $M_p \simeq 10^{14}\,\text{GeV}$, $M_{W_R} = M_C \simeq 10^5\text{-}10^7\,\text{GeV}$ with $M_{Z'} \leq 1\,\text{TeV}$, leading to a prediction of $\sin^2 \theta_W \simeq .227$. The important aspect of this symmetry breaking chain relevant to neutrino mass is the value of $M_{Z'}$ (or the scale of $B - L$ breaking) near the TeV scale.

- The second chain of symmetry breaking is given by

$$SO(10) \xrightarrow[\{45\}]{M_U=M_C=M_p} G_{3221} \xrightarrow[\{126\}]{M_{W_R},M_{Z'}} G_{321} \, . \qquad (7.30)$$

This chain is highly predictive and implies $M_U \simeq 10^{15.4}\,\text{GeV}$ for $\sin^2 \theta_W \simeq 0.23$ and $M_{W_R} \simeq M_{B-L} \simeq 10^{12}\,\text{GeV}$.

7.4 Neutrino mass in $SO(10)$ models

Let us now discuss fermion masses in $SO(10)$ theories. To see the Higgs bosons which lead to gauge invariant Yukawa couplings, we note that

$$\{16\} \otimes \{16\} = \{10\} \oplus \{120\} \oplus \{126\} \, . \qquad (7.31)$$

The Higgs bosons that give mass to quarks and leptons must belong to $\{10\}$, $\{120\}$ and $\{126\}$ dimensional representations of $SO(10)$. We can write the Yukawa couplings as:

$$\begin{aligned}
\mathcal{L}_Y = & \sum_{a,b} f^{ab}_{\{10\}} \psi_a^T \mathcal{B}C^{-1} \Gamma^i \psi_b \{10\}_i + f^{ab}_{\{120\}} \psi_a^T \mathcal{B}C^{-1} \Gamma^i \Gamma^j \Gamma^k \psi_b \{120\}_{ijk} \\
& + f^{ab}_{\{126\}} \psi_a^T \mathcal{B}C^{-1} \Gamma^i \Gamma^j \Gamma^k \Gamma^l \Gamma^m \psi_b \{126\}_{ijklm} + \text{h.c.} \qquad (7.32)
\end{aligned}$$

where Γ_i are analogs of Dirac gamma matrices for $SO(10)$ and $\mathcal{B}$ is the analog of the conjugation matrix for the spinor of $SO(10)$; a and b are generation indices.

It turns out that $\{120\}$ couplings are antisymmetric under the interchange of the $\{16\}$ fermions, whereas $\{10\}$ and $\{126\}$ are symmetric under same interchange. Hence, the Yukawa coupling matrices f have the following properties:

$$
\begin{aligned}
f^{ab}_{\{10\}} &= f^{ba}_{\{10\}} \\
f^{ab}_{\{120\}} &= -f^{ba}_{\{120\}} \\
f^{ab}_{\{126\}} &= f^{ba}_{\{126\}} \, .
\end{aligned}
\tag{7.33}
$$

The fermion masses arise when any of $\{10\}_H$, $\{120\}_H$ or $\{126\}_H$ acquire nonzero vev's, with appropriate symmetry conditions satisfied by the mass matrices. Let us consider first the case where only one $\{10\}$-dimensional Higgs boson is used to generate fermion masses. In this case, we have at grand-unification scale:

$$
M^{(u)}_{ab} = M^{(d)}_{ab} = M^{(\ell)}_{ab} = M^{(0)}_{ab} \, .
\tag{7.34}
$$

When these relations are extrapolated to the weak scale, different corrections apply to quark and leptons and we get,

$$
M^{(u)}(M_W) \simeq M^{(d)}(M_W) \simeq 3M^{(\ell)}(M_W) \simeq 3M^{(0)}(M_W) \, .
\tag{7.35}
$$

The degeneracy between up and down quark masses is clearly unrealistic. However, if we employ two $\{10\}$-dimensional Higgs bosons, the up-down mass degeneracy is split. However, we still have the unphysical relation

$$
M^{(u)}(M_W) \simeq 3M^{(0)}(M_W) \, ,
\tag{7.36}
$$

which is clearly unacceptable since it implies Dirac neutrinos with very large masses.

This problem is solved by adding a $\{126\}$-dimensional Higgs boson and assigning vev to the $SU(5)$ singlet component of it. It gives Majorana mass to the right-handed neutrino exactly as in the case of the left-right symmetric model, leading to a Majorana mass for the right handed neutrinos:

$$
\mathcal{L}^{N_R}_{\text{Maj-mass}} = f_{\{126\}} \cdot \left(\frac{M_{BL}}{g} \right) N_R^T C^{-1} N_R \, ,
\tag{7.37}
$$

where we have denoted the vev of the $SU(5)$ singlet component of $\{126\}$ by (M_{BL}/g). Again as the case of left-right symmetric theories, an $SO(10)$ invariant coupling of the form $\{126\} \cdot \{126\} \cdot \{10\} \cdot \{10\}$ appears in the Higgs potential, which leads to a non-zero vev for the $\nu_L \nu_L$ component of $\{126\}$ and a direct Majorana mass for ν_L is induced as follows:

$$\mathcal{L}^{\nu_L}_{\text{Maj-mass}} = \lambda f_{\{126\}} \frac{\kappa^2}{M_{BL}/g} \nu_L^T C^{-1} \nu_L \,, \tag{7.38}$$

where κ denotes the $SU(2)_L$-breaking scale. This leads to a neutrino mass matrix of the form

$$\begin{pmatrix} \lambda f_{\{126\}} \frac{\kappa^2}{M_{BL}/g} & \frac{1}{3} f_{\{10\}} \kappa \\ \frac{1}{3} f_{\{10\}} \kappa & f_{\{126\}} M_{BL}/g \end{pmatrix} \,. \tag{7.39}$$

It gives the following expression for the light neutrino masses:

$$m_\nu^{\text{light}} \simeq \lambda f_{\{126\}} \frac{\kappa^2}{M_{BL}/g} - \frac{1}{9} f_{\{10\}} f_{\{126\}}^{-1} f_{\{10\}} \frac{\kappa^2}{M_{BL}/g} \,. \tag{7.40}$$

This formula is very similar to Eq. (6.66); we see that as $M_{BL} \to \infty$, $m_\nu \to 0$. In models where $B - L$ and $SU(2)_R$ are broken together, we obtain the analytic connection of m_ν and suppression of $V + A$ interactions as advocated in the context of the left-right model [9]. However, the presence of the first term destroys the see-saw formula. But since in a minimal Higgs scenario, this term is bound to appear, we will first discuss its effect on neutrino masses.

Looking at the orders of magnitude, it is clear that the ratio of the second term to the first is $\delta \approx \frac{f_{\{10\}}^2}{9\lambda f_{\{126\}}^2}$. But $f_{\{10\}}$ is proportional to $M^{(u)}/M_W$. Neglecting inter-generational mixings for the moment, we expect its value to be a small number of order 10^{-4} for the first generation, about $\frac{1}{2} \times 10^{-2}$ for second generation and $\frac{1}{4}$ or greater for the third generation. Therefore, if $f_{\{126\}} \simeq 10^{-1}$ and $\lambda \simeq 10^{-1}$-10^{-2}, we expect that

$$\begin{aligned} \delta &\approx 10^{-5}\text{--}10^{-6} && \text{for the 1st generation} \\ &\approx 10^{-2} && \text{for the 2nd generation.} \end{aligned} \tag{7.41}$$

Therefore the first term is likely to dominate in this "minimal" version of the $SO(10)$ model. In this case, we expect neutrinos of all three generations to be almost degenerate. Their masses however depend on M_{BL}. Existing upper limits [10] on m_{ν_e} of 12 eV therefore imply $M_{BL} \geq 10^{10}$ to 10^{11} GeV. If indeed the scale is near that limit, assuming the elements of $f_{\{126\}}$ to be all of the same order of magnitude, we would expect

$$m_{\nu_e} \approx m_{\nu_\mu} \approx m_{\nu_\tau} \approx \text{few eV}. \tag{7.42}$$

Let us next turn to the case where the direct ν_L Majorana mass vanishes (e.g. by assuming $\lambda = 0$ or by a D-parity breaking mechanism to be described below). In this case, if we ignore quark mixing, we have for neutrino masses the following formula:

$$m_{\nu_e} \approx \frac{1}{9} g \frac{m_u^2}{f_{\{126\}} M_{BL}}$$

$$m_{\nu_\mu} \approx \frac{1}{9} g \frac{m_c^2}{f_{\{126\}} M_{BL}}$$

$$m_{\nu_\tau} \approx \frac{1}{9} g \frac{m_t^2}{f_{\{126\}} M_{BL}} \tag{7.43}$$

We have of course assumed the $f_{\{126\}}$ coupling to be roughly generation independent. An important improvement of the $SO(10)$ model over the left-right symmetric model is that here the neutrino Dirac masses are predicted to be equal to the up-quark masses of each generation. The factor $1/9$ is the effect of renormalization group extrapolation down to the weak scale. This first of all predicts the following quadratic mass formula for neutrinos:

$$m_{\nu_e} : m_{\nu_\mu} : m_{\nu_\tau} = m_u^2 : m_c^2 : m_t^2 . \tag{7.44}$$

Therefore, using [11] $m_t = 180$ GeV and the upper limit of 24 MeV on ν_τ mass given in Eq. (3.1), we predict

$$m_{\nu_\mu} \leq 2 \text{ keV} . \tag{7.45}$$

This is to be compared with the existing experimental upper limit of 170 keV.

At this stage, depending on the symmetry breaking chain one chooses, the neutrino mass spectrum can be different. Some interesting scenarios are described below.

Low M_{BL} [Eq. (7.29)]

If M_{BL} is in the TeV range (say 5 TeV), then choosing $m_u \simeq 7\,\text{MeV}$, we get

$$
\begin{aligned}
m_{\nu_e} &\approx 0.65\,\text{eV}/f_{\{126\}}\,, \\
m_{\nu_\mu} &\approx 30\,\text{keV}/f_{\{126\}}\,, \\
m_{\nu_\tau} &\approx 70\,\text{MeV}/f_{\{126\}}\,.
\end{aligned}
\tag{7.46}
$$

To be consistent with experiments, either $f_{\{126\}} \geq 3$ or $M_{BL} \geq 15\,\text{TeV}$ if $f_{\{126\}} \simeq 1$. In this case, one of course has to extend the model to let the heavy neutrinos decay so that they can be acceptable cosmologically. One such mechanism has been discussed within an $SO(10)$ context where ν_μ and ν_τ decay via a Goldstone boson emission [12].

Intermediate M_{BL} scenario [Eq. (7.30)]

An interesting scenario is the two step $SO(10)$ breaking scenario advocated in Ref. [13], where $SO(10)$ breaks to the standard model via a left-right intermediate symmetry. Here, $M_{BL} \simeq 10^{12}\,\text{GeV}$ from constraints of $\sin^2 \theta_W$. In this case, simply scaling the Eq. (7.46) by a factor of 10^7, we predict (roughly, since $f_{\{126\}}$ is unknown)

$$
\begin{aligned}
m_{\nu_e} &\approx 10^{-7}\,\text{eV}\,, \\
m_{\nu_\mu} &\approx 10^{-3}\,\text{eV}\,, \\
m_{\nu_\tau} &\approx 10^{-1}\,\text{eV}\,.
\end{aligned}
\tag{7.47}
$$

The interesting point about this scenario is that this kind of mass spectrum is required to explain the solar neutrino puzzle via an MSW type matter oscillation effect, which will be discussed in Ch. 15. This model therefore deserves special attention. An additional theoretical argument in favor of this model is that $U(1)_{\text{PQ}}$-symmetry needed to solve the strong CP-problem in the class of invisible axion models [14] also has a symmetry breaking scale around 10^{12} GeV. The models of Ref. [13] emphasize this point by identifying the $U(1)_{\text{PQ}}$ symmetry breaking scale. The proton lifetime in this model is predicted to be around 10^{34} yrs and may be within reach of dedicated next generation experimental searches.

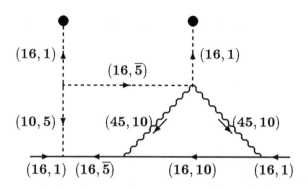

Figure 7.2: Two-loop diagram giving rise to the Majorana mass of right handed neutrinos in Witten's model. The notation (16,1) means the $SU(5)$ singlet part of the 16 representation of $SO(10)$.

Loop-induced heavy mass for N_R

An interesting scenario which does not fall into either of the above cases was proposed by Witten [15] where the actual $B - L$ breaking scale is high (i.e. perhaps of order M_G) but the effective N_R mass that goes into the see-saw matrix is actually much lower. The way that happens is that, even though the $B - L$ symmetry is broken at the tree level, it can be broken by a properly chosen $SO(10)$ representation which has no Yukawa coupling to the fermions. As a result, no tree level mass for N_R will be present. However N_R mass can in general arise from radiative corrections and will therefore be much smaller. An explicit realization of this idea [15] is to break $B - L$ symmetry by a 16-dimensional Higgs multiplet. In this case the N_R mass arises at the two-loop level from the diagram of Fig. 7.2. We get

$$M_N = \frac{f_{\{10\}} g^4}{(16\pi^2)^2} \frac{v_{BL}^2}{M_G} \qquad (7.48)$$

This leads to $M_N \sim 10^{-6}$ to $10^{-7} M_G \approx 10^8$ GeV, which in turn yields neutrino masses much larger than given in Eq. (7.47). Notice that, since the heavy neutrino mass is proportional to the Yukawa coupling $f_{\{10\}}$, the see-saw mechanism now gives light neutrino masses proportional to the single power of up quark masses.

In concluding this section, we mention how the D-parity breaking scenario of $SO(10)$ models helps to restore the see-saw picture for neutrino masses by suppressing the direct ν_L-Majorana mass term in Eq. (7.39). The basic idea is the same as in Ch. 6 except here we break the $SO(10)$ symmetry either by a $\{210\}$-dimensional representation or the $\{45\}$-dimensional representation. Writing the $\{210\}$ dimensional representation as a 4-index antisymmetric tensor $\{210\}_{ijkl}$, the D-parity and $SO(10)$ breaking is achieved by giving the vev

$$\langle\{210\}_{7890}\rangle = M_U/g. \tag{7.49}$$

This lowers the $SO(10)$ symmetry to $SU(2)_L \times SU(2)_R \times SU(4)_C$. If instead we use the antisymmetric rank-2 tensor representation $\{45\}$, we can give vev to

$$\langle\{45\}_{12}\rangle = \langle\{45\}_{34}\rangle = \langle\{45\}_{56}\rangle = M_U/g. \tag{7.50}$$

This breaks D-parity and reduces the $SO(10)$ symmetry to $SU(3)_c \times SU(2)_L \times SU(2)_R \times U(1)_{B-L}$. The consequence of D-parity breaking is that it asymmetrizes the Higgs spectrum: for instance, if we denote M_p to be the D-parity breaking scale, then the masses of Δ_L and Δ_R in $\{126\}$ become different, i.e.,

$$M_{\Delta_L}^2 \simeq M_p^2, \tag{7.51}$$

whereas

$$M_{\Delta_R}^2 \simeq M_{BL}^2. \tag{7.52}$$

As a result, we have

$$\begin{aligned}
\langle\Delta_L^0\rangle &\simeq \lambda\frac{\kappa^2 M_{BL}}{M_p^2} \\
&= \lambda\left(\frac{M_{BL}}{M_p}\right)^2\left(\frac{\kappa^2}{M_{BL}}\right).
\end{aligned} \tag{7.53}$$

Eq. (7.41) then implies that if $M_{BL}/M_p \leq 10^{-3}$, the true see-saw picture emerges [16].

7.5 Predictive $SO(10)$ scenarios for neutrino masses

In the previous section, we noted that the see-saw mechanism for neutrino masses finds its natural embedding in the $SO(10)$ models. While this provides a convincing way to understand the smallness of the neutrino masses compared to the corresponding charged fermion masses, it falls far short of predicting the detailed mass and mixing patterns for neutrinos in general. The reason for this is that, while in the $SO(10)$ model the Dirac masses for the neutrinos are related to the up quark mass matrix, the Majorana mass matrix for the right handed neutrinos involves unknown Yukawa couplings. There is however a class of minimal $SO(10)$ models, where the Majorana mass for the heavy right handed neutrino gets related to the quark and lepton masses and therefore gets predicted. This model therefore leads to firm predictions for the mixings as well as masses for neutrinos [17]. The model is also consistent with all other phenomenological constraints and could therefore be the first $SO(10)$ model to be tested by the current generation of neutrino experiments. Let us briefly describe the model and explain why it has the predictive power in the neutrino sector. We will present only the non-supersymmetric version although a supersymmetric version has been constructed in the second paper of [17].

The Higgs sector of the model is assumed to consist of a $\{210\}$-dimensional representation to break the $SO(10)$ symmetry down to $SU(2)_L \times SU(2)_R \times SU(4)_c$ model at the GUT scale. The only other Higgs fields of the theory are a complex $\{10\}$ and a $\{126\}$ dimensional (denoted by **H** and Δ respectively). To start with we will assume that as usual the $\{126\}$-dimensional multiplet breaks the $SU(2)_L \times SU(2)_R \times SU(4)_c$ group down to the standard model gauge group. It however turns out that the Higgs potential of this model consists of terms of the form $\Delta \overline{\Delta} \Delta H$. In the G_{224} decomposition, this coupling is of the form $(1, 3, \overline{10})(1, 3, 10)(2, 2, 15)(2, 2, 1)$. It is then clear that after electroweak symmetry breaking via the $\{10\}$-dimensional multiplet, the $(2, 2, 15)$ sub-multiplet of the $\{126\}$-Higgs acquires a vev. This simple observation has profound implications for the fermion masses. To see this, let us further assume that the Yukawa couplings of the fermions have only the following terms:

$$\mathcal{L}_Y = h_{ab}\psi_a\psi_b H + f_{ab}\psi_a\psi_b\overline{\Delta} \qquad (7.54)$$

Note that we have not included a coupling of the form $\psi\psi H^*$ which is forbidden if the model is supersymmetrized. The same can also happen if the model has a Peccei-Quinn symmetry. Now we see that the coupling f_{ab} plays a role in heavy neutrino Majorana masses after $B - L$ breaking and also contributes to quark and lepton masses due to the induced $(\mathbf{2,2,15})$ vev. The important point to note is that the Yukawa Lagrangian of the model has a total of twelve parameters which can be counted as follows. We can choose a basis so that h is diagonal, so that there are three parameters there. The coupling of Δ, being symmetric, has six parameters. As a result the Yukawa couplings give nine parameters. Three parameters come from the four Higgs vev's (two from H and two from from Δ) after the W-mass constraint is used. Note that in the fermion sector there are nine masses (up, down and charged leptons) and three CKM mixing angles which are known. Thus all parameters in the Yukawa sector of the theory are predicted. Thus the f_{ab} which give the right handed neutrino masses are now predicted yielding a predictive scenario for neutrino masses and mixings. The detailed predictions of the model are given in Ref. [17] and we do not reproduce them here. Let us merely display the formulae for the fermion masses that arises in this model:

$$
\begin{aligned}
M_u &= h\kappa_u + fv_u \\
M_d &= h\kappa_d + fv_d \\
M_\nu^D &= h\kappa_u - 3fv_u \\
M_\ell &= h\kappa_d - 3fv_d \\
M_{\nu_R} &= fv_R
\end{aligned}
\qquad (7.55)
$$

where $\kappa_{u,d}$ and $v_{u,d}$ are the vev's of the standard model Higgs doublets in $\{10\}$ and $\{126\}$ Higgs multiplets. Note that the same Yukawa couplings are involved in all mass matrices which is at the heart of the predictive power of this model. One important point to note is that one actually obtains several sets of predictions for the neutrino masses and mixings depending on the choice of the signs for the quark masses and one can actually compare those numbers with data. In fact one takes the MSW interpretation of the solar neutrino data, most of the predictions get eliminated.

Several other procedures for predicting the neutrino masses and mixings have also been considered in recent literature by postulating specific forms for the quark mass matrices and the right handed neutrino masses [18]. We do not discuss them here.

7.6 Neutrino masses in E_6

In this last section of this chapter, we turn to E_6 grand-unified theories and study the neutrino masses. Let us first discuss a few facts about the E_6 theories [19]. There are several maximal subgroups of E_6:

$$E_6 \supset \begin{cases} SO(10) \times U(1) \\ SU(3) \times SU(3) \times SU(3) \\ SU(6) \times SU(2) \,. \end{cases} \tag{7.56}$$

The fundamental representation of E_6 is $\{27\}$-dimensional. Under the above subgroups $\{27\}$ decomposes as follows:

$$\{27\} \overset{SO(10) \times U(1)}{\longrightarrow} \{16\}_{+1} \oplus \{10\}_{-2} \oplus \{1\}_{+4}$$
$$\overset{[SU(3)]^3}{\longrightarrow} (3, \overline{3}, 1) \oplus (\overline{3}, 1, 3) \oplus (1, 3, \overline{3})$$
$$\overset{SU(6) \times SU(2)}{\longrightarrow} (15, 1) \oplus (\overline{6}, 2) \,. \tag{7.57}$$

A look at the $SO(10) \times U(1)$ reduction of $\{27\}$ makes it clear that $\{27\}$ is suitable for assigning the fermion fields. Clearly it contains additional fermion fields than are present in the standard model. Their discovery, direct or indirect, will provide evidence for E_6 grand-unification. We will use the $[SU(3)]^3$ decomposition and assign the fermions as follows:

$$\{27\} \supset \begin{pmatrix} u \\ d \\ g \end{pmatrix} \quad : \quad (3, \overline{3}, 1)$$

$$\oplus \begin{pmatrix} \widehat{u} \\ \widehat{d} \\ \widehat{g} \end{pmatrix} \quad : \quad (\overline{3}, 1, 3)$$

$$\oplus \begin{pmatrix} E_u^0 & E_u^+ & e^+ \\ E_d^- & E_d^0 & \widehat{\nu} \\ e^- & \nu & n \end{pmatrix} : (1, 3, \overline{3}) \tag{7.58}$$

Among the new fermions required by E_6 are: a new color triplet isosnglet quark g and its antiparticle $\hat{g}$, and in addition five new leptons: E_u^0, E_u^+, E_d^-, E_d^0 and n, the last one being a singlet under $SO(10)$. The model has got five neutral leptons in each generation. Therefore its neutral lepton sector promises to be rich in physics. With so many neutral particles in a single generation, we choose to ignore generational mixing for the rest of this chapter and consider just the first generation.

Turning now to the gauge bosons, the adjoint representation of E_6 is $\{78\}$-dimensional. Under $SO(10) \times U(1)$ and $[SU(3)]^3$ it decomposes as follows:

$$\{78\} \xrightarrow{SO(10) \times U(1)} \{45\}_0 \oplus \{16\}_{+3} \oplus \{\overline{16}\}_{-3} \oplus \{1\}_0$$
$$\{78\} \xrightarrow{[SU(3)]^3} (8,1,1) \oplus (1,8,1) \oplus (1,1,8) \oplus (3,3,3) \oplus (\overline{3},\overline{3},\overline{3}) \,. \tag{7.59}$$

In order to study neutral lepton masses, let us briefly touch on the question of symmetry breaking.

The breaking of $E_6 \to SO(10)$ can be easily achieved by means of a $\{27\}$-dimensional Higgs boson. On the other hand to break $E_6 \to [SU(3)]^3$, one needs to include $\{650\}$ dimensional representation. It turns out that in superstring inspired models, the flux breaking mechanism [20] breaks E_6 down to $[SU(3)]^3$ automatically.

Even in a single generation, the neutrino mass matrix in this model is a 5×5 matrix and is in general very involved. One simple possibility is to use $\{27\}$-dimensional Higgs boson and $\{351'\}$ Higgs boson denoted by H_1 and Δ respectively. Denoting the $\{27\}$-dimensional fermion by ψ, we have the following Yukawa couplings:

$$\mathcal{L}_Y = h_1 \psi \psi H_1 + h_2 \psi \psi \Delta \,. \tag{7.60}$$

If we give a vev to $\langle n(H) \rangle$, then it gives a Dirac mass coupling $E_u^0 E_d^0$ in Eq. (7.58). If we choose $\langle n(H) \rangle$ heavy enough, two neutral leptons decouple from the low energy spectrum. We may then try to employ a see-saw type mechanism to get a light Majorana neutrino. The right-handed neutrino Majorana mass can be gotten from the Δ-coupling in Eq. (7.60) if we consider the $\{126\}$ dimensional representation of $SO(10)$

that is contained in the $SO(10) \times U(1)$ decomposition of E_6:

$$\{351'\} \xrightarrow{SO(10) \times U(1)} \{1\}_{-8} \oplus \{10\}_{-2} \oplus \{\overline{16}\}_{-5} \oplus \{54\}_4$$
$$\oplus \{\overline{126}\}_{-2} \oplus \{\overline{144}\}_1 . \qquad (7.61)$$

The Dirac masses of the neutrinos can be obtained by using another $\{27\}$-dimensional Higgs H_2 multiplet and giving vev to the following components:

$$\left\langle E_u^0(H_2) \right\rangle = \kappa_1 , \quad \left\langle E_d^0(H_2) \right\rangle = \kappa_2 . \qquad (7.62)$$

In this case, one has in the symmetry limit the following mass relations:

$$m_u = m_D = h_2 \kappa_1$$
$$m_d = m_e = h_2 \kappa_2 . \qquad (7.63)$$

Then, a see-saw scenario similar to the case of $SO(10)$ emerges. In this case, the fifth neutral lepton n_0 remains fully decoupled from other matter.

On the other hand, if the $\{351'\}$ is not included in the Higgs spectrum, one can use an extra singlet fermion, S and use an extra $\{27\}$-dimensional Higgs boson H_3 whose vev is $\langle \hat{\nu}(H_3) \rangle = V_{BL}$ and whose Yukawa coupling with fermions is h_3. In this case, at low energies $\mu \ll \langle n(H_2) \rangle$, we obtain a 3×3 neutrino mass matrix [21]:

$$\begin{pmatrix} 0 & m_D & 0 \\ m_D & 0 & h_3 V_{BL} \\ 0 & h_3 V_{BL} & 0 \end{pmatrix} \qquad (7.64)$$

Since this matrix has zero determinant, one of the eigenvalues must be zero. This eigenstate is given by

$$\nu + \epsilon \hat{\nu} \quad \text{where } \epsilon \simeq \frac{m_D}{h_3 V_{BL}} . \qquad (7.65)$$

The other two eigenvalues must be equal and opposite since the trace of the matrix in Eq. (7.64) is zero. In such a case, as was discussed in §4.5, they form a Dirac neutrino with mass $|h_3 V_{BL}|$. This is a rather interesting scenario that has been used in superstring models.

Chapter 8

Neutrino mass in supersymmetric models

8.1 Introduction

One of the fundamental new symmetries of nature that has been the subject of intense discussion in particle physics of the past decade is the symmetry between bosons and fermions, known as supersymmetry. This symmetry was introduced in the early 1970's by Golfand, Likhtman, Akulov, Volkov, Wess and Zumino. In addition to the obvious fact that it provides the hope of an unified understanding of the two known forms of matter, the bosons and fermions, it has also provided a mechanism to solve two conceptual problems of the standard model, viz. the possible origin of the weak scale as well as its stability under quantum corrections. The recent developments in strings, which embody supersymmetry in an essential way also promise the fulfillment of the eternal dream of all physicists to find an ultimate theory of everything. It would thus appear that a large body of contemporary particle physicists have accepted that theory of particles and forces must incorporate supersymmetry. This will of course have important implications for the physics of the neutrinos and the related phenomenon of lepton number violation, which we will consider in this chapter. Let us begin with a brief overview of the basic notions of supersymmetry.

There are many ways to introduce supersymmetric field theories and we refer to several texts [1, 2] and reviews [3] for more thorough discussion. Here we will content ourselves with a brief outline only.

Since supersymmetry transforms a boson to a fermion and vice versa, an irreducible representation of supersymmetry will contain in it both fermions and bosons. Therefore in a supersymmetric theory, all known particles are accompanied by a superpartner which is a fermion if the known particle is a boson and vice versa. For instance, the electron (e) super-multiplet will contain its superpartner $\tilde{e}$, (called the selectron) which has spin zero. The photon (γ) super-multiplet will contain its superpartner, $\tilde{\gamma}$, known as the photino, which is a Majorana field of spin-$\frac{1}{2}$. We will adopt the notation that the superpartner of a particle will be denoted by the same symbol as the particle with a 'tilde' as above. The name of the superpartner of any known fermion will have an extra letter 's' at the beginning, and the names of known bosons will end with the letters '-ino'. Furthermore, while supersymmetry does not commute with the Lorentz transformations, it commutes with all internal symmetries; as a result, all non-Lorentzian quantum numbers for both the fermion and boson in the same super-multiplet are the same.

A convenient mathematical way to discuss supersymmetry is to extend space-time to a superspace where one augments the normal space-time with the addition of anticommuting Grassmanian coordinates denoted by θ, where θ is a Majorana spinor. As elaborated in Ch. 4, such spinors have two independent complex components, or four independent real components. Equivalently, one can use the chiral projections θ_L and θ_R, each of which has two independent real components. Operator valued functions in the superspace will be called superfields and denoted by $\Phi(x, \theta_L, \theta_R)$. The Grassmanian property of the θ coordinates (i.e., $\theta_1 \theta_2 = -\theta_2 \theta_1$ and $(\theta_i)^2 = 0$) then implies that an expansion of the superfields Φ in the θ coordinate terminates after a few terms. The coefficients of such an expansion will be functions of x only, and are therefore fields of the ordinary field theory. Those fields will create and destroy the known particles and their superpartners. The superfields can of course have any spin but for the description of the particles of the standard model, it is enough to consider only superfields with spin zero.

There exist three kinds of spin-zero superfields which form irreducible representations of supersymmetry: (i) the chiral and anti-chiral super-fields, (ii) vector superfields and (iii) linear multiplets. Again for our purpose, we only need the first two kinds of multiplets. The chiral (and anti-chiral) multiplets contain a spin zero and a spin half particle and will be used to describe matter as well as Higgs fields. On the other hand the vector multiplets contain a spin one and a spin half fields and will be used to describe gauge fields. It turns out that real vector fields allow transformations, which can be identified with the gauge transformations as is needed if they are to describe gauge fields.

8.2 The Lagrangian for supersymmetric field theories

In order to write down the action for a supersymmetric field theory, let us start by considering generic chiral fields denoted by $\Phi(x, \theta_L)$ with component fields given by (ϕ, ψ) and gauge fields denoted by $V(x, \theta_L, \theta_R)$ with component gauge and gaugino fields given by (A^μ, λ). The action in the superfield notation is

$$
\begin{aligned}
S = & \int d^4x \int d^2\theta_L d^2\theta_R \; \Phi^\dagger e^V \Phi \\
& + \int d^4x \int d^2\theta_L \; (P(\Phi) + W(V)W(V) + \text{h.c.}) .
\end{aligned} \tag{8.1}
$$

In the above equation, the first term gives the gauge invariant kinetic energy term for the matter fields Φ; $P(\Phi)$ is a holomorphic function of Φ and is called the superpotential; it leads to the Higgs potential of the usual gauge field theories. Secondly, $W(V) \equiv \mathcal{D}^2 \overline{\mathcal{D}} V$ where $\mathcal{D} \equiv \partial\theta - i\sigma.\partial x$, and the term involving $W(V)$ leads to the gauge invariant kinetic energy term for the gauge fields as well as for the gaugino fields. In terms of the component fields the Lagrangian can be written as

$$
\mathcal{L} = \mathcal{L}_g + \mathcal{L}_{\text{matter}} + \mathcal{L}_Y - V(\phi) \tag{8.2}
$$

where

$$
\begin{aligned}
\mathcal{L}_g &= -\frac{1}{4} F^{\mu\nu} F_{\mu\nu} + \frac{1}{2}\overline{\lambda}\gamma^\mu i D_\mu \lambda \\
\mathcal{L}_{\text{matter}} &= |D_\mu \phi|^2 + \overline{\psi}\gamma^\mu i D_\mu \psi \\
\mathcal{L}_Y &= \sqrt{2}g\overline{\lambda}\psi\phi^\dagger + \psi_a \psi_b P_{ab} \\
V(\phi) &= |P_a|^2 + \frac{1}{2}\mathcal{D}_\alpha \mathcal{D}_\alpha
\end{aligned} \tag{8.3}
$$

where D_μ stands for the covariant derivative with respect to the gauge group and $\mathcal{D}_\alpha$ stands for the so-called $\mathcal{D}$-term and is given by $\mathcal{D}_\alpha = g\phi^\dagger T_\alpha \phi$ (g is the gauge coupling constant and T_α are the generators of the gauge group). P_a and P_{ab} are the first and second derivative of the superpotential P with respect to the superfield with respect to the field Φ_a, where the index a stands for different matter fields in the model.

Several new features of a supersymmetric field theories (as compared to a non-supersymmetric one) are evident from the Eq. (8.3). First, note that the Yukawa couplings and the first contribution to the Higgs potential arise from the same function $P(\Phi)$ and therefore their parameters are related. The second term in the Higgs potential involves only the gauge coupling as the free parameter. Thus the number of parameters in a SUSY field theory are expected to be fewer than the usual gauge theories. Secondly, there is a new kind of Yukawa coupling involving the gaugino, matter fermion ψ and the superpartner of the matter field with the Yukawa coupling given by the gauge coupling constant g. These generic features have many important implications for the phenomenology of these models.

To see one immediate phenomenological implication of supersymmetry, consider a superpotential $P(\Phi) = m\Phi^2$. Using Eq. (8.3), it is very easy to see that the scalar field ϕ and the fermion field ψ both have the same mass. This is again a generic prediction of the SUSY models, i.e. the particle and its superpartner have the same mass in the supersymmetric limit. This is clearly against observations for all known particles implying that in a realistic model, supersymmetry must be broken. As in the case of other global symmetries, supersymmetry breaking can be explicit or spontaneous. In the latter case, the analogs of Nambu-Goldstone theorem and the Higgs-Kibble mechanism imply that if supersymmetry is global and is spontaneously broken (i.e., $Q_{\text{susy}}|0\rangle \neq 0$), then there must exist a massless fermion in the particle spectrum (to be called the Goldstino) and it will obey low energy theorems — the analog of Adler's zeros for the pion. Since no particle with these properties is known to exist in nature, we have to assume that supersymmetry is either explicitly broken or more elegantly, supersymmetry is a local symmetry which is spontaneously broken. In the latter case, the analog of the Higgs-Kibble mechanism for this case leads to the conclusion that the Goldstino becomes the longitudinal mode of the gauge particle corresponding to the local supersymmetry. A very exciting aspect of local supersymmetry is that the corresponding gauge super-multiplet contains the graviton (which has spin 2) and its superpartner called the gravitino (which has spin 3/2). The Higgs mechanism for local SUSY leads to the Goldstino being the longitudinal mode of the massive spin 3/2 gravitino. It turns out that in this process, the theory also generates SUSY breaking terms involving the matter and other fields [1, 2]. It is beyond the scope of this book to provide derivation of these results.

Exercise 8.1 *Using the superpotential $m\Phi^2 + \lambda\Phi^3$ for a chiral superfield Φ, show that the component scalar field ϕ and the fermion field ψ both have the same mass. What is the relation between the Yukawa coupling and the quartic scalar coupling?*

8.3   Soft breaking of supersymmetry

One other very important property of supersymmetric field theories is their ultraviolet behavior. They have the extremely important property that in the exact supersymmetric limit, the parameters of the superpotential $P(\Phi)$ do not receive any (finite or infinite) corrections from Feynman diagrams involving the loops. In other words, if the value of a superpotential parameter is fixed at the classical level, it remains unchanged to all orders in perturbation theory. This is known as the non-renormalization theorem [4].

This observation was realized as the key to solving the Higgs mass problem of the standard model as follows: the radiative corrections to the Higgs mass in the standard model are quadratically divergent and admits the Planck scale as a natural cutoff if there is no new physics upto that level. Since the Higgs mass is directly proportional to the mass of the W-boson, the loop corrections would push the W-boson mass to the Planck scale destabilizing the standard model. On the other hand in the supersymmetric version of the standard model (to be called MSSM), in the limit of exact supersymmetry, there are no radiative corrections to any mass parameter and therefore to the Higgs boson mass which is a parameter of the superpotential. Thus if the world could be supersymmetric at all energy scales, the weak scale stability problem would be easily solved. However, since supersymmetry must be a broken symmetry, one has to ensure that the terms in the Hamiltonian that break supersymmetry do not spoil the non-renormalization theorem in a way that infinities creep into the self mass correction to the Higgs boson. This is precisely what happens if effective supersymmetry breaking terms are "soft" which means they are of the following type:

1. $m_a^2 \phi_a^\dagger \phi_a$, where ϕ is the bosonic component of the chiral superfield Φ_a;

2. $m \int d^2\theta_L \theta_L^2 \left(A P^{(3)}(\Phi) + B P^{(2)}(\Phi) \right)$, where $P^{(3)}(\Phi)$ and $P^{(2)}(\Phi)$ are the second and third order polynomials in the superpotential.

3. $\frac{1}{2} m_\lambda \lambda^\mathsf{T} C^{-1} \lambda$, where λ is the gaugino field.

It can be shown that the soft breaking terms only introduce finite loop corrections to the parameters of the superpotential. Since all the soft breaking terms require couplings with positive mass dimension, the loop corrections to the Higgs mass will depend on this mass and we must keep these masses less than a TeV so that the weak scale remains stabilized. This has the interesting implication that superpartners of the known particles are accessible to the ongoing and proposed collider experiments. For a recent survey of the experimental situation, see Ref. [5].

The mass dimensions associated with the soft breaking terms depend on the particular way that supersymmetry is broken. It is usually assumed that supersymmetry is broken in a sector that involves fields which do not have any quantum numbers under the standard model group. This is called the hidden sector. The supersymmetry breaking is transmitted to the visible sector either via the gravitational interactions [6] or via the gauge interactions of the standard model [7] or via anomalous U(1) $\mathcal{D}$-terms [8]. There is one particular scenario using gravity to transmit the supersymmetry breaking which has been very widely discussed and forms much of the basis for the discussion in current supersymmetry phenomenology. This was suggested by Polonyi and is based upon the following assumptions: first, the superpotential describing the model is a sum of two terms: $P = P_H + P_{\text{vis}}$. The P_H, the hidden sector potential consists of a gauge singlet field which we will denote by z and P_{vis} consists of the known fields of the standard model. Let us first give P_H:

$$P_H = \mu^2 (z + \beta) \qquad (8.4)$$

where μ and β are mass parameters to be fixed by various physical considerations. For instance, requiring the cosmological constant to vanish fixes $\beta = (2 - \sqrt{3}) M_{\text{Pl}}$. Given this potential, supergravity calculus predicts that the generic soft breaking parameters m are given by

$m \sim \mu^2/M_{\text{Pl}}$. Requiring m to be in the TeV range implies that $\mu \sim 10^{11}$ GeV. This model also predicts A and B given above. It is also important to point out that the super-Higgs mechanism operates at the Planck scale. Therefore all parameters derived at the tree level of this model need to be extrapolated to the electroweak scale. So after the soft-breaking Lagrangian is extrapolated to the weak scale, it will look like:

$$\mathcal{L}^{SB} = m \sum_{i,j,k} A_{ijk} \phi_i \phi_j \phi_k + \sum_{i,j} B_{ij} \phi_i \phi_j$$

$$+ \sum_i \mu_i^2 |\phi_i^2| + \sum_a \frac{1}{2} M_a \lambda_a^{\mathsf{T}} C^{-1} \lambda_a + \text{h.c.} \qquad (8.5)$$

8.4 Supersymmetric standard model

We will now apply the discussions of the previous section to construct the supersymmetric extension of the standard model so that the goal of stabilizing the Higgs mass is indeed realized in practice. In Table 8.1, we give the particle content of the model.

First note that an important difference between the standard model and its supersymmetric version apart from the presence of the super-partners is the presence of a second Higgs doublet. This is required both to give masses to quarks and leptons as well as to make the model anomaly-free. The gauge interaction part of the model is easily written down following the rules laid out in Ch. 2. In the weak eigenstate basis, weak interaction Lagrangian for the quarks and leptons is exactly the same as in the standard model. As far as the weak interactions of the squarks and the sleptons is concerned, the generation mixing angles are very different from those in the corresponding fermion sector due to supersymmetry breaking. This has the phenomenological implication that the gaugino-fermion-sfermion interaction changes generation leading to potentially large flavor changing neutral current effects such as K^0-$\widehat{K}^0$ mixing, $\mu \to e\gamma$ decay etc unless the sfermion masses of different generations are chosen to be very close in mass.

Let us now proceed to a discussion of the superpotential of the model. It consists of two parts:

$$W = W_1 + W_2, \qquad (8.6)$$

Table 8.1: The particle content of the supersymmetric standard model. For matter and Higgs fields, we have shown the left-chiral fields only. The right-chiral fields will have a conjugate representation under the gauge group.

Superfield	Particles	Superpartners	gauge transformation
Quarks Q	(u, d)	$(\tilde{u}, \tilde{d})$	$(3, 2, \frac{1}{3})$
Antiquarks $\hat{U}$	$\hat{u}$	$\tilde{\hat{u}}$	$(3^*, 1, -\frac{4}{3})$
Antiquarks $\hat{D}$	$\hat{d}$	$\tilde{\hat{d}}$	$(3^*, 1, \frac{2}{3})$
Leptons L	(ν, e)	$(\tilde{\nu}, \tilde{e})$	$(1, 2 - 1)$
Antileptons $\hat{E}$	$\hat{e}$	$\tilde{\hat{e}}$	$(1, 1, 2)$
Higgs Boson $\mathbf{H_u}$	(H_u^+, H_u^0)	$(\tilde{H}_u^+, \tilde{H}_u^0)$	$(1, 2, +1)$
Higgs Boson $\mathbf{H_d}$	(H_d^0, H_d^-)	$(\tilde{H}_d^0, \tilde{H}_d^-)$	$(1, 2, -1)$
Color Gauge Fields	G_a	$\tilde{G}_a$	$(8, 1, 0)$
Weak Gauge Fields	$W^\pm, Z$	$\tilde{W}^\pm, \tilde{Z}$	
Photon	γ	$\tilde{\gamma}$	

where

$$W_1 = h_\ell^{ij}\hat{E}_i L_j \mathbf{H_d} + h_d^{ij}Q_i\hat{D}_j\mathbf{H_d} + h_u^{ij}Q_i\hat{U}_j\mathbf{H_u} + \mu\mathbf{H_u}\mathbf{H_d},$$
$$W_2 = \lambda_{ijk}L_iL_j\hat{E}_k + \lambda'_{ijk}L_iQ_j\hat{D}_k + \lambda''_{ijk}\hat{U}_i\hat{D}_j\hat{D}_k, \qquad (8.7)$$

i, j, k being generation indices. We first note that the terms in W_1 conserve baryon and lepton number whereas those in W_2 do not. The latter are known as the R-parity breaking terms where R-parity is defined as

$$R = (-1)^{3(B-L)+2S}, \qquad (8.8)$$

where S is the spin of the particle. It is interesting to note that the R-parity symmetry defined above assigns even R-parity to known particles of the standard model and odd R-parity to their superpartners. This has the important experimental implication that for theories that conserve R-parity, the super-partners of the particles of the standard model must always be produced in pairs and the lightest superpartner must be a stable particle. This is generally called the LSP. If the LSP turns out to be neutral, it can be thought of as the dark matter particle of the universe.

We now embed this model into the minimal $N = 1$ supergravity model with a Polonyi type hidden sector. As a result, we get the mass splitting for the squarks and sleptons from the quarks and the leptons. We also get trilinear scalar interactions among the sfermions as follows:

$$\mathcal{L}^{SB} = m_{3/2}[A_{e,ab}\widetilde{\bar{E}}_a\widetilde{L}_b H_d + A_{d,ab}\widetilde{Q}_a H_d \widetilde{\bar{d}}_b + A_{u,ab}\widetilde{Q}_a H_u \widetilde{\bar{u}}_b]$$

$$+ B\mu m_{3/2} H_u H_d + \sum_{i=\text{scalars}} \mu_i^2 \phi_i^\dagger \phi_i + \sum_a \frac{1}{2} M_a \lambda^\mathsf{T} C^{-1} \lambda_a \qquad (8.9)$$

There will also be the corresponding terms involving the R-parity breaking terms, which we omit here for simplicity.

The subject of supersymmetry is now a vast topic which touches many areas of particle physics starting from physics just beyond the standard model to physics of grand unification and beyond to superstrings. In this chapter, we will discuss only those aspects that have a bearing on neutrino mass. This discussion can be neatly split into two parts: (*i*) the MSSM with R-parity breaking [2] and (*ii*) supersymmetric left-right model which leads to automatic R-parity conservation [9]. Both kinds of theories lead to non-vanishing neutrino masses in two very different ways. Related lepton number violating phenomena also have different complexions as we will discuss in some of the later chapters.

Before proceeding to the discussion of neutrino masses in the supersymmetric models, we wish to point out that one reason for the popularity of the supersymmetric models is the observation that unlike the non-supersymmetric $SU(5)$, in SUSY $SU(5)$ models, the three gauge couplings unify at a scale of about 2×10^{16} GeV. The difference arises from the fact that the beta functions receive additional contributions from the superpartners of various standard model particles. As a result the b_n's defined in Eq. (7.10) are now given by $2N + N_H - 3n$ where N is the number of generations and n is the $SU(n)$ group in question and N_H denotes the Higgs contribution. It is easy to see that in this case the three gauge coupling starting at their values at M_Z nicely unify at the scale mentioned above.

8.5 Neutrino mass in MSSM

It is clear from Eq. (8.7) that in the absence of the R-parity violating superpotential W_2, MSSM conserves lepton and baryon number to all or-

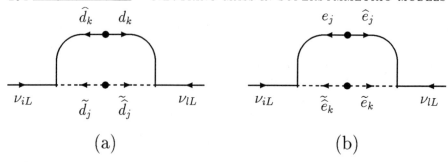

Figure 8.1: One-loop diagrams giving neutrino masses in R-parity violating MSSM.

ders in perturbation theory. Therefore as in the standard model, MSSM without R-parity violation implies $m_\nu = 0$. Thus any evidence for non-vanishing neutrino masses will imply either the presence of R-parity violating terms (either arising from explicit breaking or spontaneous breaking via the nonzero vacuum expectation values of the sneutrino field $\langle \tilde{\nu} \rangle \neq 0$ [10] or new physics beyond MSSM, such as SUSY left-right model or SUSY grand unification.

Exercise 8.2 *Suppose the R-parity violating terms are absent in the superpotential. In this case, the Lagrangian would conserve $B - L$. If one of the sneutrino fields obtain a vev v_ν alongwith the vevs of H_u and H_d, $B - L$ will be broken spontaneously and a Majoron will result [10]. Identify the Majoron field.*

Within the framework of supersymmetry, there is however no good theoretical reason to omit W_2 from the MSSM since all symmetries of the theory allow it. Once W_2 is included, it is clear that both lepton and baryon numbers are violated. Thus they lead to non-zero Majorana mass for the left-handed neutrinos without the need for a right handed neutrino. Indeed such mass terms for neutrinos arise at the one loop level from the Feynman diagrams in Fig. 8.1. The contribution of Fig. 8.1a to the Majorana mass matrix of left-handed neutrinos can now be easily estimated from this diagram. First, notice that both the vertices are of the λ'-type. The blob on the internal fermion line represents just the fermion mass, which gives a factor of m_{d_k}. The blob on the internal scalar line, however, is somewhat more complicated because it can come from two sources. These are shown in Fig. 8.2. The contribution in Fig. 8.2a comes from the soft supersymmetry breaking term in Eq. (8.9),

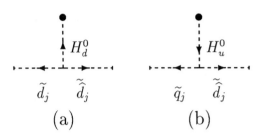

$$(a) \qquad\qquad (b)$$

Figure 8.2: Two contributions to the blob on the sfermion line which appears in Fig. 8.1a.

and this can be written as $m_{3/2}A_d v_d$, where $v_d = \langle H_d^0 \rangle$. We will denote this contribution by $A m_{d_k}$ with a suitably defined dimensionless constant A. The other contribution, represented in Fig. 8.2b, comes from the supersymmetric scalar potential whose general form was given in Eq. (8.3). The first term in that equation is $|P_a|^2$, where the subscript meant differentiation with respect to the superfield Φ_a. Applying it to Eq. (8.7), we find one term which is of the form

$$\left| h_d^{ij} Q_i \widehat{D}_j + \mu \mathbf{H_u} \right|^2 , \qquad (8.10)$$

which is obtained by taking the derivative with respect to the superfield $\mathbf{H_d}$. The contribution shown in Fig. 8.2b is the cross term in this expression, obtained after putting the vev of H_u^0. This contribution has a factor of μ, and is proportional to the down-type quark masses due to the first term of Eq. (8.10). However, the vev appearing in the down-type mass matrix is v_d, not v_u. Thus, denoting v_u/v_d by $\tan\beta$, the contribution of this diagram to the blob is $\mu m_{d_j} \tan\beta$. Collecting all the terms together, we can now write

$$m_{il} \simeq \frac{\lambda'_{ijk}\lambda'_{lkj}}{16\pi^2} \frac{m_{d_k} m_{d_j} (A + \mu \tan\beta)}{M^2} \qquad (8.11)$$

where M denotes a typical mass of the squarks and the factor of $16\pi^2$ in the denominator comes from loop integration.

The contribution of the Fig. 8.1b is given by a similar formula with m_d in the above formula replaced by m_l of the corresponding generation. For a typical supersymmetric theory, we can expect the parameters

Table 8.2: Examples of bounds on the R-parity breaking couplings.

Coupling	Upper limit	Source
λ_{12k}	$0.04 \times \left(M_{\tilde{e}_k}/100 \text{ GeV} \right)$	Beta decay and muon decay universality
$\lambda_{13k,23k}$	$0.1 \times \left(M_{\tilde{e}_k}/100 \text{ GeV} \right)$	Leptonic tau decay
λ'_{11k}	$0.05 \times \left(M_{\tilde{d}_k}/100 \text{ GeV} \right)$	$\pi \to e\nu_e$ decay

A, μ, M to be in the 100 GeV range. Choosing them to be 100 GeV and keeping the dominant contribution from the b-quark mass, we estimate

$$m_{\nu_{il}} \simeq 50 \,\text{eV} \left(\frac{\lambda'_{ijk}\lambda'_{lkj}}{10^{-4}} \right) . \tag{8.12}$$

The presence of the couplings in W_2 leads to corrections to many standard model processes as well as processes that are forbidden in the standard model. In particular, the couplings λ_{ijk} and λ'_{ijk}, being lepton number violating, contributes not only to Majorana neutrino mass but also to other lepton number violating effects such as neutrinoless double beta decay, muonium-antimuonium transition etc., some of which will be discussed in the later chapters. Present low energy data therefore lead to upper bounds on various λ_{ijk} and λ'_{ijk} [11] some typical examples of which are summarized in Table 8.2.

In this table, we have only given a small subset of possible constraints on the R-parity violating parameters, but they are enough to show that these couplings must be small. Thus we see that for present upper limits for the R-parity violating couplings, the typical contributions to the neutrino masses are in the eV range. One could therefore use the limits on neutrino masses to put constraints on the parameters of the supersymmetric models.

Exercise 8.3 *The presence of the λ'' term also leads to baryon number violation. Show that if λ'' were nonzero, in combination with λ', it would lead to rapid proton decay at the tree level. Estimate the decay rate from this, and use it to show that the present lower bounds on proton lifetime imply that $\lambda''\lambda' \leq 10^{-24}$ for squark masses of order 100 GeV.*

Another point worth emphasizing about the role of R-parity violating couplings in the discussion of neutrino masses is that one can generate interesting patterns for the neutrino masses by imposing symmetries on the couplings. As an example, if we impose $L_e - L_\mu$ invariance, all λ''s vanish and the only nonzero λ's are: λ_{123}, λ_{131} and λ_{232}. Thus in this case only Fig. 8.1b contributes and it is easily checked that only $m_{\nu_{12}} \neq 0$ and all other contributions vanish. In particular, even though we would have expected on the basis of lepton number symmetry considerations that ν_τ could have a mass, the antisymmetry of the couplings λ makes it vanish.

8.6 Supersymmetric Left-Right model

It is well-known that one of the attractive features of the supersymmetric models is its ability to provide a candidate for the cold dark matter of the universe. This however relies heavily on the theory obeying R-parity conservation. In the MSSM, this is achieved by imposing global baryon and lepton number conservation on the theory as additional requirements. First of all, this takes us one step back from the non-supersymmetric standard model where the conservation B and L arise automatically from the gauge symmetry and the field content of the model. Secondly, there is a prevalent lore supported by some calculations that in the presence of nonperturbative gravitational effects such as black holes or worm holes, any externally imposed global symmetry must be violated by Planck suppressed operators [12]. In this case, the R-parity violating effects again become strong enough to cause rapid decay of the lightest R-odd neutralino so that again there is no dark matter particle in supersymmetric theories. It is therefore desirable to seek supersymmetric theories where, like the standard model, R-parity conservation (hence baryon and lepton number conservation) becomes automatic i.e. guaranteed by the field content and gauge symmetry. It was realized in mid-80's [9] that such is the case in the supersymmetric version of the left-right model that implements the see-saw mechanism. We briefly discuss this model in the section.

The gauge group for this model is $SU(2)_L \times SU(2)_R \times U(1)_{B-L} \times SU(3)_c$. The chiral superfields denoting left-handed and right-handed quark superfields are denoted by $Q \equiv (u, d)$ and $\widehat{Q} \equiv (\widehat{u}, \widehat{d})$ respectively and similarly the lepton superfields are given by $L \equiv (\nu, e)$ and $\widehat{L} \equiv (\widehat{\nu}, \widehat{e})$. The Q and L transform as left-handed doublets with the obvious values for the $B - L$ and the $\widehat{Q}$ and $\widehat{L}$ transform as the right-handed doublets with opposite $B - L$ values. The symmetry breaking is achieved by the following set of Higgs superfields: $\phi_a(2, 2, 0, 1) \, (a = 1, 2)$; $\Delta(3, 1, +2, 1); \overline{\Delta}(3, 1, -2, 1); \Delta'(1, 3, -2, 1)$ and $\overline{\Delta}'(1, 3, +2, 1)$. Unlike in the MSSM, the allowed terms in the superpotential are very limited in this case:

$$\begin{aligned} W &= h_a Q \phi_a \widehat{Q} + h_a' L \phi_a \widehat{L} + f(LL\Delta + \widehat{L}\widehat{L}\overline{\Delta}) \\ &\quad + \mu_{ab} \, \mathrm{Tr} \, \phi_a \phi_b + M(\Delta\overline{\Delta} + \Delta'\overline{\Delta}') \end{aligned} \quad (8.13)$$

It is clear from the above equation that this theory has no baryon or lepton number violating terms and it allows for a dark matter particle. In the second paper [9] a subtle point about the model was noted. If we take the minimal version of this model, parity violation requires that R-parity must be spontaneously violated. However, one can add superfields such as $\delta(3, 1, 0, 1) + \overline{\delta}(1, 3, 0, 1)$ in which case spontaneous R-parity violation is not required to generate a parity violating global minimum. Such extra fields are always present if the model emerges from a grand unified theory or an underlying composite model.

The phenomenology of this model has been extensively studied [13] in recent papers. The neutrino masses in this model arise from the same see-saw mechanism as in the usual left-right models discussed in the Ch. 6. The associated phenomenology is quite similar and we do not go into it any further.

Part III

Phenomenology

Chapter 9

Kinematic tests of neutrino mass

In Ch. 3, we discussed two kinds of tests of neutrino mass. One kind involves looking for processes which cannot occur in absence of neutrino masses. Such processes will be discussed in the subsequent chapters of this part of the book. In this chapter, we discuss some processes of the other kind which occur even in the standard model, but whose rates are nevertheless modified if the neutrinos are massive. To keep the discussion simple, we assume for the most part that there is no neutrino mixing, so that ν_e, ν_μ and ν_τ are the physical particles. Exception is made only in §9.1.3, where the effects of neutrino mixing has been discussed.

Our discussion mainly concerns the theoretical aspects of the problem and the experimental results. For details on the experimental techniques, see some excellent reviews on the subject [1, 2].

9.1 Beta decay and the mass of the ν_e

In a nuclear beta decay, a neutron inside a nucleus decays into a proton, an electron and a $\hat{\nu}_e$, inflicting the following process at the nuclear level:

$$(A, Z) \rightarrow (A, Z + 1) + e^- + \hat{\nu}_e \,. \tag{9.1}$$

The electron and the antineutrino escape from the nucleus. Although the $\hat{\nu}_e$ is hard to detect, measurements on the escaping electron (historically called a beta-particle) can provide information about the $\hat{\nu}_e$. We start by outlining how to calculate the spectrum of the electrons.

161

9.1.1 The electron spectrum

At the hadronic level, a neutron decays in a beta decay:

$$n \to p + e^- + \hat{\nu}_e \, . \tag{9.2}$$

The transition amplitude for this process can be written down, using the 4-Fermi form of interaction, as

$$T_{fi} = -i \frac{4 G_F \cos \theta_C}{\sqrt{2}} \int d^4 x \, \left[\overline{\psi}_e(x) \gamma_\lambda \mathsf{L} \psi_{\nu_e}(x) \right] \left[\overline{\psi}_p(x) \gamma^\lambda \mathsf{L} \psi_n(x) \right] \, , \tag{9.3}$$

where the subscripts denote the type of the fermion field. The Fermi constant G_F is accompanied with a factor $\cos \theta_C$. This is because at the quark level, the process involves a transition from a down quark to an up quark, and the relevant element of the quark mixing matrix is the cosine of the Cabibbo angle θ_C.

To evaluate this amplitude, we specialize to the case where the initial and the final nucleus both have the same parity. Thus, in the hadronic part of the matrix element, there is no change of parity. Hence, the axial vector part does not contribute.

The second important point is that the energy released in any beta decay process is of the order of a few MeV at most, which is negligible compared to the masses of the nuclei. Thus the final nucleus is very non-relativistic. Looking back at the spinor solutions of Eq. (4.32), we see that the lower components of the u-spinors are negligible in this case, and therefore the only non-negligible term in the bilinear comes from γ^0.

Extracting the matrix element of the leptonic part, we can rewrite Eq. (9.3) as

$$T_{fi} = -i \frac{2 G_F \cos \theta_C}{\sqrt{2}} \left[\overline{u}_e(p) \gamma_0 \mathsf{L} v_{\nu_e}(k) \right] \int d^4 x \, \Psi_p^\dagger(x) \Psi_n(x) e^{-i(p+k)\cdot x} \, , \tag{9.4}$$

where p and k are the momenta of the outgoing electron and the antineutrino, and Ψ denotes the non-relativistic wave-functions for the nucleons.

The electron and the antineutrino are emitted with energies of the order of 1 MeV at most, so their de Broglie wavelengths are at least of order 10^{-11} cm. This is much larger than the nuclear radius. Thus, we can set

$$e^{i(p+k)\cdot x} \simeq 1 \tag{9.5}$$

to an excellent approximation. This is called the *allowed approxima-tion*. Performing the spatial integration in Eq. (9.4) now, we obtain the following expression for the invariant amplitude for the process:

$$\mathcal{M} = \frac{2G_F \cos\theta_C}{\sqrt{2}} \left[\bar{u}_e(p)\gamma_0 L v_{\nu_e}(k)\right] 2m_N \times I, \qquad (9.6)$$

where the factor $2m_N$ comes from the normalization of the hadronic wave-functions, m_N being the nucleon mass. And I is an isospin factor which depends on the isospin properties of the initial and the final nuclei.

In the expression for the decay rate, the initial state factor for the neutron and the final state factor of the proton cancel with the factor $(2m_N)^2$ coming from the square of the matrix element of Eq. (9.6). Integrating over the 3-momentum of the final proton, we can then write down the decay rate in the form

$$d\Gamma = 2G_F^2 \cos^2\theta_C I^2 \left| \sum_{\text{spin}} \bar{u}_e(p)\gamma_0 L v_{\nu_e}(k) \right|^2$$

$$\frac{d^3p}{(2\pi)^3 2E} \frac{d^3k}{(2\pi)^3 2\omega} 2\pi\, \delta(E_0 - E - \omega). \qquad (9.7)$$

Here, E_0 denotes the total energy available to the lepton pair. Since the nucleus is much heavier than the electron, the kinetic energy of the recoil nucleus is negligible. Thus, to a good approximation, E_0 is given by the mass difference of the initial and the final nuclei.

The spin sum is given by

$$\left| \sum_{\text{spin}} \bar{u}_e(p)\gamma_0 L v_{\nu_e}(k) \right|^2 = \frac{1}{2}\text{Tr}\left[\not{k}\gamma_0\not{p}\gamma_0(1 - \gamma_5)\right]$$

$$= 2\left[E\omega + \boldsymbol{p}\cdot\boldsymbol{k}\right]$$

$$= 2E\omega\left[1 + v_e v_\nu \cos\theta\right], \qquad (9.8)$$

where v_e and v_ν are the magnitudes of the velocities of the outgoing leptons and θ is the angle between their directions. Putting this in Eq. (9.7) and integrating over the angular variables other than θ, we obtain

$$d\Gamma = \frac{2G_F^2 \cos^2\theta_C I^2}{(2\pi)^3} |\boldsymbol{p}|\, E\, |\boldsymbol{k}|\, \omega\, \delta(E_0 - E - \omega)$$

$$[1 + v_e v_\nu \cos\theta]\, dE\, d\omega\, d(\cos\theta), \qquad (9.9)$$

where we have used $p^2 d\,|p| = |p|\,E\,dE$, and the similar expression for the $\hat{\nu}_e$-momentum. Now, from the integration over $\cos\theta$, the term proportional to $v_e v_\nu$ drops out, and the other term gives 2. Finally, performing the integration over ω, we obtain

$$\frac{d\Gamma}{dE} = \frac{G_F^2 \cos^2\theta_C I^2}{2\pi^3} \, |p|\, E\,(E_0 - E)\sqrt{(E_0 - E)^2 - m_{\nu_e}^2} \,. \qquad (9.10)$$

Exercise 9.1 *Integrate the above expression to obtain the total decay rate in terms of E_0. Neglect the electron mass and the neutrino mass for this purpose, assuming E_0 is large. For the transition $^{14}O \to {}^{14}N^* + e^+ + \nu_e$, $E_0 = 1.81\,\mathrm{MeV}$ and the lifetime is $102\,\mathrm{s}$ (half-life is $71\,\mathrm{s}$). If $I = \sqrt{2}$, find the value of $G_F \cos\theta_C$ from this data.*

Customarily, one uses $Q \equiv E_0 - m_e$ in place of E_0. Then we can write the number of electrons emitted with energy between E and $E + dE$ as

$$
\begin{aligned}
n(E)dE \;=\; & \frac{G_F^2 \cos^2\theta_C I^2}{2\pi^3} F(Z, R, E)\,|p|\,E \\
& \times (Q - T_e)\sqrt{(Q - T_e)^2 - m_{\nu_e}^2}\, dE \,.
\end{aligned}
\qquad (9.11)
$$

where $T_e = E - m_e$ is the kinetic energy of the electron. Notice that we have also put in a factor $F(Z, R, E)$. This factor, called the Coulomb correction factor, appears because of the electrostatic interaction between the charged nucleus and the escaping electron, which modifies the outgoing electron energy. This factor depends on the nuclear charge Z, the nuclear radius R and the electron energy.

If $m_{\nu_e} = 0$, Eq. (9.11) immediately shows that

$$K(E) \equiv \left[\frac{n(E)}{F(Z, R, E)\,|p|\,E}\right]^{1/2} \propto Q - T_e \,. \qquad (9.12)$$

Thus, if we plot the quantity $K(E)$ vs T_e, we would obtain a straight line. Such a plot is called the Kurie plot, as shown in Fig. 9.1 by the solid line. Since the quantity $K(E)$ must be non-negative, it follows that the maximum value of the electron kinetic energy in this case would be Q.

However, the linearity of the Kurie plot is lost if the neutrino has a non-zero mass. The effect of this mass becomes appreciable only near the end point of the plot where $Q - T_e$ is comparable to m_{ν_e}. The resulting nature of the plot is shown by a dashed line in Fig. 9.1. Notice that in

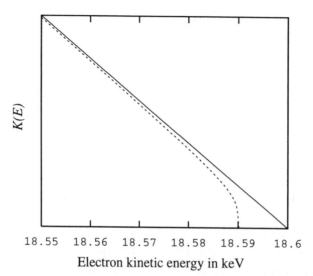

Electron kinetic energy in keV

Figure 9.1: Plot of $K(E)$ vs the electron kinetic energy in tritium beta decay near the end point. The solid line denotes the plot in absence of neutrino mass. The dashed line is the plot if the ν_e has a mass of $10\,\text{eV}$.

this case, the maximum kinetic energy of the electron is not given by Q, but rather by the vanishing of the quantity inside the square root sign in Eq. (9.11), i.e.,

$$(T_e)_{\text{max}} = Q - m_{\nu_e}. \qquad (9.13)$$

Thus, in principle, if one measures the electron energy near the end point of the spectrum, one can determine the mass of the electron neutrino.

9.1.2 Discussion of experimental efforts

The practical difficulty in carrying out this program arises from the fact that near the end point of the spectrum, $K(E)$ itself is very low, so that any deviation of $K(E)$ from the straight-line nature is hard to detect. To get a quantitative feeling, consider a small energy range $\Delta E \ll Q$ near the endpoint. The number of decays having electron energy in this range is given by

$$\int_{Q-\Delta E}^{Q} dE\, n(E) = \Delta E\, n(Q) - \frac{(\Delta E)^2}{2} n'(Q) + \frac{(\Delta E)^3}{6} n''(Q) + \ldots \qquad (9.14)$$

where the primes on n denote the number of derivatives with respect to the variable E. Now, in absence of neutrino mass, the function $n(E)$ has a factor of $(Q - T_e)^2$, so that the first two terms of the last equation vanish. We thus get for the fraction of decays having electron energy in the range from $Q - \Delta E$ to Q to be given by

$$\frac{\int_{Q-\Delta E}^{Q} dE\, n(E)}{\int_0^Q dE\, n(E)} \propto \left(\frac{\Delta E}{Q}\right)^3. \tag{9.15}$$

It is therefore extremely important that we choose a beta decay process with as small value of Q as possible. The lowest known Q value, 18.6 keV, occurs for tritium. However, even in this case, if we want to be sensitive to a ν_e-mass in the range of 10 eV, we are aided by only a small fraction (of order 10^{-9}) of the total decay events, viz the ones occurring with $Q - T_e$ of order 10 eV. Such small branching fraction means that one must be very careful about the background. Also, since the important part of the measurement concerns the highest few tens of eV within the total range of 18.6 keV, one must have a spectrometer that is sensitive to minute changes in energy.

Apart from the problems mentioned above, there are other effects which make the analysis of experimental data very difficult. Firstly, the decaying nucleus is in general surrounded by atomic electrons, and this affects the β-spectrum in many ways [1]. The Coulomb factor $F(Z, R, E)$ is modified since the atomic electrons screen the nuclear charge. There can also be energy exchange between the β-particle and the atomic electrons. In particular, the β-particle can even impart excitation energy to atomic electrons. Another important factor is the presence of atomic energy levels. For example, the tritium atom turns into a ^{3}He atom after the β-decay; but it is not clear if the final ^{3}He atom is in its ground state. Calculations show that about 30% of the time, it ends up in an excited state which is about 43 eV above the ground state. All these issues make the analysis complicated, particularly if the parent nucleus is a heavy one. Fortunately, tritium has only three nucleons in its nucleus and only one electron orbiting outside, so the analysis can be handled with a reasonable amount of confidence. Tritium molecule, however, is somewhat more involved.

These questions assumed paramount importance when the ITEP group announced evidence for a non-zero neutrino mass [3]. They looked

Table 9.1: Negative $m^2_{\nu_e}$ values reported by various tritium decay experiments.

Experiment	Ref.	$m^2_{\nu_e}$ in eV^2
Los Alamos	[8]	$-147 \pm 68 \pm 41$
Zurich	[6]	$-24 \pm 48 \pm 61$
INS Tokyo	[9]	$-65 \pm 85 \pm 65$
Mainz	[10]	$-39 \pm 34 \pm 15$
Livermore	[11]	$-130 \pm 20 \pm 15$

at the β-decay spectrum from valine molecule ($C_5H_{11}NO_2$). About 18% of the hydrogen nuclei in valine are tritium, and analysis of their β-spectrum yielded a value of m_{ν_e} between 14 and 46 eV. This initial estimate was revised in subsequent publications by the same group [4], but no other group could reproduce their result. The analysis of the ITEP data was criticized by several people on various grounds [5].

Meanwhile, a group in Zurich performed a similar experiment [6] with tritium implanted in carbon and obtained the result $m_{\nu_e} < 18\,$eV. This cast further skepticism on the ITEP analysis, although the analysis of this new experiment was also not beyond criticism [7].

More recently, new results have been obtained from a group at Los Alamos [8]. They use a source of gaseous tritium. Although it makes the experimental setup more complicated, the analysis becomes easier since the final state spectrum is very easy to understand. The recent result of this experiment gives

$$m_{\nu_e} < 12\,\text{eV}\,, \tag{9.16}$$

Other experiments using the same atom are from the INS in Tokyo [9] which found an upper limit of 13 eV, from Mainz [10] which gave an upper limit of 7.2 eV and Livermore [11] which reported an upper limit of 5 eV. There is also a recent experiment from Troitsk [12] which gives an upper limit of 4.35 eV.

A very curious feature of all the neutrino experiments is that one obtains a negative (mass)^2 for ν_e. We list the results from different experiments in Table 9.1.

It is of course quite premature to say what the negative value for $m^2_{\nu_e}$ could be due to. If it is assumed to be a genuine effect, it is very hard to

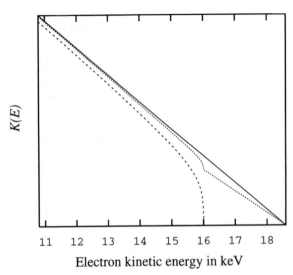

$K(E)$

Electron kinetic energy in keV

Figure 9.2: Plot of $K(E)$ in tritium beta decay, showing the effect of neutrino mixing. The solid and the dashed lines denote the plots in absence of mixing with neutrino masses of $10\,\mathrm{eV}$ and $2.6\,\mathrm{keV}$. The dotted line denotes the plot if neutrinos with both these masses are emitted, the heavier one having a probability of 0.2.

understand without postulating a very new kind of physics beyond the standard model. Several interpretations suggested are: (i) the existence of anomalous new long range force between neutrinos and matter [13] and (ii) neutrino as a tachyon [14]. Since these interpretations are indeed quite speculative, we do not elaborate on them here.

9.1.3   Effect of neutrino mixing

In presence of neutrino mixing, the final state of beta decay should not always contain the same neutrino eigenstate. Rather, one will obtain all the eigenstates which mix with the electron and which have masses in the kinematically allowed range. To see how the electron spectrum will be modified in this case, let us assume, for the sake of simplicity, a two-generation case. Let the neutrino eigenstates have masses m_1 and m_2, and let the mixing angle be θ. In this case, one should obtain

$$K(E) = \cos^2\theta K(E, m_1) + \sin^2\theta K(E, m_2), \qquad (9.17)$$

where $K(E, m)$ is a function which equals the Kurie function for a neutrino of mass m provided $T_e \leq Q - m$, and zero otherwise. The plot of this function is shown in Fig. 9.2 with some arbitrarily chosen values of the masses and the mixing angle. The plot shows a dip from the straight line behavior at the value $T_e = Q - m_2$, where a kink appears. The presence of such a kink would then indicate neutrino mixing in the final state of beta decay.

From time to time, various experimentalists have reported observing such a dip corresponding to a heavier neutrino mass around 17 keV. Further analysis, however, revealed that these experiments were faulty, and by now it is agreed upon that no such indications exist [15].

9.2 Pion decay and the mass of the ν_μ

The muon neutrino appears in the final state of muon decay. However, this final state is a 3-body state, and so kinematical analysis is complicated. On the other hand, the charged pion, π^+, decays almost entirely through the 2-body final state:

$$\pi^+(p) \to \mu^+(k) + \nu_\mu(q) \,, \tag{9.18}$$

where we have put the notations for the 4-momenta of different particles, which will be used in what follows. The momentum conservation law can be written as

$$q = p - k \,. \tag{9.19}$$

If the decaying pion is at rest, we can square both sides of this equation to obtain

$$m_{\nu_\mu}^2 = m_\pi^2 + m_\mu^2 - 2m_\pi E_\mu \,. \tag{9.20}$$

Thus, if one knows the muon and the pion masses and measures the muon energy in the decay, one can determine the mass of the ν_μ.

The problem in this apparently simple program stems from the fact that m_{ν_μ} is very small compared to m_π and m_μ. As a result, E_μ is very insensitive to the actual value of m_{ν_μ}. Take, as an example, two extreme possible allowed values of m_{ν_μ}, i.e., zero and 170 keV. From Eq. (9.20), the fractional change of the corresponding values of E_μ comes out to be about 3×10^{-5}. This means that E_μ has to be measured with extreme accuracy.

Another serious limitation of the procedure is the knowledge of the masses of the π^+ and the μ^+. Clearly, a small uncertainty in these masses can inflict a large error on the mass of the m_{ν_μ}. With the presently known values [16] of the pion mass

$$m_\pi = 139.56995 \pm 0.00035 \, \mathrm{MeV} \,, \qquad (9.21)$$

and the muon mass

$$m_\mu = 105.658389 \pm 0.000034 \, \mathrm{MeV} \,, \qquad (9.22)$$

the present experimental bound on the muon neutrino mass is found to be [17]

$$m_{\nu_\mu} < 170 \, \mathrm{keV} \,. \qquad (9.23)$$

It is interesting that in this case also the central value obtained for the squared mass of ν_μ is negative: $m_{\nu_\mu}^2 = -0.016 \pm 0.023 \, \mathrm{MeV}^2$.

> **Exercise 9.2** π^+ *also decays to* $e^+ + \nu_e$. *Since* π^+ *has the quark struc-ture of* $u\widehat{d}$, *at the quark level this implies a transition* $u\widehat{d} \to e^+ + \nu_e$. *Write down the quark-level amplitude for this process in the 4-Fermi interaction limit. If*
>
> $$\langle 0 \, | \overline{d}\gamma^\lambda \gamma_5 u | \, \pi^+(p) \rangle = i f_\pi p^\lambda \qquad (9.24)$$
>
> *whereas the matrix element for the vector current vanishes, calculate the decay rate in terms of* f_π *and the masses of the particles involved. Assuming the neutrinos to be massless, show that*
>
> $$\frac{\Gamma(\pi^+ \to e^+ \nu_e)}{\Gamma(\pi^+ \to \mu^+ \nu_\mu)} = \frac{m_e^2}{m_\mu^2} \frac{(m_\pi^2 - m_e^2)^2}{(m_\pi^2 - m_\mu^2)^2} \,. \qquad (9.25)$$

9.3   Tau decay and the mass of the ν_τ

The tau lepton has a large mass (1.777 GeV) so that it can even have semileptonic decay modes. This is advantageous in the determination of the ν_τ mass because one can consider a τ decay into several hadrons (say pions) and the ν_τ. Since a lot of the decay energy goes into the pions, the neutrino gets a small share of the energy, so that the energy is more sensitive to the mass. In actual experiment, one produces a $\tau^+\tau^-$ pair from e^+e^- collision. The produced pair is identified by the observation of a simple decay mode of one member of the pair — typically to a single charged particle and neutrinos. The multi-pion decay mode of the other member is then measured, and the missing energy and missing momentum are reconstructed to determine the ν_τ-mass. To date, the best limits are derived from a two dimensional fit to the decay mode of τ containing five charged pions in the final state:

$$\tau^- \rightarrow 2\pi^+ 3\pi^- \pi^0 \nu_\tau \,, \tag{9.26}$$

which gives [18]

$$m_{\nu_\tau} < 24 \, \text{MeV} \,. \tag{9.27}$$

Experiments are under way to improve this bound.

9.4 Other processes

9.4.1 Electron capture: mass of the ν_e

It was suggested [19] that one can measure the mass of the electron neutrino in the following process:

$$(A, Z) + e^- \rightarrow (A, Z-1) + \gamma + \nu_e \,. \tag{9.28}$$

The electron emits a photon in the vicinity of a nucleus. The resulting virtual electron is absorbed by the nucleus, triggering an inverse beta decay. The spectrum of the photon depends on the mass of the neutrino. Measurements were done using the ^{163}Ho nucleus which has the smallest known Q-value. So far, however, the limits are very weak [20], the best upper limit being $m_{\nu_e} < 225 \, \text{eV}$.

9.4.2 Kaon decay: mass of the ν_τ

Some time ago it was suggested [21] that the pion spectrum in the decay $K^+ \rightarrow \pi^+ \nu \hat{\nu}$ can be used to find information on neutrino masses. The advantage of this suggestion is that the neutrinos are produced via neutral currents in this reaction, so all neutrino flavors can appear in the final state. However, since the K-π mass difference is about $360 \, \text{MeV}$, the method can be useful only if some neutrino has mass in the $100 \, \text{MeV}$ range. Moreover, in that case, one could distinguish between the Dirac and Majorana nature of the neutrinos from this process [22]. At the time the process was proposed, the upper limit of ν_τ mass was known to be 164 MeV, so this suggestion offered some hope. However, with improvements of the upper bound of the ν_τ mass, any further insight from this process would be hard to obtain experimentally.

Chapter 10

Neutrino oscillations

In almost all experiments that are performed with neutrinos, the neutrinos are produced by charged current weak interactions. If neutrino mixing exists, the charged current can produce any physical neutrino in conjunction with a charged lepton, as shown, e.g., in Eq. (5.9). Thus the neutrino beam produced is a superposition of different particle eigenstates. As the beam propagates, different components of this beam evolve differently, so that the probability of finding different eigenstates in the beam varies with time. This consequence of neutrino mixing, first suggested by Pontecorvo [1], is called neutrino oscillation.

The physics is exactly similar to that of strangeness oscillation in neutral kaons. Kaons are produced in strong interactions, which produces either K^0 or $\widehat{K^0}$. But none of these is a physical particle. Rather, they are superpositions of the physical states K_L and K_S, which do not have any well-defined value of strangeness as K^0 and $\widehat{K^0}$ do. As the original K^0 (or $\widehat{K^0}$) beam evolves, the proportion of K^0 and $\widehat{K^0}$ changes in the superposition. In fact, since K_S has a very short lifetime, the K_S component of the original beam decays quickly. Thus, no matter whether one starts with a K^0 beam or $\widehat{K^0}$, after a while one obtains almost a pure K_L beam, which contains about the same amount of K^0 and $\widehat{K^0}$.

10.1 Neutrino oscillations in vacuum

In this section, we discuss the mathematical formulation of neutrino propagation through the vacuum. Quite apart from the theoretical beauty of it, the formalism is also important for experimentalists because there has been great many number of experiments looking for neutrino oscillations. We discuss the results obtained so far. Even if they are null results, they help us constrain the neutrino masses and mixings.

10.1.1 Theory of neutrino oscillations

Consider a neutrino beam created in a charged current interaction along with the antilepton $\hat{\ell}$. By definition, the neutrino created is called ν_ℓ. In general, this is not a physical particle, but rather is a superposition of the physical fields ν_α with different masses m_α:

$$|\nu_\ell\rangle = \sum_\alpha U_{\ell\alpha} |\nu_\alpha\rangle , \qquad (10.1)$$

where U is a unitary matrix as obtained, e.g., in the models of Ch. 5, which signifies neutrino mixing.

For a simple-minded approach to the propagation of this state, we assume that the 3-momentum $\boldsymbol{p}$ of the different components in the beam are the same. However, since their masses are different, the energies of all these components cannot be equal. Rather, for the component ν_α, the energy is given by the relativistic energy-momentum relation

$$E_\alpha = \sqrt{\boldsymbol{p}^2 + m_\alpha^2} . \qquad (10.2)$$

After a time t, the evolution of the initial beam of Eq. (10.1) gives

$$|\nu_\ell(t)\rangle = \sum_\alpha e^{-iE_\alpha t} U_{\ell\alpha} |\nu_\alpha\rangle . \qquad (10.3)$$

In writing this, we assume that the neutrinos ν_α are stable particles. If that is not the case, the entire analysis below is modified, as is discussed in §10.3.

Since all E_α's are not equal if the masses are not, Eq. (10.3) represents a different superposition of the physical eigenstates ν_α compared to Eq. (10.1). In general, this state has not only the properties of a ν_ℓ, but also of other flavor states. The amplitude of finding a $\nu_{\ell'}$ in the original ν_ℓ beam is

$$
\begin{aligned}
\langle \nu_{\ell'} | \nu_\ell(t) \rangle &= \sum_{\alpha,\beta} \left\langle \nu_\beta \left| U^\dagger_{\beta\ell'} e^{-iE_\alpha t} U_{\ell\alpha} \right| \nu_\alpha \right\rangle \\
&= \sum_\alpha e^{-iE_\alpha t} U_{\ell\alpha} U^*_{\ell'\alpha} \qquad (10.4)
\end{aligned}
$$

using the fact that the mass eigenstates are orthonormal:

$$\langle \nu_\beta \mid \nu_\alpha \rangle = \delta_{\alpha\beta} \, . \tag{10.5}$$

At $t = 0$, as expected, the amplitude is just $\delta_{\ell\ell'}$, using the unitarity of the matrix U. At any time t, the probability of finding a $\nu_{\ell'}$ in an originally ν_ℓ beam is

$$
\begin{aligned}
P_{\nu_\ell \nu_{\ell'}}(t) &= |\langle \nu_{\ell'} \mid \nu_\ell(t)\rangle|^2 \\
&= \sum_{\alpha,\beta} \left| U_{\ell\alpha} U^*_{\ell'\alpha} U^*_{\ell\beta} U_{\ell'\beta} \right| \cos[(E_\alpha - E_\beta)t - \varphi_{\ell\ell'\alpha\beta}] \, ,
\end{aligned}
\tag{10.6}
$$

where

$$\varphi_{\ell\ell'\alpha\beta} = \arg(U_{\ell\alpha} U^*_{\ell'\alpha} U^*_{\ell\beta} U_{\ell'\beta}) \, . \tag{10.7}$$

In all practical situations, neutrinos are extremely relativistic, so that we can approximate the energy-momentum relation as

$$E_\alpha \simeq |\boldsymbol{p}| + \frac{m_\alpha^2}{2|\boldsymbol{p}|} \, , \tag{10.8}$$

and can also replace t by the distance x traveled by the beam. Thus we obtain

$$P_{\nu_\ell \nu_{\ell'}}(x) = \sum_{\alpha,\beta} \left| U_{\ell\alpha} U^*_{\ell'\alpha} U^*_{\ell\beta} U_{\ell'\beta} \right| \cos\left(\frac{2\pi x}{L_{\alpha\beta}} - \varphi_{\ell\ell'\alpha\beta} \right) \, , \tag{10.9}$$

where, writing $|\boldsymbol{p}| = E$ for the sake of brevity, we defined

$$L_{\alpha\beta} \equiv \frac{4\pi E}{\Delta_{\alpha\beta}} \, , \tag{10.10}$$

with

$$\Delta_{\alpha\beta} \equiv m_\alpha^2 - m_\beta^2 \, . \tag{10.11}$$

The quantities $|L_{\alpha\beta}|$ are called the *oscillation lengths*, which give a distance scale over which the oscillation effects can be appreciable.

Again, in Eq. (10.9), notice that if the distance x is an integral multiple of all $L_{\alpha\beta}$, we obtain $P_{\nu_\ell \nu_{\ell'}} = \delta_{\ell\ell'}$, as in the original beam. But at distances where that condition is not satisfied, we can see nontrivial effects, which are sought for in the experiments.

Exercise 10.1 *In deriving the oscillation probabilities in this section, we assumed that the 3-momentum of all constituent eigenstates are equal. Show that, when Eq. (10.8) is used, the same probabilities are obtained if instead we assume that the energies of all constituent eigenstates are equal.*

Exercise 10.2 *Consider N generations of neutrinos. If the energies are such that the oscillatory terms average to zero, show that*

$$P_{\nu_e \nu_e} \geq \frac{1}{N} \, . \tag{10.12}$$

10.1.2 Experimental searches of neutrino oscillations

The goal of neutrino oscillation experiments is to look for the effects when the distance x in Eq. (10.9) is far from any integral multiple of all $L_{\alpha\beta}$. If it is so, there will be two kinds of consequences.

Firstly, we will obtain

$$P_{\nu_\ell \nu_\ell}(x) < 1 \, , \tag{10.13}$$

so that we will conclude that some of the ν_ℓ of the original beam have disappeared. Looking for such effects are thus usually called *disappearance experiments*. In a typical experiment of this sort, one takes a $\widehat{\nu}_e$ beam, let it travel for a certain distance, and let it hit a target. Owing to charged current weak interaction, a $\widehat{\nu}_e$-beam will produce positrons by inverse beta-decay processes:

$$\widehat{\nu}_e + X \to e^+ + Y \, , \tag{10.14}$$

where X and Y are two nuclei. If there is no neutrino oscillation, one can calculate the flux of positrons thus produced. If the observed flux is less, the implication is that some $\widehat{\nu}_e$ has disappeared from the original beam, thus providing evidence for neutrino oscillation.

Of course such disappearance must be compensated by the appearance of other flavor, i.e., by the fact that

$$P_{\nu_\ell \nu_{\ell'}}(x) > 0 \qquad \text{for } \ell' \neq \ell \, . \tag{10.15}$$

This raises the possibility of a second kind of experiments, called *appearance experiments* where, for example, one starts with a $\widehat{\nu}_\mu$ beam, let it travel through some distance, and let it hit a target to see if it can produce any e^+. In the absence of neutrino oscillation, it would be impossible.

From this discussion, it might seem that the appearance experiments are much simpler, since the observation of even one event of the kind sought for would provide evidence of neutrino oscillation, whereas in a disappearance experiment one must meticulously observe a significant deviation of a certain probability from unity. But there is another side to it. An appearance experiment can look only for a specific channel, e.g., a ν_μ oscillating to a ν_e or something like that. If ν_μ mixes very little with ν_e but very strongly with some other state, the $\nu_\mu \to \nu_e$ appearance experiment would be in bad shape. On the other hand, suppose we are doing a disappearance experiment on a ν_e beam. If there is an observable effect, it will show no matter whether the ν_e oscillates to ν_μ, or ν_τ, or to anything else.

The last point is more interesting for another reason. As we saw in Part II of this book, many models of neutrino mass predict some *sterile* neutrinos which do not have interactions mediated by the gauge bosons of the standard model. If the ν_e, for example, mixes significantly with one of these sterile neutrinos, an appearance experiment would be fruitless, since the cross section for the production of a charged lepton must be extremely small if one uses a sterile neutrino. On the other hand, a disappearance experiment can still register the depletion in the probability of a ν_e beam. In short, we can say that an appearance experiment measures oscillations $\nu_\ell \to \nu_{\ell'}$ for specific neutrinos, whereas a disappearance experiment is sensitive to $\nu_\ell \to \nu_X$ oscillations where ν_X can be any flavor including the sterile ones.

The sensitivities of all oscillation experiments depend on a particular parameter [2],

$$\overline{m}^2 \equiv \frac{E}{x}, \qquad (10.16)$$

which we will call the *figure of merit* of the experiment. Notice that the elements of the mixing matrix U and the mass eigenvalues m_α, m_β come from parameters in the fundamental Lagrangian and we have no handle over them. Thus, the amplitudes of the cosine terms in Eq. (10.9) are fixed for all experiments, and so are the values of $m_\alpha^2 - m_\beta^2$ and $\varphi_{\ell\ell'\alpha\beta}$ in the argument of the cosine. However, one can vary the momentum of the produced neutrinos as well as the distance through which they

Table 10.1: Order of magnitude estimates of the figure of merit of neutrino oscillation experiments.

Source	Typical values of		
	x in cm	E in MeV	$\overline{m}^2$ in eV^2
Reactor	10^2	1	10^{-2}
Meson factory	10^2	10	10^{-1}
Accelerators	10^3	10^3	1
Atmosphere	10^7	10^4	10^{-3}
Solar core	10^{11}	1	10^{-11}

are let to travel, which appear only through their ratio $\overline{m}^2$. Thus, Eq. (10.9) can be rewritten as

$$P_{\nu_\ell \nu_{\ell'}}(x) = \sum_{\alpha,\beta} \left| U_{\ell\alpha} U_{\ell'\alpha}^* U_{\ell\beta}^* U_{\ell'\beta} \right| \cos \left(\frac{m_\alpha^2 - m_\beta^2}{2\overline{m}^2} - \varphi_{\ell\ell'\alpha\beta} \right) , \quad (10.17)$$

using the figure of merit. From this formula, it is clear that if the figure of merit of a particular experiment is much bigger than the values of $\left| m_\alpha^2 - m_\beta^2 \right|$, any oscillation effect would be hard to observe. A neutrino oscillation experiment is thus sensitive to the values of the mass-square difference satisfying the relation

$$\overline{m}^2 \lesssim \left| m_\alpha^2 - m_\beta^2 \right| . \quad (10.18)$$

The values of $\overline{m}^2$ are listed in Table 10.1 for different types of neutrino sources. These include laboratory sources as well as atmospheric and solar neutrinos. In this chapter, we will discuss the experimental results of laboratory and atmospheric experiments. The implication of neutrino oscillation on solar neutrinos will be discussed in Ch. 15.

10.1.3 Understanding the experimental results

There have been many experiments to search for neutrino oscillation and there are many ongoing ones. So far, the only experiment which has given a positive indication for neutrino oscillation is the one from

Los Alamos Liquid Scintillation Neutrino Detector (LSND) [3], which we summarize below after a summary of the other experimental results. The situation upto 1987 has been nicely summarized in the review article by Bilenky and Petcov [4].

It is convenient to analyze the oscillation data in terms of the simplest assumption, viz., there is oscillation between two Dirac neutrinos only. In this case, the matrix U defined in §10.1.1 takes a particularly simple form:

$$U = \begin{pmatrix} \cos\theta & \sin\theta \\ -\sin\theta & \cos\theta \end{pmatrix}. \tag{10.19}$$

Eq. (10.9) now reduces to

$$P_{\text{conv}}(x) = \sin^2 2\theta \sin^2\left(\frac{\Delta}{4E} x\right)$$

$$P_{\text{surv}}(x) = 1 - \sin^2 2\theta \sin^2\left(\frac{\Delta}{4E} x\right) \tag{10.20}$$

where $\Delta = |m_1^2 - m_2^2|$, and the subscripts 'conv' and 'surv' on the left sides of these equations denote the conversion and the survival probabilities of a particular flavor of neutrino. The experimental data thus restricts Δ as a function of $\sin^2 2\theta$ from the limits of the observed probabilities.

Exercise 10.3 *For two generations of Majorana neutrinos, the most general form for the mixing matrix is*

$$U = \begin{pmatrix} \cos\theta & e^{-i\varrho}\sin\theta \\ -e^{i\varrho}\sin\theta & \cos\theta \end{pmatrix}. \tag{10.21}$$

Show that the phase ϱ does not appear in oscillation probabilities, so that the probabilities are still given by Eq. (10.20).

To understand the nature of these limits, let us first realize that in a given experiment, the neutrinos can never come with a well-defined value of E, a fact which will be further emphasized in §10.1.4. Thus, the conversion probability obtained in a real experiment should be given by

$$P_{\text{conv}}(x) = \sin^2 2\theta \int dE \, \Phi(E) \sin^2\left(\frac{\Delta}{4E} x\right), \tag{10.22}$$

where $\Phi(E)$ is the normalized energy spectrum of the incoming beam,

$$\int dE \, \Phi(E) = 1. \tag{10.23}$$

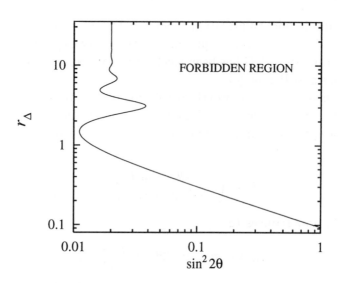

Figure 10.1: The constraints imposed by a neutrino oscillation experiment with $r_\sigma = 0.2$ which obtains $P_{\text{conv}} < 0.01$, or, equivalently, $P_{\text{surv}} > 0.99$.

To proceed further, we need to know the shape of this energy spectrum. For the sake of definiteness, let us assume that the shape is a Gaussian wave packet with an average energy $\overline{E}$ and standard deviation σ_E. Then, using the dimensionless parameters

$$r_\Delta = \frac{\Delta}{4\overline{E}} x \,, \qquad r_\sigma = \frac{\sigma_E}{\overline{E}} \,, \qquad (10.24)$$

one can rewrite Eq. (10.22) in the form

$$P_{\text{conv}}(x) = \sin^2 2\theta \int \frac{dy}{\sqrt{2\pi}} \sin^2 \left(\frac{r_\Delta}{y r_\sigma + 1} \right) e^{-y^2/2} \,, \qquad (10.25)$$

where y is the reduced energy variable $(E - \overline{E})/\sigma_E$. If now in an experiment one obtains, say, $P_{\text{conv}} < P_0$ for some value of P_0, one can constrain the possible values in the r_Δ vs $\sin^2 2\theta$ plane using the value of r_σ. Substituting the values of x and $\overline{E}$ appropriate for the experiment, one can then convert it to constraints in the Δ vs $\sin^2 2\theta$ plane. The results of a typical experiment are presented in Fig. 10.1. Some of the features of this figure can be easily understood. For example, if r_Δ is large, the argument of the sine function oscillates violently as a function of energy in the region where the Gaussian energy distribution is appreciably different from zero. Thus, the integral in Eq. (10.25) gives the value 0.5, and we obtain the constraint $\sin^2 2\theta < 2P_0$. This is the vertical part of the line at the top of Fig. 10.1. On the other hand, if r_Δ is small, we can replace the sine function by its argument, which yields a constraint of the form

$$r_\Delta^2 \sin^2 2\theta < P_0 f(r_\sigma) \tag{10.26}$$

where $f(r_\sigma)$ is a function of r_σ whose form can be easily read off from Eq. (10.25). In a log-log plot of r_Δ vs $\sin^2 2\theta$, this part therefore has a slope of $-\frac{1}{2}$, as seen in the lower part of Fig. 10.1.

These characteristics of the asymptotic behavior do not depend on the shape of the energy spectrum. In Table 10.2, we give the asymptotes of the excluded regions for various appearance and disappearance experiments.

Exercise 10.4 *Show that, irrespective of the incoming energy spectrum, the allowed region in a null experiment on neutrino oscillation will be bounded by the lines* $\sin^2 2\theta = 2P_0$ *for large* r_Δ, *and by* $r_\Delta^2 \sin^2 2\theta = constant$ *for small* r_Δ.

The basic message conveyed by Table 10.2 is the following: the known neutrinos can have appreciable mixing only if their masses are almost degenerate, at the level of about $1\,\text{eV}$ or less. This is a very strong constraint on neutrino mixing.

Let us now turn to the Los Alamos LSND experiment where positive indications for neutrino oscillations seem to have appeared[3]. In this experiment, 97% of the π^+'s produced in proton reaction at LAMPF are stopped and decay to μ^+'s $(\pi^+ \to \mu^+ + \nu_\mu)$ and the μ^+'s subsequently decay to $e^+ + \nu_e + \bar{\nu}_\mu$. The resulting $\bar{\nu}_\mu$'s have a maximum energy of 52.8 MeV and the ν_e's have a maximum energy of 36 MeV. The $\hat{\nu}_\mu$'s are then allowed to scatter off Carbon in mineral oil. What has been observed are electrons (or positrons) of energy higher than 36 MeV in coincidence with a 2.2 MeV gamma ray. Such energetic $e^\pm$'s cannot come from the ν_e's from muon decay and are interpreted as arising from $\hat{\nu}_\mu$-$\hat{\nu}_e$ oscillation and subsequent scattering of $\hat{\nu}_e$ off proton to give $e^+ + n$ and the neutron subsequently scatters in the $np \to d\gamma$ reaction giving the 2.2 MeV gamma rays. They have found 13 events against a possible background of about 4. If one takes this indication seriously, then it implies

Table 10.2: Summary of experimental results on neutrino oscillations. The experiments are of disappearance type when the initial and the final beams are the same, and of appearance type when they are different.

Beam			Limit on one parameter for extreme values of the other	
Initial	Final	Reference	Δ in eV^2 for $\sin^2 2\theta = 1$	$\sin^2 2\theta$ for large Δ
$\hat{\nu}_e$	$\hat{\nu}_e$	[5]	< 0.014	< 0.14
ν_e	ν_e	[6]	< 2.3	< 0.07
ν_μ	ν_e	[7]	< 0.09	< 0.0034
$\hat{\nu}_\mu$	$\hat{\nu}_e$	[6]	< 0.9	< 0.004
ν_μ	ν_τ	[8]	< 0.9	< 0.004
$\hat{\nu}_\mu$	$\hat{\nu}_\tau$	[9]	< 2.2	< 0.044
ν_μ	ν_μ	[10]	< 0.23	< 0.02
$\hat{\nu}_\mu$	$\hat{\nu}_\mu$	[10]	< 7.0 (or > 1200)	< 0.02
ν_e	ν_τ	[8]	< 9.0	< 0.12

a Δ between .1 to 10 eV^2. As far as the mixing angle is concerned, in the higher Δ range, $\sin^2 2\theta \approx .005$ whereas in the lower range for Δ, $\sin^2 2\theta$ can go upto one. There are two other experiments which have searched for neutrino oscillation in the overlapping Δ and $\sin^2 \theta$ with negative results by E776 group at Brookhaven [11] and KARMEN [12] group at Rutherford Laboratory. But the results are mutually consistent so that optimistically one could take this as an evidence for neutrino oscillations.

10.1.4 Wave packet treatment of neutrino oscillations

In §10.1.1, we derived the oscillation probabilities assuming that the neutrino state produced as a definite value of 3-momentum, which is shared by all the constituent eigenstates. This, of course, cannot be strictly true in any quantum theory because of the uncertainty principle. In particular, if p is very well determined, the production point of the neutrino has a large uncertainty. If this uncertainty is larger than any of the oscillation lengths, the quantity x, which was the distance traveled by the beam, becomes meaningless. Therefore, a correct treatment should use a wave packet of the neutrino, which would have a spread in momentum.

However, one can argue that in cases of practical interest, the wave packet treatment would make no difference in the prediction of the probabilities [13]. To show this, we restrict ourselves to the 2-generation case for the sake of notational simplicity. The oscillation length in this case will be denoted by L. Let us say that the uncertainty in the production point is δx. As argued above, we must demand $\delta x \ll L$ so that the arguments of the cosine in Eq. (10.9) makes sense. The uncertainty relation would then imply that the momentum uncertainty δp satisfies

$$\frac{\delta p}{p} \gg \frac{1}{Lp} = \frac{\Delta}{4\pi E^2}, \qquad (10.27)$$

where we have used the definition of the oscillation length L from Eq. (10.10) for the last step. As seen from Table 10.1, the typical values of E are in the range of $1\,\mathrm{MeV}$ or higher, whereas one typically searches for values of Δ which are smaller than a few eV^2. Thus, the right side of Eq. (10.27) is of order 10^{-12} or less. This implies that, although the momentum uncertainty of the initial wave packet cannot be zero, it can be very small. Hence the formulas for oscillation probabilities derived by assuming a fixed momentum hold with very good accuracy [13].

10.2 Atmospheric neutrinos

10.2.1 Summary of experimental data

The interaction of cosmic ray protons with the nuclei in upper atmosphere leads to production of both electron and muon type neutrinos. During the past decade, these neutrinos have been observed and analyzed in detail in several experiments [14, 15, 16, 17]. The observations have the potential to provide sensitive information on the nature of neutrino oscillations. As we discuss below, the present observations can be interpreted as positive indications of oscillations between different neutrino species. In order to see how one arrives at this conclusion, let us first summarize the observations and theoretical expectations for the case of massless neutrinos, for which there are no oscillations.

The atmospheric neutrinos are supposed to originate in the following chain of reactions:

$$p + X \to \pi^{\pm} + Y$$
$$\hookrightarrow \mu^{\pm} + \text{``}\nu_{\mu}\text{''}$$
$$\hookrightarrow e^{\pm} + \text{``}\nu_{e}\text{''} + \text{``}\nu_{\mu}\text{''} \tag{10.28}$$

Here $\pi^{\pm}$'s could be replaced by $K^{\pm}$'s. The notation "ν_{μ}", for example, means that we have not distinguished between ν_{μ}'s and $\widehat{\nu}_{\mu}$'s as they are not distinguished in the experiments. Naive counting argument then says that the ratio r of ν_{e}-type to ν_{μ}-type events in the underground detectors (not distinguishing between particles and antiparticles) is 0.5. The experiments count the number of e- and μ-type events produced in the collisions of the neutrinos with the water detector. The events may either be fully contained or partially contained. We will consider mostly the contained events.

In this class of events, the neutrino-nucleus interaction vertex is located inside the detector as are all final state particles. The charged lepton energies for these events range from a few hundred MeV to 1.2 GeV. Using the contained events of electron- and muon- type, one constructs the following double ratio:

$$R \equiv \frac{(\mu/e)}{(\mu/e)_{\text{MC}}} \tag{10.29}$$

where MC stands for Monte-Carlo predictions. The denominator gives the theoretical expectation for (μ/e) ratio assuming neutrinos are massless and all their interactions are given by the standard model. An advantage of considering μ/e is that the cosmic ray flux uncertainties do not effect this result. If there are no neutrino oscillations, R should be 1 [18]. The values for this double ratio obtained by various experiments, as of now, are given below:

Kamiokande [14]	$.60 \pm .06 \pm .05$
IMB [15]	$.54 \pm .05 \pm .12$
Sudan2 [16]	$.69 \pm .19 \pm .09$
FREJUS [17]	$.87 \pm .21$
NUSEX [17]	$.99 \pm .40$
Super-Kamiokande [19]	$0.63 \pm 0.03 \pm 0.05$ (sub-GeV)
	$0.60^{+0.07}_{-0.06} \pm 0.07$ (multi-GeV)

The first two experiments use water Cerenkov detectors whereas the next three use iron-calorimeters. The last three have (as of this writing) much lower statistics than the first two. In the water Cerenkov detectors, the muons are distinguished from the electrons in two ways: by the size of the rings of Cerenkov light or by observing decay products from the muons.

The first three sets of data imply that there has either been a depletion of muon neutrinos or an excess of electron-type events due to some non-standard property of the neutrinos.

10.2.2 Neutrino masses and mixings implied by data

The simplest way to understand the atmospheric neutrino puzzle is to assume that the neutrinos oscillate into other species, thereby leading to observed values of the fluxes in the underground than what would be expected on the basis of the standard model. If $\nu_e \to \nu_\mu$ oscillation is to provide the solution to the solar neutrino puzzle which will be discussed in Ch. 15, then the most appealing scenario for the atmospheric neutrino puzzle is to assume that the atmospheric ν_μ to oscillate into ν_τ leaving the ν_e flux unchanged. The observed energy spectra for ν_μ and ν_e by the Kamiokande as well as super-Kamiokande collaboration would seem to indicate support for this hypothesis. We will therefore consider [20] $\nu_\mu \to \nu_\tau$ oscillation to explain the atmospheric neutrino deficit. The allowed masses and mixing angles are constrained by observations prior to the start of super-Kamiokande to be in the range [21]:

$$\Delta_{\mu\tau} \simeq 0.025 \text{ to } 0.005 \, \text{eV}^2$$
$$\sin^2 2\theta_{\mu\tau} \simeq 0.6 \text{ to } 1. \tag{10.30}$$

As data from super-Kamiokande are analyzed, these numbers will most likely change.

To understand the above conclusion regarding the masses and mixing angles, let us start with the probability for $\nu_\mu \to \nu_\tau$ oscillation. Assuming for the sake of simplicity a two-flavor case, Eq. (10.20) can be re-written as:

$$P_{\nu_\mu \nu_\tau} = \sin^2 2\theta_{\mu\tau} \sin^2 \left(\frac{1.27 \Delta_{\mu\tau}(\text{eV}^2) L(\text{km})}{E(\text{GeV})} \right). \tag{10.31}$$

Roughly speaking, for the contained events, we would like $P_{\nu_\mu \nu_\tau} \simeq 1/2$, which can be satisfied if $\sin^2 2\theta_{\mu\tau} \simeq .5-1$ and $\Delta_{\mu\tau} \geq 10^{-3}$ eV^2 since $E_\nu \simeq 1$ GeV and $L \simeq 1000$ km. On the other hand, there now appears to be evidence for zenith angle dependence in the multi-GeV data. Since this dependence implies that the sine function in the above formula is varying with distance, its argument can at most be of order one. Assuming the energy of the multi-GeV events to be roughly in the 10 GeV range and requiring the argument of the sine function in the above equation to be of order one, we get $\Delta_{\mu\tau} \leq 10^{-2}$ eV^2. A more detailed preliminary analysis from the Super Kamiokande collaboration gives an allowed range between 2×10^{-4} eV$^2 \leq \Delta_{\mu\tau} \leq 10^{-2}$ eV^2 [22].

These experimental results are under active review, discussion and analysis by the experimentalists as well as theorists. If they stand the test of time, they would provide evidence for neutrino masses and mixings, separate from the solar neutrino experiments and will thus pioneer the new era of neutrino physics.

10.3   Oscillation with unstable neutrinos

The analysis of neutrino oscillation experiments becomes quite different from that described in §10.1.1 if the neutrinos are unstable [23]. For clarity of presentation, we restrict ourselves to the case of two neutrino flavors which we will call ν_e and ν_μ. The mass eigenstates ν_1 and ν_2 are related as usual to these flavor states through Eq. (10.1), with the mixing matrix U given in Eq. (10.19). Let ν_2 be the heavier of the two neutrinos, and let us say that it decays via the mode $\nu_2 \to \nu_1 X$, where X can be anything. Let the lifetime of this decay be Γ^{-1}.

Starting from the initial beam

$$|\nu_{\text{in}}\rangle = \cos\alpha \, |\nu_e\rangle + \sin\alpha \, |\nu_\mu\rangle \, , \qquad (10.32)$$

one then obtains at time t the state

$$
\begin{aligned}
|\psi(t)\rangle = {} & e^{-iE_1 t}\cos(\theta+\alpha)\,|\nu_1\rangle + e^{-iE_2 t - \Gamma t/2}\sin(\theta+\alpha)\,|\nu_2\rangle \\
& + \sum_k c_k \, |\nu_1(k)X\rangle \, ,
\end{aligned}
\qquad (10.33)
$$

where $|\nu_1(k)X\rangle$ denotes the final states of ν_2 decay where ν_1 has a momentum k. From this, we obtain

$$
\begin{aligned}
P_{\nu_{\text{in}} \nu_e}(t) &= \cos^2(\theta + \alpha) \cos^2\theta + e^{-\Gamma t} \sin^2(\theta + \alpha) \sin^2\theta \\
&\quad + \frac{1}{2} e^{-\Gamma t/2} \sin 2\theta \sin 2(\theta + \alpha) \cos \frac{t\Delta}{2E} \\
P_{\nu_{\text{in}} \nu_\mu}(t) &= \cos^2(\theta + \alpha) \sin^2\theta + e^{-\Gamma t} \sin^2(\theta + \alpha) \cos^2\theta \\
&\quad - \frac{1}{2} e^{-\Gamma t/2} \sin 2\theta \sin 2(\theta + \alpha) \cos \frac{t\Delta}{2E} .
\end{aligned}
\tag{10.34}
$$

The interesting point is that, even for fast decays, i.e., $\Gamma t \gg 1$, we get

$$
\begin{aligned}
P_{\nu_{\text{in}} \nu_e}(t) &= \cos^2(\theta + \alpha) \cos^2\theta \\
P_{\nu_{\text{in}} \nu_\mu}(t) &= \cos^2(\theta + \alpha) \sin^2\theta ,
\end{aligned}
\tag{10.35}
$$

so that none of these probabilities is zero. Despite their fast decays, the ν_μ's can survive after a long journey.

Since the sum of all probabilities must be unity, we get

$$
\sum_k P_{\nu_{\text{in}} \to \nu_1 X} = (1 - e^{-\Gamma t}) \sin^2(\theta + \alpha) .
\tag{10.36}
$$

The ν_1's produced in the decay of ν_2 are incoherent with respect to the ν_1's in the original beam. Therefore,

$$
\begin{aligned}
\sum_k P_{\nu_{\text{in}} \to \nu_e X} &= (1 - e^{-\Gamma t}) \sin^2(\theta + \alpha) \cos^2\theta \\
\sum_k P_{\nu_{\text{in}} \to \nu_\mu X} &= (1 - e^{-\Gamma t}) \sin^2(\theta + \alpha) \sin^2\theta .
\end{aligned}
\tag{10.37}
$$

To point out the difference of this case with that of stable neutrinos, let us take the specific example of an initial ν_e beam, i.e., the one for which $\alpha = 0$. Specializing now to the case of fast decay, we obtain

$$
P_{\nu_e \nu_e}(t) = \cos^4\theta ,
\tag{10.38}
$$

which can be very small if θ is close to $\pi/2$. However, without decay, this probability would be given by Eq. (10.20). Averaging over the distance, we obtain in that case

$$
P_{\nu_e \nu_e}(t) = 1 - \frac{1}{2} \sin^2 2\theta ,
\tag{10.39}
$$

which is greater than $\frac{1}{2}$ for all values of θ. Thus, the physical implications of the instability of neutrinos can be dramatic.

10.4 Neutrino oscillations in matter

Wolfenstein [24] pointed out that the patterns of neutrino oscillation
might be significantly affected if the neutrinos travel through a material
medium rather than through the vacuum. The basic reason for this is
simple. Normal matter has electrons but no muons or taus at all. Thus,
if a ν_e beam goes through matter, it encounters both charged and neutral
current interactions with the electrons. But a ν_μ or a ν_τ interacts with
the electron only via the neutral current, so their interaction is different
in magnitude than that of the ν_e.

Interactions modify the effective mass that a particle exhibits while
traveling through a medium. An well-known example is that of the photon, which is massless in the vacuum but develops an effective mass in a
medium. As a result, electromagnetic waves do not travel with speed c
through a medium. The effective masses of neutrinos are similarly modified in a medium by their interactions. Since ν_e has different interaction
than the other neutrinos, the modification is different for ν_e than for the
other flavored neutrinos.

This fact can have dramatic consequences if the neutrinos mix in the
vacuum. In this case, a physical eigenstate can have components of ν_e,
ν_μ, ν_τ and other possible states. When such a state travels through a
medium, the modulation of its ν_e component is different from the same
modulation in the vacuum. This leads to changes in the oscillation
probabilities compared to their values in the vacuum.

For a quantitative treatment of the above ideas, let us again stick to
the simplest case of two flavors ν_e and ν_μ. We first recapitulate the case
of vacuum oscillation in a way that is most suitable for the generalization
to matter oscillation. If the mass eigenstates are ν_1 and ν_2, the evolution
equation for these states can be written as

$$i\frac{d}{dt}\begin{pmatrix} \nu_1(t) \\ \nu_2(t) \end{pmatrix} = H \begin{pmatrix} \nu_1(t) \\ \nu_2(t) \end{pmatrix}, \tag{10.40}$$

where H is diagonal in this basis:

$$H = \begin{pmatrix} E_1 & 0 \\ 0 & E_2 \end{pmatrix} \simeq E + \begin{pmatrix} m_1^2/2E & 0 \\ 0 & m_2^2/2E \end{pmatrix}, \tag{10.41}$$

using the approximate expressions of Eq. (10.8). Recalling the form for
the matrix U that connects mass eigenstates to flavor states, we can

rewrite Eq. (10.40) as

$$i\frac{d}{dt}\begin{pmatrix} \nu_e(t) \\ \nu_\mu(t) \end{pmatrix} = H' \begin{pmatrix} \nu_e(t) \\ \nu_\mu(t) \end{pmatrix} , \qquad (10.42)$$

where

$$\begin{aligned} H' &= UHU^\dagger \\ &= E + \frac{m_1^2 + m_2^2}{4E} + \frac{\Delta}{4E}\begin{pmatrix} -\cos 2\theta & \sin 2\theta \\ \sin 2\theta & \cos 2\theta \end{pmatrix} \qquad (10.43) \end{aligned}$$

From this, we can see that the diagonalizing angle θ is given by

$$\tan 2\theta = \frac{2H'_{12}}{H'_{22} - H'_{11}} \qquad (10.44)$$

in terms of the elements of the matrix H', as is usual for a 2×2 matrix.

We now consider the same problem when the neutrinos are traveling through matter. For simplicity, we assume that the density of the background matter is uniform, with n_e, n_p and n_n denoting the number of electrons, protons and neutrons per unit volume. Elastic scattering off these particles change the effective masses of the neutrinos, whose magnitude we now estimate.

Consider first elastic scattering through charged current interactions. The only interaction of this sort is the $\nu_e e$ scattering, since there are no μ's or τ's in the background at normal temperatures. The diagram for this process has been given in Fig. 2.1b. The contribution of this diagram to the effective Lagrangian is given by

$$i\mathcal{L}_{\text{eff}} = \left(\frac{ig}{\sqrt{2}}\right)^2 \left\{\overline{e}(p_1)\gamma^\lambda L\nu_e(p_2)\right\} \frac{ig_{\lambda\rho}}{M_W^2} \left\{\overline{\nu_e}(p_3)\gamma^\rho Le(p_4)\right\} , \quad (10.45)$$

where we have neglected the momentum dependence in the W-propagator since we are interested in the leading order contribution in the Fermi constant. Using the relation $g^2/8M_W^2 = G_F/\sqrt{2}$, we thus obtain

$$\begin{aligned} \mathcal{L}_{\text{eff}} &= -\frac{4G_F}{\sqrt{2}}\left\{\overline{e}(p_1)\gamma_\lambda L\nu_e(p_2)\right\}\left\{\overline{\nu_e}(p_3)\gamma^\lambda Le(p_4)\right\} \\ &= -\frac{4G_F}{\sqrt{2}}\left\{\overline{\nu_e}(p_3)\gamma_\lambda L\nu_e(p_2)\right\}\left\{\overline{e}(p_1)\gamma^\lambda Le(p_4)\right\} , \quad (10.46) \end{aligned}$$

where the second form is obtained via Fierz transformation. For forward scattering where $p_2 = p_3 = p$, this gives the following contribution that affects the propagation of the ν_e:

$$-\sqrt{2}G_F \bar{\nu}_{eL}(p)\gamma_\lambda\nu_{eL}(p)\left\langle \bar{e}\gamma^\lambda(1-\gamma_5)e\right\rangle , \qquad (10.47)$$

averaging the electron field bilinear over the background. To calculate that average, note that the axial current reduces to spin in the non-relativistic approximation, which is negligible for a non-relativistic collection of electrons. The spatial components of the vector current give the average velocity, which is negligible as well. So the only non-trivial average is

$$\left\langle \bar{e}\gamma^0 e\right\rangle = \left\langle e^\dagger e\right\rangle = n_e , \qquad (10.48)$$

which gives a contribution to the effective Lagrangian

$$-\sqrt{2}G_F n_e \bar{\nu}_{eL}\gamma_0\nu_{eL} . \qquad (10.49)$$

Exercise 10.5 *Write down the equation of motion of a neutrino, using the free Lagrangian augmented by the term in Eq. (10.49). Show that the effect of this term is to change the effective energy of the neutrino, viz.,*

$$E = \sqrt{|\mathbf{p}|^2 + m^2} + \sqrt{2}G_F n_e . \qquad (10.50)$$

One can interpret the density dependent term as a contribution to the potential energy:

$$V_{cc} = \sqrt{2}G_F n_e . \qquad (10.51)$$

Turning now to neutral current contributions, we find in an exactly similar way the following contributions to effective energies of both ν_e and ν_μ:

$$V_{nc} = \sqrt{2}G_F \sum_f n_f \left[I_{3L}^{(f)} - 2\sin^2\theta_W Q^{(f)} \right] \qquad (10.52)$$

where f stands for the electron, the proton or the neutron, $Q^{(f)}$ is the charge of f and $I_{3L}^{(f)}$ is the third component of weak isospin of the left-chiral projection of f. Thus, for the proton, $Q = 1$ and $I_{3L} = \frac{1}{2}$, whereas for the electron, $Q = -1$ and $I_{3L} = -\frac{1}{2}$. Also, for normal neutral matter,

$n_e = n_p$ to guarantee charge neutrality. Therefore the contributions of the electron and the proton cancel each other and we are left with the neutron contribution, which is

$$V_{\text{nc}} = -\sqrt{2}G_F n_n/2 \,. \tag{10.53}$$

We emphasize that this neutral current contribution is the same for all flavors of neutrinos whereas the charged current contribution affects ν_e only. Thus, the evolution equation of neutrino beams should now be given by Eq. (10.42) with H' replaced by

$$\tilde{H} = E + \frac{m_1^2 + m_2^2}{4E} - \frac{1}{\sqrt{2}}G_F n_n$$
$$+ \begin{pmatrix} -\frac{\Delta}{4E}\cos 2\theta + \sqrt{2}G_F n_e & \frac{\Delta}{4E}\sin 2\theta \\ \frac{\Delta}{4E}\sin 2\theta & \frac{\Delta}{4E}\cos 2\theta \end{pmatrix}. \tag{10.54}$$

The effective mixing angle in matter, $\tilde{\theta}$, would accordingly be given by

$$\tan 2\tilde{\theta} = \frac{2\tilde{H}_{12}}{\tilde{H}_{22} - \tilde{H}_{11}} = \frac{\Delta \sin 2\theta}{\Delta \cos 2\theta - A} \,, \tag{10.55}$$

where for the sake of convenience, we defined

$$A = 2\sqrt{2}G_F n_e E \,. \tag{10.56}$$

The effective mixing angle thus changes inside matter. The change is most dramatic if $A = \Delta \cos 2\theta$, i.e., if the electron number density is given by

$$n_e = \frac{\Delta \cos 2\theta}{2\sqrt{2}G_F E} \,. \tag{10.57}$$

Then, even if the vacuum mixing angle θ is small, we obtain $\tilde{\theta} = \pi/4$, which is to say that ν_e and ν_μ mix maximally.

To understand the exact nature of this maximal mixing, let us write down the eigenvalues of $\tilde{H}$. The results are

$$E_\alpha = E - \frac{1}{\sqrt{2}}G_F n_n + \frac{\tilde{m}_\alpha^2}{2E} \,, \tag{10.58}$$

where

$$\tilde{m}_{1,2}^2 = \frac{1}{2}\left[(m_1^2 + m_2^2 + A) \mp \sqrt{(\Delta \cos 2\theta - A)^2 + \Delta^2 \sin^2 2\theta}\right]. \tag{10.59}$$

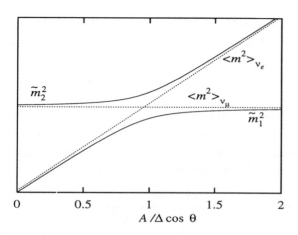

Figure 10.2: The effective masses of neutrinos in matter. The quantity A is proportional to matter density. The solid lines are mass squared values of the physical eigenstates, the dashed lines are the expectation values of squared mass for the states ν_e and ν_μ. We have taken $\theta = 0.15$ for the plot. The vertical scale is arbitrary.

The behavior of $\tilde{m}_1^2$ and $\tilde{m}_2^2$ as functions of A are shown schematically in Fig. 10.2.

Consider now the case when θ is very small. For $A = 0$ which corresponds to the vacuum, the lighter eigenstate ν_1 is almost purely ν_e whereas the heavier one, ν_2, is almost purely ν_μ. In the other extreme when $A \gg \Delta \cos 2\theta$, $\tilde{\theta} \to \pi/2$, which means that the lighter eigenstate is almost purely ν_μ whereas the heavier one is ν_e to a good approximation. Thus, if we follow the effective mass squared values of ν_e and ν_μ, as shown by the dotted lines in Fig. 10.2, we see that the effective mass squared value of ν_e starts lower than that of ν_μ but with the increase of A, the difference diminishes until finally ν_e overtakes ν_μ. This phenomenon is known as *level-crossing*.

It was realized by Mikheyev and Smirnov [25] that this phenomenon of level crossing might be crucial in understanding the neutrino flux from the Sun, a detailed account of which will be given in Ch. 15. In view of this importance, there has been a great deal of work in this field [26]. Several authors have analyzed matter effects with three generations of neutrinos instead of two [27]. A rigorous field theoretic derivation of the effective mass matrix has been performed as well, the details of which will be presented in Ch. 14.

Chapter 11

Electromagnetic properties of neutrinos

Neutrinos do not have electric charge, but that does not mean that they do not have any electromagnetic interaction. Consider the familiar example of quantum electrodynamics, where the magnetic moment of a fermion is induced by quantum loops. In the context of electroweak theories, similar quantum loops can induce electromagnetic properties of the neutrino. This is true even in the standard electroweak model, where a charge radius is induced for the neutrinos, as mentioned in Ch. 2.

The subject becomes richer if the neutrinos are massive, as we will see in this chapter. For the case of Dirac neutrinos, one can define as many as four electromagnetic form factors. On the other hand, if the massive neutrino is a Majorana particle, we will see that there is only one form factor. This brings out the difference between Dirac and Majorana properties in a striking manner, as we will see again and again in this chapter.

Moreover, since neutrino mass models usually predict neutrino mixing, one can have flavor changing processes such as $\mu \to e + \gamma$. Similarly, in the neutrino sector, one can have decays like $\nu_\alpha \to \nu_{\alpha'} + \gamma$, where ν_α and $\nu_{\alpha'}$ are two different neutrinos. There have been many suggestions from time to time that this process might have important cosmological implications. In view of those suggestions, some of which will be discussed in Ch. 17, here we give a detailed discussion of the rate of this process in different models.

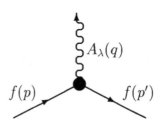

Figure 11.1: The effective electromagnetic vertex of a fermion.

11.1 Electromagnetic form factors of a neutrino

The electromagnetic properties of any fermion shows up, in a quantum field theory, as its interaction with the photon, and is described by the interaction vertex shown in Fig. 11.1.

For charged fermions, there is a diagram of this form even at the tree level, since the basic interaction Lagrangian contains a term

$$\mathcal{L}_{\text{int}} = eQ\overline{\psi}\gamma_\lambda\psi A^\lambda \,, \tag{11.1}$$

where e is the charge of the positron and Q is the charge of the relevant particle in units of e. For an uncharged fermion like the neutrino, this term is absent. The interactions arise only from loops, and are therefore momentum dependent in general. We can summarize these interactions by writing the effective interaction term in analogy with the direct term in Eq. (11.1):

$$\mathcal{L}_{\text{eff}} = \overline{\psi}\mathcal{O}_\lambda\psi A^\lambda \equiv j_\lambda A^\lambda \,, \tag{11.2}$$

where the form of j_λ is the subject of our present discussion. The form depends on the Dirac or Majorana nature of the neutrino, as commented earlier.

11.1.1   Form factors of a Dirac neutrino

Let us first focus on a Dirac neutrino. In this case, taking the matrix element of $\mathcal{L}_{\text{eff}}$ between two one-particle states, we get

$$\langle \boldsymbol{p}', s' | j_\lambda | \boldsymbol{p}, s \rangle = \overline{u}_{s'}(\boldsymbol{p}')\Gamma_\lambda(p, p')u_s(\boldsymbol{p}) \,, \qquad (11.3)$$

using the plane wave expansion of Eq. (4.1). Hermiticity of the effective Lagrangian implies that Γ_λ must satisfy

$$\Gamma_\lambda(p, p') = \gamma_0 \Gamma_\lambda^\dagger(p', p)\gamma_0 \,. \qquad (11.4)$$

In addition, electromagnetic current conservation implies

$$\begin{aligned} 0 &= \left\langle \boldsymbol{p}', s' \left| \partial^\lambda j_\lambda(x) \right| \boldsymbol{p}, s \right\rangle \\ &= iq^\lambda \, \overline{u}(\boldsymbol{p}')\Gamma_\lambda(p, p')u(\boldsymbol{p}) \,, \end{aligned} \qquad (11.5)$$

where

$$q = p - p' \,. \qquad (11.6)$$

While this is the general statement about electromagnetic gauge invariance which is valid for any fermion, for neutral fermions like neutrinos a more restricted condition can be obtained from the Ward identity:

$$q^\lambda \, \Gamma_\lambda(p, p') = 0 \,. \qquad (11.7)$$

Subject to these conditions, we can now write down the most general form for Γ_λ which is also consistent with Lorentz covariance. It is [1, 2, 3]

$$\begin{aligned} \Gamma_\lambda(p, p') &= (q^2\gamma_\lambda - q_\lambda \not{q})[R(q^2) + r(q^2)\gamma_5] \\ &\quad + \sigma_{\lambda\rho}q^\rho[D_M(q^2) + iD_E(q^2)\gamma_5]. \end{aligned} \qquad (11.8)$$

A few comments should be made here. First, notice that Γ_λ depends only on the vector q, and not on $p + p'$. This is because of the fact that all possible terms involving $p + p'$ can be converted to terms involving q by the use of Gordon identity. Secondly, the hermiticity condition in Eq. (11.4) implies that the form factors R, r, D_M and D_E are all real. Thirdly, the form factors are Lorentz invariant. Since $p^2 = p'^2 = m^2$, where m is the mass of the fermion f, there is only one independent dynamical quantity, viz. q^2, which is Lorentz invariant. The form factors thus depend on q^2 only.

Exercise 11.1 *Consider two fermion fields whose quanta have masses m and m'. The positive energy spinors corresponding to these fields are denoted by u and u'. Show that*

$$\bar{u}'(p')\left[(a+b\gamma_5)\left\{(m+m')\gamma_\lambda + i\sigma_{\lambda\rho}q^\rho - (p+p')_\lambda\right\}\right]u(p) = 0. \quad (11.9)$$

This is the generalized form of Gordon identity, which is obtained when m = m' and b = 0.

The physical significance of the form factors is easily understood by considering the non-relativistic limit. Using the explicit form of the Dirac matrices and the plane wave solution given in §4.3, we can see that the D_M term reduces to $D_M(0)\boldsymbol{\sigma}\cdot\boldsymbol{B}$ in the non-relativistic limit, where $\boldsymbol{B}$ is the magnetic field. Thus, $D_M(0)$ is the magnetic moment of the particle, which we will denote by μ. In general, $D_M(q^2)$ is called the *magnetic form factor*. Similarly, $D_E(q^2)$ is the *electric form factor* since $d \equiv D_E(0)$ is the electric dipole moment. The quantity $R(q^2)$ is called the *charge radius*, and $r(q^2)$ is the *axial charge radius*.

Exercise 11.2 *Verify explicitly, using the plane wave solution of §4.3, that in the non-relativistic limit, the operator associated with the form factor D_M gives an interaction $\boldsymbol{\sigma}\cdot\boldsymbol{B}$.*

In the above discussion, we have not made any assumption about the validity of the discrete symmetries $\mathcal{C}$, $\mathcal{P}$, $\mathcal{T}$ and their combinations. For neutrinos, Γ_λ arises entirely from loop diagrams. Such loops necessarily include weak interaction and therefore violate discrete symmetries like $\mathcal{P}$ and $\mathcal{C}$. If $\mathcal{CP}$ were conserved, it would put extra constraint on the form factors, viz., that $D_E(q^2) = 0$. However, $\mathcal{CP}$ is not conserved in the quark sector at least. At some high order loops involving virtual quark lines, this will induce a nonzero D_E for the neutrinos. In addition, if the leptonic sector contains $\mathcal{CP}$ violation in the presence of neutrino masses, that will contribute to a nonzero value of D_E as well. In any case, all four form factors are nonzero for a Dirac neutrino.

Exercise 11.3 *Using the CP-transformation properties of spinors from §4.4, show that the operator associated with D_E is CP-violating.*

11.1.2 Form factors of a Majorana neutrino

For any uncharged fermion, we can take the form of Γ_λ in Eq. (11.8) as the starting point. However, the Majorana nature of the neutrino would imply that

$$R = D_M = D_E = 0, \quad (11.10)$$

as we will see shortly.

There are two equivalent ways to see this. In the first way, one takes the definition of Γ_λ in Eq. (11.3) along-with its explicit form in Eq. (11.8) and tries to write down the explicit form for the effective Lagrangian, $\mathcal{L}_{\text{eff}}$, in the co-ordinate representation. This will not be easy since the momentum dependence of form factors will imply derivatives in the co-ordinate language. By partial integration, we can transfer all these derivatives to act on the photon field A^λ. In that case, it is clear that the R-term in Eq. (11.8) gives some interaction which involves the fermion bilinear $\overline{\psi}\gamma_\lambda\psi$ multiplied by a functional of the photon field. Similarly, the r-term will involve a bilinear $\overline{\psi}\gamma_\lambda\gamma_5\psi$, the D_M-term $\overline{\psi}\sigma_{\lambda\rho}\psi$ and the D_E-term $\overline{\psi}\sigma_{\lambda\rho}\gamma_5\psi$.

Next notice that for a Majorana field since ψ equals $\widehat{\psi}$ apart from a possible phase, we can write, for any 4×4 matrix F, the following identity

$$
\begin{aligned}
\overline{\psi}F\psi &= \overline{\widehat{\psi}}F\widehat{\psi} \\
&= -\psi^\mathsf{T}C^{-1}FC\gamma_0^\mathsf{T}\psi^* = \psi^\dagger\gamma_0\left(C^{-1}FC\right)^\mathsf{T}\psi \\
&= \overline{\psi}CF^\mathsf{T}C^{-1}\psi,
\end{aligned}
\tag{11.11}
$$

using the properties of the matrix C and interchanging the order of the spinors in the middle of this operation, as demonstrated in Eq. (4.74). For $F = \gamma_\lambda\gamma_5$, this identity is trivially satisfied since

$$
\gamma_\lambda\gamma_5 = C(\gamma_\lambda\gamma_5)^\mathsf{T}C^{-1}
\tag{11.12}
$$

which follows from Eq. (4.24). However, if F is γ_λ, or $\sigma_{\lambda\rho}$, or $\sigma_{\lambda\rho}\gamma_5$, we get

$$
F = -CF^\mathsf{T}C^{-1}
\tag{11.13}
$$

by directly using Eq. (4.24). Comparing with Eq. (11.11) just obtained, we conclude that for a Majorana neutrino, the vector current vanishes, as does the tensor and the axial tensor ones. Thus, the form factors which go with these currents are meaningless for a Majorana neutrino. The only form factor for a Majorana neutrino is the axial charge radius $r(q^2)$, which goes with the axial vector current.

The other, may be more instructive, way of deducing Eq. (11.10) is the following. Consider again any fermion bilinear $\overline{\psi}F\psi$. For Dirac fermions, the matrix element of this operator between one particle states is clearly given by

$$\left\langle p', s' \left| \overline{\psi} F \psi \right| p, s \right\rangle = \overline{u}_{s'}(p') F u_s(p). \qquad (11.14)$$

The same is not true for a Majorana field, however. To see that, we can use the plane wave expansion of a Majorana field given in Eq. (4.16). The operators $f_s(p)$ and $f_s^\dagger(p)$ satisfy the relation

$$\left\langle p', s' \left| f_{s_2}^\dagger(p_2) f_{s_1}(p_1) \right| p, s \right\rangle = \delta_{s,s_1} \delta_{s',s_2} \delta^3(p - p_1) \delta^3(p' - p_2). \quad (11.15)$$

Using the anticommutation rule

$$\left\{ f_{s_1}(p_1), f_{s_2}^\dagger(p_2) \right\} = 0, \qquad (11.16)$$

for $p_1 \neq p_2$, we can deduce the relation

$$\left\langle p', s' \left| f_{s_2}(p_2) f_{s_1}^\dagger(p_1) \right| p, s \right\rangle = -\delta_{s,s_2} \delta_{s',s_1} \delta^3(p - p_2) \delta^3(p' - p_1). (11.17)$$

Using the above matrix elements along-with the fact that the matrix elements of ff and $f^\dagger f^\dagger$ vanish, we finally get

$$\left\langle p', s' \left| \overline{\psi} F \psi \right| p, s \right\rangle = \overline{u}_{s'}(p') F u_s(p) - \overline{v}_s(p) F v_{s'}(p') \qquad (11.18)$$

for a Majorana neutrino. We can further simplify this expression by using the relation between the spinors given in Eq. (4.18). Thus,

$$\begin{aligned} \overline{v}_s(p) F v_{s'}(p') &= -u_s^\mathsf{T}(p) C^{-1} F C \gamma_0^\mathsf{T} u_{s'}^*(p') \\ &= -\overline{u}_{s'}(p') \left(C^{-1} F C \right)^\mathsf{T} u_s(p), \end{aligned} \qquad (11.19)$$

using the same steps as those used to prove Eq. (11.11) above, with the difference that no negative sign comes when one interchanges two spinors. Now, plugging back in Eq. (11.18), we obtain for a Majorana neutrino

$$\left\langle p', s' \left| \overline{\psi} F \psi \right| p, s \right\rangle = \overline{u}_{s'}(p')(F + C F^\mathsf{T} C^{-1}) u_s(p). \qquad (11.20)$$

Again, noting that Eq. (11.13) holds if F is γ_λ or $\sigma_{\lambda\rho}$ or $\sigma_{\lambda\rho}\gamma_5$, we realize that the only form factor for a Majorana neutrino is the one associated with the axial current $\gamma_\lambda \gamma_5$.

Moreover, using Eq. (11.12), we see that for a Majorana neutrino, the matrix element of $\overline{\psi} \gamma_\lambda \gamma_5 \psi$ is twice as much as it is for a Dirac neutrino.

Exercise 11.4 *Consider the decay $Z \to f\widehat{f}$ discussed in Exercise 2.4, where f can be a Dirac or a Majorana fermion. Define the matrix element for the neutral current, $\left\langle f\widehat{f} \left| K_\mu \right| 0 \right\rangle$, as in Eq. (2.32). This time one gets*

$$\Gamma(Z \to f\widehat{f}) = \frac{1}{n!} \frac{\sqrt{2}}{3\pi} (a^2 + b^2) G_F M_Z^3 , \qquad (11.21)$$

instead of Eq. (2.33), where n is the number of identical particles in the final state, which is 1 if f is a Dirac fermion, but is 2 for Majorana particles since $f \equiv \widehat{f}$.

Using the method of evaluation of matrix elements involving Majorana particles described here, show that for Majorana neutrinos, show that $a_D = b_D$, whereas $a_M = 0$, $b_M = 2a_D$, where the subscripts indicate the Dirac and the Majorana cases. Thus, that the rate is the same whether the neutrino is Majorana or Dirac. If m_f is not negligible, show that the rate depends on the Majorana or Dirac character of the neutrino.

11.1.3   Form factors for a Weyl neutrino

In Ch. 4, we explained why a Majorana neutrino is a completely different type of object compared to a Weyl neutrino, despite the fact that both are spinors with two independent components. As further justification to that analysis, here we discuss the electromagnetic form factors of a Weyl neutrino and show how they differ from those of a Majorana neutrino.

For this, let us go back again to the discussion of putting the effective vertex in the co-ordinate representation, in the form of an effective interaction. We mentioned that the dipole moment terms will then involve the fermion bilinears $\overline{\psi}\sigma_{\lambda\rho}\psi$ and $\overline{\psi}\sigma_{\lambda\rho}\gamma_5\psi$. Using the left and right chiral projection operators, we find

$$\overline{\psi}\sigma_{\lambda\rho}\psi = \overline{\psi}_R\sigma_{\lambda\rho}\psi_L + \overline{\psi}_L\sigma_{\lambda\rho}\psi_R , \qquad (11.22)$$

and similarly

$$\overline{\psi}\sigma_{\lambda\rho}\gamma_5\psi = \overline{\psi}_R\sigma_{\lambda\rho}\gamma_5\psi_L + \overline{\psi}_L\sigma_{\lambda\rho}\gamma_5\psi_R . \qquad (11.23)$$

However, for Weyl neutrinos, there is no ψ_R, so these operators, being chirality changing operators, do not make sense. Hence, D_M and D_E must vanish for a Weyl neutrino.

Regarding the charge radius and the axial charge radius form factors, we notice that

$$\overline{\psi}\gamma_\lambda[R + r\gamma_5]\psi = \overline{\psi}\gamma_\lambda[(R - r)\mathsf{L} + (R + r)\mathsf{R}]\psi$$
$$= \overline{\psi}_L\gamma_\lambda[R - r]\psi_L + \overline{\psi}_R\gamma_\lambda[R + r]\psi_R. \quad (11.24)$$

Once again, the last term is meaningless for a Weyl neutrino, so that $R - r$ is left as the only form factor, as was mentioned in Ch. 2.

This discussion brings out some important points. We see that although a Weyl neutrino has the same number of electromagnetic form factors as a Majorana neutrino, the form factors are different combination of the general Lorentz covariant and gauge invariant form factors.

A second point is that the dipole moments of a neutrino vanish in the massless limit. For neutrinos with small masses, the dipole moment must then be small in some sense. In model calculations below, we will see that the dipole moment comes out to be proportional to the masses.

11.2   Kinematics of radiative decays

In this section, we consider a process closely related to that of §11.1, viz., that of the possibility of the decay of a neutrino ν_α in the channel

$$\nu_\alpha \to \nu_{\alpha'} + \gamma, \quad (11.25)$$

where $\nu_{\alpha'}$ is a lighter neutrino. Schematically, the process is represented by diagram like the one in Fig. 11.1, with the modification that now the incoming and the outgoing fermions are not the same. For Dirac neutrinos, we can write down the T-matrix element of the process as

$$T = \overline{u'}(\boldsymbol{p'})\Gamma_\lambda u(\boldsymbol{p})\epsilon^{*\lambda}, \quad (11.26)$$

where ϵ^λ is the photon polarization. For off-shell photons, the most general form for Γ_λ would still contain four form factors as in Eq. (11.8). However, for physical photons the on-shell condition and the Lorentz gauge condition read

$$q^2 = 0, \quad \epsilon^\lambda q_\lambda = 0, \quad (11.27)$$

so that the most general form for Γ_λ is given by

$$\Gamma_\lambda = \left[F(q^2) + F_5(q^2)\gamma_5\right]\sigma_{\lambda\rho}q^\rho, \quad (11.28)$$

where F and F_5 are Lorentz invariant form factors whose values are model-dependent. They are often called the *transition magnetic and electric dipole moments* between the two neutrinos involved.

The decay rate of ν_α can be obtained in a straightforward manner from the matrix element given above. In the rest frame of the decaying neutrino, one obtains

$$\Gamma = \frac{(m_\alpha^2 - m_{\alpha'}^2)^3}{8\pi m_\alpha^3} \left(|F|^2 + |F_5|^2 \right) . \tag{11.29}$$

Exercise 11.5 *Verify Eq. (11.29).*

In general, not much can be said about F and F_5. Hermiticity alone cannot restrict them. In particular, hermiticity relates form factors for the process in Eq. (11.25) to those for the process $\nu_{\alpha'} \to \nu_\alpha + \gamma$. However, in conjunction with CP invariance, hermiticity implies that F and F_5 must be relatively real.

If in addition, CP invariance can be assumed, nothing new is learned in the case of Dirac neutrinos. However, if both the initial and final neutrinos are Majorana particles, then one gets non-trivial constraints on the form factors. Recall, from the discussion of §4.4, that a physical Majorana neutrino is a CP eigenstate if CP is not violated by interactions. The eigenvalue can be $+i$ or $-i$. If both ν_α and $\nu_{\alpha'}$ have the same eigenvalues, then it follows that in Eq. (11.28), $F = 0$. On the other hand, if their CP eigenvalues are opposite, then $F_5 = 0$. Thus the transition is either purely of electric dipole type of purely of magnetic dipole type [3, 4]. Any mixture of these two kinds would mean CP violation for Majorana neutrinos.

11.3 Model calculations of dipole moments and radiative lifetime

In this section, we will illustrate the calculation of the electromagnetic vertex, defined in Eq. (11.3), in some models. To be specific, we consider the photon to be a real one, so that the restriction in Eq. (11.27) apply. Thus, we can determine only the magnetic and the electric form factors.

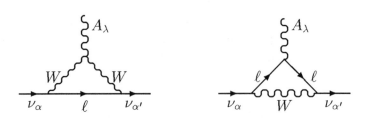

Figure 11.2: One-loop Feynman diagrams for the decay $\nu_\alpha \to \nu_{\alpha'} + \gamma$. In renormalizable gauges, there are extra diagrams where the W lines are replaced by the unphysical Higgs fields.

11.3.1 $SU(2)_L \times U(1)_Y$ model with Dirac neutrinos

Let us first take the $SU(2)_L \times U(1)_Y$ model of §5.1.1 which gives rise to Dirac neutrinos. The one-loop diagrams that contribute to the electromagnetic vertex of neutrinos have been shown in Fig. 11.2. These diagrams contain the W boson as internal lines. In general, this would give rise to infinities which would be canceled by accompanying diagrams where one or more of the internal W^+ lines are replaced by the unphysical Higgs H^+ which is eaten up by the W^+ in the process of symmetry breaking. The Feynman rules for the calculation are summarized in Fig. 11.3 using the renormalizable R_ξ gauge. The quantity ξ is a gauge parameter which must drop out from the calculation of all physical quantities.

> **Exercise 11.6** *Take the R_ξ gauge propagators of the W gauge boson and the associated unphysical Higgs boson. Consider the elastic scattering $\nu_\alpha \ell \to \nu_\alpha \ell$ at the tree level. Take the W coupling to fermions as given in Fig. 11.3. The W-mediated diagram will have some unphysical, i.e., ξ-dependent poles. Show that the unphysical Higgs mediated diagram cancels these poles if the unphysical Higgs coupling to fermions is as given in Fig. 11.3.*

Even without the detailed calculation, a few observations can be made. First, the magnetic form factor term comes with the dimension-5 operator $\overline{\psi}\sigma_{\lambda\rho}\psi q^\rho A^\lambda$. Since no such term was present in the original Lagrangian, no parameter can absorb any infinities of this term by a renormalization procedure. The renormalizability of the theory then guarantees that the magnetic form factor comes out to be finite, as it does in pure quantum electrodynamics. The same argument ensures the finiteness of the electric form factor.

Secondly, notice that both tensor and pseudo-tensor bilinears are chirality changing operators. This means that they connect left-chiral states with right-chiral ones, as shown in Eq. (11.22) and Eq. (11.23). For the $SU(2)_L \times U(1)_Y$ models, this has an interesting consequence. Consider the diagrams in Fig. 11.2 as flavor diagrams. Since the right chiral projections of fermions are all $SU(2)_L$ singlets, they do not have any interaction with $W^\pm$. Hence, it seems that only left chiralities are flowing on the fermion line. To obtain a chirality changing contribution, we must then put a mass insertion on one of the external legs. Thus, the magnetic and electric form factors must involve at least one factor of the mass of the external fermion(s).

It is important to realize that one power from the internal fermion line will not work. Starting with a state of left chirality, an internal mass insertion will produce a right chiral state. But this cannot interact with $W^\pm$, so one needs another chirality flip before the end of the internal fermion line. So, one can have even powers of internal fermion masses in the amplitude, but external masses must be there to cause the net chirality flip.

The calculation of the diagrams is straightforward. In order to extract as much information as possible, we do not make any assumption about whether ν_α and $\nu_{\alpha'}$ are, in fact, the same neutrino. Making the reasonable assumptions that

$$m_\alpha \ll M_W, \quad m_\alpha \ll m_\ell, \tag{11.30}$$

one obtains

$$\Gamma_\lambda = -\frac{eG_F}{4\sqrt{2}\pi^2}(m_\alpha \mathsf{R} + m_{\alpha'}\mathsf{L})\sigma_{\lambda\rho}q^\rho \sum_\ell U_{\ell\alpha}U^*_{\ell\alpha'}f(r_\ell). \tag{11.31}$$

Here, we have used the shorthands

$$r_\ell \equiv (m_\ell/M_W)^2, \tag{11.32}$$

and

$$f(r) \equiv \frac{3}{4(1-r)^2}\left[-2 + 5r - r^2 + \frac{2r^2 \ln r}{1-r}\right] \simeq -\frac{3}{2} + \frac{3}{4}r + \dots, \tag{11.33}$$

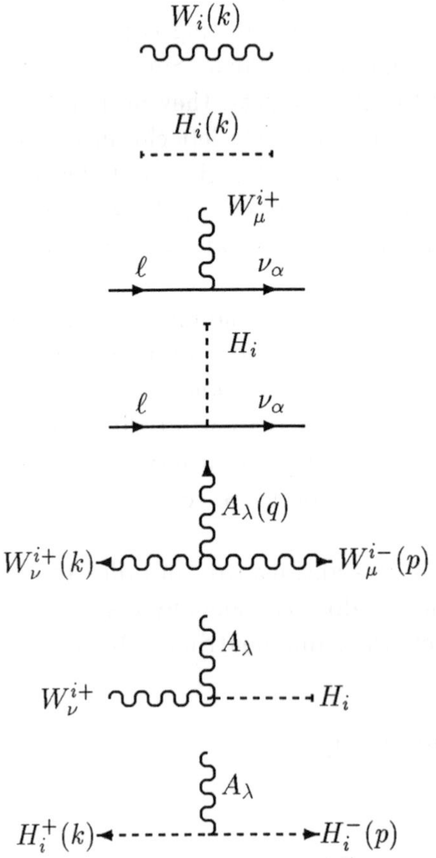

$$W_i(k)$$

$$\frac{-i}{k^2-M_{W_i}^2}\left[g_{\mu\nu}-\frac{(1-\xi)k_\mu k_\nu}{k^2-\xi M_{W_i}^2}\right]$$

$$H_i(k)$$

$$\frac{i}{k^2-\xi M_{W_i}^2}$$

$$W_\mu^{i+}$$
$$\ell \qquad \nu_\alpha$$

$$\frac{ig}{\sqrt{2}}\gamma_\mu(O_{Li}U_{\ell\alpha}\mathsf{L}+O_{Ri}V_{\ell\alpha}\mathsf{R})$$

$$H_i$$
$$\ell \qquad \nu_\alpha$$

$$\frac{ig}{\sqrt{2}M_{W_i}}[O_{Li}U_{\ell\alpha}(m_\alpha\mathsf{R}-m_\ell\mathsf{L})$$
$$+O_{Ri}V_{\ell\alpha}(m_\alpha\mathsf{L}-m_\ell\mathsf{R})]$$

$$A_\lambda(q)$$
$$W_\nu^{i+}(k)\qquad W_\mu^{i-}(p)$$

$$ie\left[g_{\nu\lambda}(q-k)_\mu+g_{\mu\nu}(k-p)_\lambda\right.$$
$$\left.+g_{\lambda\mu}(p-q)_\nu\right]$$

$$A_\lambda$$
$$W_\nu^{i+}\qquad H_i$$

$$ieg_{\nu\lambda}$$

$$A_\lambda$$
$$H_i^+(k)\qquad H_i^-(p)$$

$$ie(p-k)_\lambda$$

Figure 11.3: Feynman rules for calculations of various neutrino processes in renormalizable R_ξ gauges. For $SU(2)_L \times U(1)_Y$ models, disregard the index i on gauge boson W and unphysical Higgs H, set $O_{L1}=1$ and $O_{R1}=0$. For left-right symmetric models, the index i can take values of 1 and 2. ξ is a gauge parameter which must vanish in calculations for physical quantities.

where the last form gives the leading terms of $f(r)$ for small values of r.

If now we consider the diagonal matrix element for $\nu_\alpha = \nu_{\alpha'}$, the factor $(m_\alpha \mathsf{R} + m_{\alpha'}\mathsf{L})$ in the amplitude becomes m_α, so that we see that there is no $\sigma_{\lambda\rho}\gamma_5$ term. This means that, whether CP is conserved or not, the electric dipole moment vanishes at the one loop level. The magnetic dipole moment of ν_α, on the other hand, is given by [5]

$$\mu_\alpha = -\frac{eG_F}{4\sqrt{2}\pi^2}m_\alpha \sum_\ell U_{\ell\alpha}^* U_{\ell\alpha} f(r_\ell)$$

$$\simeq \frac{3eG_F}{8\sqrt{2}\pi^2}m_\alpha, \tag{11.34}$$

where in the last step, we have used the unitarity of the matrix U and the leading term of $f(r)$ from Eq. (11.33) for $r \ll 1$, an approximation which is valid for all known charged leptons. Thus, Eq. (11.34) shows that for Dirac neutrinos having usual weak interactions, the magnetic moment is given by [5, 6]

$$\mu_\alpha \simeq 3.1 \times 10^{-19} \mu_B \cdot \left(\frac{m_\alpha}{1\,\mathrm{eV}}\right), \tag{11.35}$$

μ_B being the Bohr magneton, $e/(2m_e)$.

Next we consider the case $\nu_\alpha \neq \nu_{\alpha'}$. Now Eq. (11.31) tells us that

$$F = K(m_\alpha + m_{\alpha'}) \quad, \quad F_5 = K(m_\alpha - m_{\alpha'}), \tag{11.36}$$

in the language of Eq. (11.28), where

$$K = -\frac{eG_F}{8\sqrt{2}\pi^2}\sum_\ell U_{\ell\alpha}U_{\ell\alpha'}^* f(r_\ell). \tag{11.37}$$

Thus, using Eq. (11.29), we obtain the decay rate to be [7, 8]

$$\Gamma = \frac{\alpha G_F^2}{128\pi^4}\left(\frac{m_\alpha^2 - m_{\alpha'}^2}{m_\alpha}\right)^3 (m_\alpha^2 + m_{\alpha'}^2)\left|\sum_\ell U_{\ell\alpha}U_{\ell\alpha'}^* f(r_\ell)\right|^2. \tag{11.38}$$

Since $\alpha \neq \alpha'$ now, the leading constant term of $f(r)$ sums up to zero due to the unitarity of the mixing matrix U. The next leading term proportional to r contributes to the sum over intermediate charged leptons in Eq. (11.38) and results in the expression

$$\Gamma = \frac{\alpha}{2}\left(\frac{3G_F}{32\pi^2}\right)^2\left(\frac{m_\alpha^2 - m_{\alpha'}^2}{m_\alpha}\right)^3 (m_\alpha^2 + m_{\alpha'}^2)\left|\sum_\ell U_{\ell\alpha}U_{\ell\alpha'}^* r_\ell\right|^2. \tag{11.39}$$

Since $r \ll 1$ for known leptons, this results in an extra suppression in the rate. In hadronic processes, the same phenomenon is called GIM suppression after the names of Glashow, Iliopoulos and Maiani [9] who first proposed that an unitary mixing matrix in the quark sector produces extra suppression in flavor changing neutral currents. This realization was crucial to include hadrons in the standard model.

If we assume that the tau-lepton term contributes most to the sum over internal leptons and that the elements of U are of order unity, Eq. (11.39) gives a lifetime of order 10^{29} years if $m_\alpha = 30\,\text{eV}$ and $m_{\alpha'} \ll m_\alpha$. Since the age of the universe is no more than a few times 10^{10} years, this means that neutrinos are practically stable towards a radiative decay. If there is a fourth lepton heavier than the tau then it can increase the decay rate. However, analysis of the Z decay width renders the existence of a fourth generation rather improbable.

11.3.2 $SU(2)_L \times U(1)_Y$ models with Majorana neutrinos

We now turn our attention to the models of §5.2, which yield Majorana neutrinos within the context of the gauge group $SU(2)_L \times U(1)_Y$. They involve extra scalar fields. Three things, therefore, must be kept in mind while calculating the electromagnetic vertex.

First, the charged unphysical Higgs scalar eaten up by the W^+ boson is, in general, not just ϕ_+, the charged component of the usual doublet Higgs representation. Hence it is not obvious that the Feynman rules given in Fig. 11.3 involving the unphysical Higgs still hold. However, it is possible to avoid such couplings altogether by going to the unitary gauge where diagrams involving unphysical Higgs fields do not contribute. In this gauge, the calculation of these two diagrams are obviously unaffected by the details of the Higgs sector. Hence, by gauge invariance, one can show that the couplings of the unphysical Higgs is still given by Fig. 11.3 so that the calculation of the diagrams involving W or the unphysical Higgs bosons is unaffected in the Feynman-'tHooft gauge also by the introduction of extra scalar multiplets.

Secondly, there are, in general, physical charged scalars present in the model. They will contribute through diagrams which are similar to those in Fig. 11.2 with the W-lines replaced by the physical scalars. The importance of these diagrams, evidently, depends on the mass of these scalars and therefore on the details of the scalar content.

Thirdly, one must remember that for each diagram in Fig. 11.2 there exists a second diagram in which all the internal lines shown are replaced by their conjugate lines [10, 4]. Such a contribution is absent in the Dirac case because the right handed neutrinos have no weak interaction, but occurs in the Majorana case since the right handed components of the Majorana fields are superpositions of $\hat{\nu}_{\ell R}$ which have nontrivial gauge interaction.

To determine the contribution from these conjugate diagrams, we notice that the charged current interaction in the leptonic sector can be written down as follows:

$$
\begin{aligned}
\mathcal{L}_{\text{cc}} &= \frac{g}{\sqrt{2}} \sum_{\ell} \left(\overline{\nu}_{\ell} \gamma^{\mu} \mathsf{L} \ell W_{\mu}^{+} + \overline{\ell} \gamma^{\mu} \mathsf{L} \nu_{\ell} W_{\mu}^{-} \right) \\
&= \frac{g}{\sqrt{2}} \sum_{\ell} \sum_{\alpha} \left(U_{\ell\alpha}^{*} \overline{\nu}_{\alpha} \gamma^{\mu} \mathsf{L} \ell W_{\mu}^{+} + U_{\ell\alpha} \overline{\ell} \gamma^{\mu} \mathsf{L} \nu_{\alpha} W_{\mu}^{-} \right) .
\end{aligned}
\tag{11.40}
$$

Using the definition of a conjugate field from Eq. (4.10), we can now cast it in the form

$$
\mathcal{L}_{\text{cc}} = \frac{g}{\sqrt{2}} \sum_{\ell} \sum_{\alpha} \left(U_{\ell\alpha}^{*} \overline{\nu}_{\alpha} \gamma^{\mu} \mathsf{L} \ell W_{\mu}^{+} - U_{\ell\alpha} \overline{\hat{\nu}}_{\alpha} \gamma^{\mu} \mathsf{R} \hat{\ell} W_{\mu}^{-} \right) .
\tag{11.41}
$$

While the above equation is valid in general, for Majorana neutrinos we can use the proportionality of $\hat{\nu}_{\alpha}$ and ν_{α}, as in Eq. (4.17), to write

$$
\mathcal{L}_{\text{cc}} = \frac{g}{\sqrt{2}} \sum_{\ell} \sum_{\alpha} \overline{\nu}_{\alpha} \gamma^{\mu} \left(U_{\ell\alpha}^{*} \mathsf{L} \ell W_{\mu}^{+} - \lambda_{\alpha} U_{\ell\alpha} \mathsf{R} \hat{\ell} W_{\mu}^{-} \right) ,
\tag{11.42}
$$

λ_{α} being the creation phase factor for ν_{α}.

The diagrams of Fig. 11.2 involve the first term in Eq. (11.42), whereas the conjugate diagrams involve the second term. Thus, clearly the amplitude of the conjugate diagrams can be obtained from the previous ones if we make the following substitutions:

$$
\begin{aligned}
W^{+} &\to W^{-} & , & \qquad \ell \to \hat{\ell} \\
\gamma_{5} &\to -\gamma_{5} & , & \qquad U_{\ell\alpha}^{*} \to -\lambda_{\alpha} U_{\ell\alpha} .
\end{aligned}
\tag{11.43}
$$

In addition to the factors coming from these substitutions, there will be an extra minus sign since in the conjugate diagram, the photon connects

to opposite sign particle compared to the original diagram. Thus, referring back to Eq. (11.31), we find the contribution in the Majorana case,
including both kinds of diagrams, is

$$
\begin{aligned}
\Gamma_\lambda \;=\; &-\frac{eG_F}{4\sqrt{2}\pi^2}\sigma_{\lambda\rho}q^\rho \sum_\ell [U_{\ell\alpha}U_{\ell\alpha'}^*(m_\alpha \mathsf{R} + m_{\alpha'}\mathsf{L}) \\
&- \lambda_{\alpha'}\lambda_\alpha^* U_{\ell\alpha'}U_{\ell\alpha}^*(m_\alpha \mathsf{L} + m_{\alpha'}\mathsf{R})]\,f(r_\ell)\,.
\end{aligned}
\tag{11.44}
$$

Clearly, if $\alpha = \alpha'$, the two terms in the square bracket cancel each
other, showing that the dipole moments of a Majorana neutrino vanish,
as discussed in §11.1.2.

For $\alpha \neq \alpha'$, the interpretation of Eq. (11.44) becomes clear if we
assume CP invariance. In this case, we can easily show that

$$
\Xi U_{\ell\alpha}^* \bar{\nu}_\alpha \gamma^\mu \mathsf{L}\ell W_\mu^+ \Xi^{-1} = U_{\ell\alpha}^* \eta_{\Xi(\alpha)}^* \bar{\ell}\gamma^\mu \mathsf{L}\nu_\alpha W_\mu^-\,,
\tag{11.45}
$$

where $\eta_{\Xi(\alpha)}$ is the CP eigenvalue of the neutrino ν_α as defined in §4.4.
Similar phases for ℓ and W^+ can always be chosen to be unity by suitably
defining the states $\hat{\ell}$ and W^-, but for a Majorana neutrino the phase
is physically meaningful. Looking at the charged current interaction in
Eq. (11.40), we now see that CP invariance of these terms is obtained if
the second term is the CP conjugate of the first term, i.e., if

$$
U_{\ell\alpha} = U_{\ell\alpha}^* \eta_{\Xi(\alpha)}^*\,.
\tag{11.46}
$$

The quantity appearing in square brackets in Eq. (11.44) can now be
written as

$$
\begin{aligned}
U_{\ell\alpha}U_{\ell\alpha'}^* &\left[(m_\alpha \mathsf{R} + m_{\alpha'}\mathsf{L}) - \frac{\lambda_{\alpha'}U_{\ell\alpha'}}{U_{\ell\alpha'}^*}\frac{U_{\ell\alpha}^*}{\lambda_\alpha U_{\ell\alpha}}(m_\alpha \mathsf{L} + m_{\alpha'}\mathsf{R}) \right] \\
&= U_{\ell\alpha}U_{\ell\alpha'}^* \left[(m_\alpha \mathsf{R} + m_{\alpha'}\mathsf{L}) - \frac{\tilde{\eta}_{\Xi(\alpha)}}{\tilde{\eta}_{\Xi(\alpha')}}(m_\alpha \mathsf{L} + m_{\alpha'}\mathsf{R}) \right]
\end{aligned}
\tag{11.47}
$$

using Eq. (11.46) as well as the definition of the CP eigenvalue of a state
from Eq. (4.58).

Thus, if the Majorana neutrinos ν_α and $\nu_{\alpha'}$ have same CP eigenvalues, we obtain, in the notation of Eq. (11.28),

$$
F = 0\,,\quad F_5 = 2K(m_\alpha - m_{\alpha'})\,,
\tag{11.48}
$$

where K has been defined in Eq. (11.37). The transition is thus purely electric dipole type, as mentioned earlier, and the decay rate is

$$\Gamma = \frac{\alpha G_F^2}{64\pi^4} \left(\frac{m_\alpha^2 - m_{\alpha'}^2}{m_\alpha} \right)^3 (m_\alpha - m_{\alpha'})^2 \left| \sum_\ell U_{\ell\alpha} U_{\ell\alpha'}^* f(r_\ell) \right|^2. \quad (11.49)$$

On the other hand, if the CP eigenvalues of the two neutrinos are opposite, we obtain a purely magnetic transition since

$$F = 2K(m_\alpha + m_{\alpha'}), \quad F_5 = 0. \quad (11.50)$$

This gives

$$\Gamma = \frac{\alpha G_F^2}{64\pi^4} \left(\frac{m_\alpha^2 - m_{\alpha'}^2}{m_\alpha} \right)^3 (m_\alpha + m_{\alpha'})^2 \left| \sum_\ell U_{\ell\alpha} U_{\ell\alpha'}^* f(r_\ell) \right|^2. \quad (11.51)$$

Note that if $m_{\alpha'} \ll m_\alpha$, the rates for Majorana neutrino decay is twice that of a Dirac neutrino. However, it has to be remembered that so far, we have not calculated the diagrams with physical charged scalars in the internal lines.

This last contribution depends on the details of the scalar sectors of a model. In the model with a triplet Δ discussed in §5.2.1, the contribution from diagrams involving physical scalars can enhance the decay rate by upto two orders of magnitude [4] assuming that the mass of the charged scalar is above 80 GeV. For Zee's model in §5.2.2, the enhancement from physical Higgs diagrams can be enormous [11] and can result in lifetimes as low as 10^{19} yr for $m_\alpha = 30$ eV.

11.3.3 Left-right symmetric model

The left-right symmetric model was discussed in Ch. 6. The model was based on the gauge group $SU(2)_L \times SU(2)_R \times U(1)_{B-L}$. Apart from the charged gauge boson W_L that interacts with $V - A$ currents of fermions, here we have another charged gauge boson W_R, coming from the $SU(2)_R$ part of the group. Thus, in addition to the W_L diagrams discussed so far, we will have similar diagrams involving the W_R which will contribute to the effective electromagnetic vertex. The contributions from these diagrams can be estimated by changing γ_5 to $-\gamma_5$ (i.e., interchanging L with R) in the expressions obtained in §11.3.1 and §11.3.2, and simultaneously replacing the W_L mass inherent in G_F by the W_R mass.

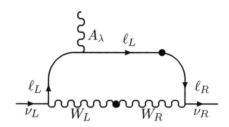

Figure 11.4: Flavor diagram to show that the left-right mixing contribution to the amplitude is proportional to the internal charged lepton mass.

However, since W_R has to be heavier than the W_L, as expressed in Eq. (6.27), we conclude that the W_R mediated diagrams are negligible.

The situation becomes more interesting if W_L and W_R are not the physical eigenstates in the gauge boson sector, but are rather linear combinations of the particles W_1 and W_2:

$$W_L = \sum_{i=1}^{2} O_{Li}W_i, \quad W_R = \sum_{i=1}^{2} O_{Ri}W_i. \qquad (11.52)$$

It can be shown that the elements of the matrix O can be taken to be real without any loss of generality, as mentioned in Eq. (6.15). In this case one can obviate the argument, given in §11.3.1, concluding that the electric or magnetic form factors are always proportional to neutrino mass. The point is that now the right handed fermions have interactions with the W_R, so we can draw a flavor diagram as in Fig. 11.4 to convince us that we can obtain a term that is proportional to the mass of the internal lepton line and not of the external neutrino line. Since the charged leptons are heavier than known neutrinos, one might expect larger contributions here.

Since left-right symmetric models naturally yield Majorana neutrinos, we will consider that case first. As discussed in Ch. 6, there are now two Majorana neutrinos per generation. We can call them n_α where the index α runs from 1 to $2\mathcal{N}$ for $\mathcal{N}$ generations. We can now write the elements of the mixing matrix as

$$\nu_{\ell L} = \sum_{\alpha=1}^{2\mathcal{N}} U_{\ell\alpha}n_{\alpha L}, \quad N_{\ell R} = \sum_{\alpha=1}^{2\mathcal{N}} V_{\ell\alpha}n_{\alpha R}, \qquad (11.53)$$

where U and V are $\mathcal{N} \times 2\mathcal{N}$ matrices. In general, both U and V are complex. Notice that although these formulas look the same as in the case of an $SU(2)_L \times U(1)_Y$ model without right handed neutrinos, the range of the sum in the other case is from 1 to $\mathcal{N}$, so that U is the square matrix used in §11.3.2. The contributions from the purely left- handed currents are given by Eq. (11.44) multiplied by O_{L1}^2, neglecting the terms suppressed by $M_{W_1}^2/M_{W_2}^2$. The mixed terms give a contribution [12]

$$\Gamma_\lambda^{(LR)} = O_{L1}O_{R1}\frac{eG_F}{4\sqrt{2}\pi^2}\sigma_{\lambda\rho}q^\rho \sum_\ell m_\ell$$

$$\times \left[\left(1 - \frac{\lambda_{\alpha'}}{\lambda_\alpha}\right)U_{\ell\alpha}^*V_{\ell\alpha'}\mathsf{R} + \left(1 - \frac{\lambda_\alpha}{\lambda_{\alpha'}}\right)U_{\ell\alpha}V_{\ell\alpha'}^*\mathsf{L}\right]. \quad (11.54)$$

Here, V is the mixing matrix in the right handed neutrino sector, which contributes because the light neutrinos have a small component of right handed neutrinos, as shown in Eq. (6.52). However, in the see-saw picture, the elements of V are so small, and the value of O_{R1} is so small from Eq. (6.27) that, despite the dependence on m_ℓ rather than on neutrino masses, the contribution from Eq. (11.54) is much smaller than that of Eq. (11.44) for light neutrinos. If, on the other hand, we consider the decay $N \to \nu + \gamma$ where N is predominantly right handed, both U and V elements can be appreciable and the decay rate can be large. This case was alluded to in §6.3.2.

Under reasonable assumptions about their masses, the physical charged Higgs bosons can mediate the decay at a much larger rate [13]. Alternatively, one can obtain a larger rate at the cost of sacrificing the see-saw mechanism of neutrino mass, whereby the elements of the matrix V can also be large and therefore the contribution in Eq. (11.54) is indeed large [14, 12].

In Ch. 6, we showed that one can make left right symmetric models which give Dirac neutrinos. If this is the case, the analysis above is somewhat modified. In particular, the Dirac neutrinos can have magnetic moments. The LL part of this moment will be the same as that given in $SU(2)_L \times U(1)_Y$ models, and the RR part is small because of the heaviness of the W_R gauge boson.

To obtain the LR part, let us, for simplicity, consider one generation first. It was shown in §4.5 that a Dirac neutrino can be seen as the degenerate limit of two Majorana neutrinos. The magnetic moment operator involves a transition from left to right helicity of the Dirac fermion, which appear from the two different basis Majorana spinors, as shown in Eq. (4.79). Thus we can take $\alpha = 1$ and $\alpha' = 2$. Since there is no diagonalization to be performed in the single generation case, $U = (1\ 0)$, $V = (0\ 1)$. Also, since the relative CP eigenvalues of the two Majorana spinors involved is negative, we use $\lambda_1/\lambda_2 = -1$. Thus, using Eq. (11.54), we obtain the following terms:

$$O_{L1}O_{R1}\frac{eG_F}{2\sqrt{2}\pi^2}m_\ell\left[\overline{\psi}_1\sigma_{\lambda\rho}q^\rho\psi_2 + \overline{\psi}_2\sigma_{\lambda\rho}q^\rho\psi_1\right] . \qquad (11.55)$$

Using the identifications of the Dirac spinor states from Eq. (4.79), this can be rewritten as

$$O_{L1}O_{R1}\frac{eG_F}{\sqrt{2}\pi^2}m_\ell\overline{\Psi}\sigma_{\lambda\rho}q^\rho\Psi , \qquad (11.56)$$

which shows that

$$\Gamma_\lambda^{(LR)} = O_{L1}O_{R1}\frac{eG_F}{\sqrt{2}\pi^2}m_\ell\,\sigma_{\lambda\rho}q^\rho . \qquad (11.57)$$

The magnetic moment is therefore given by

$$O_{L1}O_{R1}\frac{eG_F}{\sqrt{2}\pi^2}m_\ell . \qquad (11.58)$$

In the more general case with generational mixing, it is easy to see that one gets the following contribution to the magnetic moment [15]:

$$O_{L1}O_{R1}\frac{eG_F}{\sqrt{2}\pi^2}\sum_\ell m_\ell \text{Re}\left(U_{\ell\alpha}V_{\ell\alpha}^*\right) , \qquad (11.59)$$

where U and V are the mixing matrices in the left and right handed neutrino sector.

11.4 Large magnetic moment and small neutrino mass

One of the main reasons for extensive discussion on magnetic moment of the neutrino is the possibility that it may provide a solution to solar neutrino puzzle [16] (see Ch. 15 for details). The required magnitude of the neutrino magnetic moment for this purpose is

$$\mu_\nu \simeq (.3 \text{ to } 1) \times 10^{-10}\mu_B . \qquad (11.60)$$

It is clear from the discussion immediately preceding this section that neither the standard model with N_R nor the left-right symmetric model can provide such a large magnetic moment for ν_e. Therefore, new extensions of the standard model have been searched for that can yield a large magnetic moment.

One interesting possibility is to adjoin, in addition to N_R, an iso-singlet singly charged Higgs boson h_+ to the standard model [17]. Then the following additional terms are allowed in the Yukawa coupling in addition to those already present in the standard model:

$$-\mathcal{L}'_Y = \sum_{a,b} \left(f^L_{\ell\ell'} \psi^{\mathrm{T}}_{\ell L} C^{-1} \epsilon \psi_{\ell L} h_+ + f^R_{\ell\ell'} N^{\mathrm{T}}_{\ell R} C^{-1} \ell_R h_+ \right) + \mathrm{h.c.}. \quad (11.61)$$

The quantum numbers of the h_+, as well as its couplings with the lepton doublets, is the same as in Zee's model presented in §5.2.2. As in that model, we have $f^L_{\ell\ell'} = -f^L_{\ell'\ell}$. The difference now is the presence of the right handed neutrinos, and the absence of two doublet Higgs fields without which $B - L$ is not broken. Thus the neutrinos are Dirac particles in this model. There is a new contribution to the magnetic moment of the ν_e arising from the diagram involving physical scalars in the loop, which is:

$$\mu_{\nu_e} = \frac{f^L_{e\tau} f^R_{e\tau} m_e m_\tau}{16\pi^2 M_h^2} \left[\ln\left(M_h^2/m_\tau^2\right) - 1 \right] \mu_B. \quad (11.62)$$

Since $f^{L,R}$ are very poorly constrained by low energy data, for $f^L_{e\tau} f^R_{e\tau} \simeq 10^{-1}$ and $M_h \simeq 100\,\mathrm{GeV}$, one easily obtains $\mu_{\nu_e} \simeq 10^{-10} \mu_B$. A notable feature of this model is that due to antisymmetry of f^L, the magnetic moment of ν_τ receives contribution only from electron intermediate state and is therefore small. However, despite its above mentioned advantage, this extension of the standard model is quite unsatisfying as a model for the neutrinos since it leaves the neutrino mass as a free parameter in the Lagrangian. Thus, the smallness of the neutrino mass is not understood.

Extensions of this model that lead to finite neutrino mass as well as finite magnetic moment have been constructed [18]. In such models, there is an intimate connection between neutrino mass and magnetic moment. This is because when neutrino mass arises in a gauge theory at the loop level involving virtual charged bosons and fermions, the magnetic moment is induced by attaching a photon line to the internal lines.

Thus, if the typical masses in the loop are denoted by M, the neutrino mass term would be given by $\mathcal{G}M$, where $\mathcal{G}$ is some combination of coupling constants and other numerical factors. The magnetic moment will roughly be given by $e\mathcal{G}/M$, so that

$$\mu_{\nu_e} \approx \frac{e}{M^2} \, m_{\nu_e} \approx \frac{2m_e m_{\nu_e}}{M^2} \, \mu_B \, . \tag{11.63}$$

For $M \simeq 100\,\text{GeV}$, one needs $m_{\nu_e} \geq 10\,\text{keV}$ in order to satisfy the requirement of Eq. (11.60). This is unacceptable because of the experimental bounds quoted in Eq. (3.1). Thus, constructing models for large magnetic moment while keeping $m_{\nu_e} \lesssim 10\,\text{eV}$ is a new challenge to theorists.

One way to answer this challenge has been proposed by Voloshin [19]. To understand his argument, let us go back to the Majorana basis of mass terms, introduced in §4.5. Consider the Majorana basis spinors ψ_a. The mass terms in this basis have the form $\psi_a^{\mathsf{T}} C^{-1} \psi_b$, which are symmetric in the interchange of the indices a and b, as shown in Eq. (4.78). However, the magnetic moment term, $\psi_a^{\mathsf{T}} C^{-1} \sigma_{\lambda\rho} \psi_b F^{\lambda\rho}$ is antisymmetric under the same interchange.

> **Exercise 11.7** *Following the steps leading to Eq. (4.74), show that $\psi_a^{\mathsf{T}} C^{-1} \sigma_{\lambda\rho} \psi_b$ is antisymmetric under the interchange $\psi_a \to \psi_b$.*

The proposal of Voloshin [19] was to consider the interactions of ν_{eL} and $\widehat{N}_{eL}$ (i.e. the antiparticle of N_{eR}) to be symmetric under an $SU(2)_\nu$ symmetry under which ν_e and $\widehat{N}_e$ form a doublet. Then, the Dirac mass of ν_e is a triplet under $SU(2)_\nu$ whereas the magnetic moment term is a singlet. So a model that respects $SU(2)_\nu$ invariance will yield a large μ_ν with $m_\nu = 0$. Gauge models implementing this symmetry have been discussed extensively [20]. They generally require the introduction of new gauge bosons with masses in the 100 GeV to 1 TeV range and have to be carefully constructed to avoid conflict with low energy phenomenology arising from the presence of the new gauge bosons.

The problem with the implementation of Voloshin symmetry is that it does not commute with the gauge symmetry of $SU(2)_L$. This is because the $SU(2)_\nu$ symmetry connects two states of different $SU(2)_L$ quantum number. To avoid these problems, one can look for other $SU(2)$ symmetries which can serve to restrict the mass while at the same time being orthogonal to $SU(2)_L$. Of particular interest is the suggestion [21] that $SU(2)_\nu$ is identified with the horizontal symmetry group between the electron and the muon generations, which is a global symmetry of the standard model in the limit $m_e = m_\mu = 0$.

More recently a new class of models have been constructed which avoid the use of new gauge degrees of freedom but use horizontal discrete symmetries present in the standard model to obtain consistency between large magnetic moment and small neutrino mass [22]. As an example, consider the discrete horizontal symmetry D operating between the first two generations [22]:

$$\begin{pmatrix} \psi_e \\ \psi_\mu \end{pmatrix} \to \begin{pmatrix} 0 & 1 \\ -1 & 0 \end{pmatrix} \begin{pmatrix} \psi_e \\ \psi_\mu \end{pmatrix} \tag{11.64}$$

where ψ_ℓ denotes the doublet whose elements are ν_ℓ and ℓ. Note that under this symmetry $m_{\nu_e \nu_\mu}$ is odd whereas $\mu_{\nu_e \nu_\mu}$ is even. So, if in this model we impose $L_e - L_\mu$ conservation in addition to D-symmetry, the model in the D-symmetric limit leads to a large transition magnetic moment between ν_e and ν_μ while keeping $m_{\nu_e \nu_\mu} = 0$. (The $m_{\nu_e \nu_e} = m_{\nu_\mu \nu_\mu} = 0$ by $L_e - L_\mu$ conservation). In the symmetry limit however, $m_e = m_\mu$. The process of lifting this degeneracy generally implies $m_{\nu_e \nu_\mu} \neq 0$ but by a controlled amount if the D-symmetry is broken below the weak scale.

Other ways to evade the relation in Eq. (11.63) have also been proposed [23]. For instance, if the neutrino mass involves not only charged virtual particles in the loop but also neutral ones, then there is no strict relation between μ_ν and m_ν [24]. Another suggestion is to construct theories where the dominant one loop graph involves necessarily a gauge boson. In these theories, the magnetic moment graph will involve the transverse component of the gauge boson whereas the neutrino mass graph will exchange only the longitudinal component [25], thereby avoiding the constraint of Eq. (11.63).

Chapter 12

Double beta decay

12.1 Introduction

It is straightforward to see that in second order in the weak Hamiltonian, the following process can arise:

$$n + n \rightarrow p + p + e^- + e^- + \widehat{\nu}_e + \widehat{\nu}_e \, . \tag{12.1}$$

Such a process is called *double beta decay* since two β-rays (or electrons) emerge in the final states[1]. Usually, this process is denoted by $\beta\beta_{2\nu}$ since it is accompanied by two (anti)neutrinos. It has a strength G_F^2 and therefore occurs very rarely. Furthermore, for double beta decay to occur naturally, the arrangement of nuclei of different neighboring Z-values must be such that single beta decay (i.e. $n \rightarrow p + e^- + \widehat{\nu}_e$) is energetically forbidden. Thus double beta decay is a very rare process indeed. In fact, only recently, nearly a hundred years after Becquerel first observed beta decay, was double beta decay process observed in a beautiful experiment by Elliot, Hahn and Moe [2] for the Selenium nucleus:

$$^{82}\text{Se} \rightarrow {}^{82}\text{Kr} + 2e^- + 2\widehat{\nu}_e \tag{12.2}$$

with a half-lifetime 1.1×10^{20} years. The observation of another $\beta\beta_{2\nu}$-process has subsequently been reported in the process

$$^{76}\text{Ge} \rightarrow {}^{76}\text{Se} + 2e^- + 2\widehat{\nu}_e \tag{12.3}$$

with a half-lifetime of 1.77×10^{21} years by the Heidelberg-Moscow group [3]. Several geochemical experiments have also reported evidence for $\beta\beta_{2\nu}$ processes, which are listed in Table 12.1.

Table 12.1: Results for experiments on $\beta\beta_{2\nu}$ decay.

Transition	Heidelberg [4]	Missouri [5]
$^{82}Se \rightarrow {}^{82}Kr$	$(1.3 \pm .05) \times 10^{20}\,yr$	$(1.0 \pm .8) \times 10^{20}\,yr$
$^{128}Te \rightarrow {}^{128}Xe$	$> 5 \times 10^{24}\,yr$	$(1.4 \pm .4) \times 10^{24}\,yr$
$^{130}Te \rightarrow {}^{130}Xe$	$(1.63 \pm .14) \times 10^{21}\,yr$	$(7 \pm 2) \times 10^{21}\,yr$

The processes described above conserve lepton numbers. Therefore, they provide a confirmation of the standard model of weak interaction. Our concern in this chapter will be to consider processes where two electrons are emitted in nuclear transmutation without being accompanied by neutrinos. Such processes violate lepton number by two units and are therefore signatures of new physics beyond the standard model. We consider three kinds of such processes:

$$(A, Z) \rightarrow (A, Z+2) + 2e^-$$
$$(A, Z) \rightarrow (A, Z+2) + 2e^- + J$$
$$(A, Z) \rightarrow (A, Z+2) + 2e^- + 2J. \tag{12.4}$$

In the above equation, J is a massless Goldstone boson called the Majoron [6], which results in models with spontaneous breaking of global $B - L$ symmetry. We can call the first of the three processes by the name $\beta\beta_{0\nu}$, the second one by $\beta\beta_{0\nu J}$ etc. As we will see below, the different double beta decay processes are identified by the different electron sum energy spectra. Several experiments [7]-[11] have searched for neutrinoless double beta decay. No evidence for it has been uncovered yet. The best limit on $\beta\beta_{0\nu}$ lifetime comes from ^{76}Ge experiments which give [7]

$$\tau_{0\nu}(^{76}Ge) \geq 9.1 \times 10^{24}\,yr. \tag{12.5}$$

The best limit on single Majoron emitting mode $\beta\beta_{0\nu J}$ is [7, 8]

$$\tau_{0\nu J}(^{76}Ge) \geq 7.91 \times 10^{21}\,yr. \tag{12.6}$$

The double beta decay transition can occur between $0^+ \rightarrow 0^+$ as well as $0^+ \rightarrow 2^+$ spin-parity states of nuclei. We will see that each one carries valuable information about theory. Before proceeding to that discussion, we present in Table 12.2 some examples of double beta decay nuclei along with energy release Q in each process.

Table 12.2: Examples of double beta decay nuclei.

Parent nucleus	Daughter nucleus	Energy release Q (in MeV)
^{48}Ca	^{48}Ti	4.27
^{76}Ge	^{76}Se	2.04
^{82}Se	^{82}Kr	3.00
^{100}Mo	^{100}Ru	3.03
^{130}Te	^{130}Xe	2.53
^{128}Te	^{128}Xe	0.87
^{136}Xe	^{136}Ba	2.48
^{150}Nd	^{150}Sm	3.37
^{232}Th	^{232}U	0.86
^{238}U	^{232}Pu	1.16

12.2 Kinematical properties

Before going into the detailed particle physics models, in this section we discuss [12] some of the kinematical properties for the various double beta decay modes. The decay width consists of a dynamical part which is the absolute square of the matrix element, $|M|^2$, and a phase space part. In this section, we will assume that the matrix element part is energy independent and will see how the lifetime depends on the Q value and the nature of electron energy spectrum. To discuss this, we also assume that the nuclei remain at rest subsequent to the decay, which is of course an excellent approximation.

To sort out the energy dependence in the phase space integrals of the various processes, note that the momentum integration involves a factor $\int d^3 p/(2E)$ for any particle with momentum p. Of course we can write $|\boldsymbol{p}| \, d \, |\boldsymbol{p}| = E \, dE$, where E is the energy. Thus, if a massless boson is emitted as in the Majoron emission case, the factors in the momentum integration is just $E \, dE$, using $|\boldsymbol{p}| = E$. On the other hand, for a fermion emitted, we cannot use $|\boldsymbol{p}| = E$, so that the momentum integration factors reduce to $p \, dp$. But now the spinors coming in the matrix element would give an extra factor which is of the order of the energy. Thus for fermions, the integral involves a factor $pE \, dE$.

To apply these rules, let us first consider the process $\beta\beta_{2\nu}$. Integrating over the momentum of the final nucleus, we obtain

$$
\Gamma_{2\nu} = 2\pi \int |\mathcal{R}_{2\nu}|^2 \, \delta(M_I - M_F - E_1 - E_2 - \omega_1 - \omega_2)
$$
$$
\frac{d^3 p_1}{(2\pi)^3 2E_1} \frac{d^3 p_2}{(2\pi)^3 2E_2} \frac{d^3 k_1}{(2\pi)^3 2\omega_1} \frac{d^3 k_2}{(2\pi)^3 2\omega_2}, \quad (12.7)
$$

where $\mathcal{R}_{2\nu}$ is the matrix element whose magnitude will not be discussed in this section, M_I and M_F denote the masses of the initial and the final nuclei, $p_i = (E_i, \boldsymbol{p}_i)$ are the four momenta of the electrons and $k_i = (\omega_i, \boldsymbol{k}_i)$ of the neutrinos. Ignoring the angular variables in momenta integrations and taking out the factor of momentum coming from the spinors, we get [13]

$$
\Gamma_{2\nu} \propto \int d\omega_1 \, k_1\omega_1 \int d\omega_2 \, k_2\omega_2 \int dE_1 \, p_1 E_1 \int dE_2 \, p_2 E_2
$$
$$
|\mathcal{R}_{2\nu}(E_1, E_2)|^2 \, \delta(M_I - M_F - E_1 - E_2 - \omega_1 - \omega_2), \quad (12.8)
$$

using the above power counting rules. Similarly we get for other final states [14, 15]:

$$
\Gamma_{0\nu} \propto \int dE_1 \, p_1 E_1 \int dE_2 \, p_2 E_2
$$
$$
|\mathcal{R}_{0\nu}(E_1, E_2)|^2 \, \delta(M_I - M_F - E_1 - E_2), \quad (12.9)
$$
$$
\Gamma_{0\nu J} \propto \int dE_1 \, p_1 E_1 \int dE_2 \, p_2 E_2 \int d\Omega \, \Omega
$$
$$
|\mathcal{R}_{0\nu J}(E_1, E_2)|^2 \, \delta(M_I - M_F - E_1 - E_2 - \Omega), \quad (12.10)
$$
$$
\Gamma_{0\nu JJ} \propto \int dE_1 \, p_1 E_1 \int dE_2 \, p_2 E_2 \int d\Omega_1 \, \Omega_1 \int d\Omega_2 \, \Omega_2
$$
$$
|\mathcal{R}_{0\nu JJ}(E_1, E_2)|^2 \, \delta(M_I - M_F - E_1 - E_2 - \Omega_1 - \Omega_2).
$$
$$
(12.11)
$$

In these equations, Ω denotes the Majoron energy.

Let us now make the following change of variables:

$$
\begin{aligned}
T &\equiv Q/m_e = (M_I - M_F - 2m_e)/m_e \\
y_1 &\equiv (E_1 - m_e)/m_e \\
y_2 &\equiv (E_2 - m_e)/m_e \\
y &= y_1 + y_2 = (E_1 + E_2 - 2m_e)/m_e. \quad (12.12)
\end{aligned}
$$

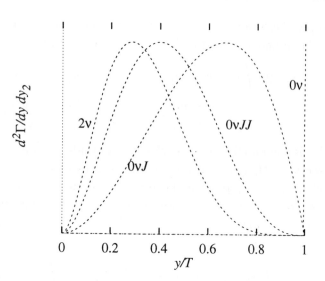

Figure 12.1: Schematic energy spectrum in different double beta processes. The normalization is arbitrary.

Using these new variables in Eq. (12.9), we find

$$\Gamma_{0\nu} \propto \int_0^T dy \int_0^y dy_2 \; |\mathcal{R}_{0\nu}|^2 \, p_1 p_2 E_1 E_2 \delta(T-y). \quad (12.13)$$

Thus in the plot of $d^2\Gamma/dy dy_2$ vs y, $\beta\beta_{0\nu}$ simply gives a spike at $y = T$. As for the other processes, we find

$$\Gamma_{2\nu} \propto \int_0^T dy \int_0^y dy_2 \; |\mathcal{R}_{2\nu}|^2 \, p_1 p_2 E_1 E_2 (T-y)^5, \quad (12.14)$$

$$\Gamma_{0\nu J} \propto \int_0^T dy \int_0^y dy_2 \; |\mathcal{R}_{0\nu J}|^2 \, p_1 p_2 E_1 E_2 (T-y), \quad (12.15)$$

$$\Gamma_{0\nu JJ} \propto \int_0^T dy \int_0^y dy_2 \; |\mathcal{R}_{0\nu JJ}|^2 \, p_1 p_2 E_1 E_2 (T-y)^3. \quad (12.16)$$

We thus see that each process has got its own characteristic sum energy spectrum which can be used to detect its presence experimentally. We show this in Fig. 12.1.

To see qualitatively how the lifetime depends on the Q-value, it is clear from Eq. (12.13) and Eq. (12.14) that, if we assume the matrix element to be constant, we get

$$\Gamma_{2\nu} \sim Q^{11} |\mathcal{R}_{2\nu}|^2$$
$$\Gamma_{0\nu} \sim Q^{5} |\mathcal{R}_{0\nu}|^2 . \tag{12.17}$$

From these formulas, we can get a rough estimate of the relative rates of the two processes. For that, we note that the amplitude $\mathcal{R}_{2\nu}$ has the spinors for the neutrino field corresponding to the outgoing antineutrinos. In general, these two antineutrinos have different momenta k_1 and k_2, and the amplitude depends on these momenta. We can represent this functional dependence by explicitly writing the $\beta\beta_{2\nu}$ amplitude as $\mathcal{R}_{2\nu}(k_1, k_2)$. However, if $k_1 = -k_2$, and if lepton number is violated, we can close the neutrino lines and take the neutrino propagator, and this will give us the amplitude for the neutrinoless double beta decay. Thus, we can write

$$\mathcal{R}_{0\nu} \simeq \lambda \int \frac{d^3 k}{(2\pi)^3} \mathcal{R}_{2\nu}(k, -k) , \tag{12.18}$$

where λ denotes the strength of lepton number violating processes, expressed as a dimensionless ratio. Now, if the intermediate neutrino typically has energies of order ω, we can approximately write the last expression as

$$\mathcal{R}_{0\nu} \simeq \lambda \omega^3 \mathcal{R}_{2\nu} . \tag{12.19}$$

Putting this back in Eq. (12.17), we finally obtain

$$\frac{\Gamma_{0\nu}}{\Gamma_{2\nu}} \simeq \lambda^2 \left(\frac{\omega}{Q} \right)^6 . \tag{12.20}$$

From Table 12.2, we see that the Q values are typically of order 2 to 3 MeV. On the other hand, the Fermi momentum of the nucleons in a nucleus are of the order of 200 MeV, so this should be an order of magnitude estimate of the energy carried by the neutrino line in a $\beta\beta_{0\nu}$ process. At present, the experimental bounds are roughly summarized as $\tau_{0\nu} \geq 10^4 \tau_{2\nu}$, i.e., as $\Gamma_{0\nu}/\Gamma_{2\nu} < 10^{-4}$. Using all these information, we can get a bound on the strength of lepton number violation, λ, from Eq. (12.20):

$$\lambda \lesssim 10^{-8} . \tag{12.21}$$

We will discuss the particle physics implications of this result in the next section.

> **Exercise 12.1** *A nucleus can be thought of as a sphere of radius $R_0 A^{1/3}$, where A is the total number of neutrons and protons in the nucleus and $R_0 = 1.2$ fm, where 'fm' means femtometer, i.e., 10^{-15} m. Considering the collection of nucleons to be a Fermi gas, calculate its Fermi momentum and Fermi energy.*

Next experimentally relevant quantity is the angular distribution, which has the following forms [13, 16] for the two neutrino decay $\beta\beta_{2\nu}$:

$$0^+ \to 0^+ \quad P_{2\nu}(\theta_{12}) \quad \propto 1 - v_1 v_2 \cos\theta_{12}$$
$$0^+ \to 2^+ \quad P_{2\nu}(\theta_{12}) \quad \propto 1 + \frac{1}{3} v_1 v_2 \cos\theta_{12}\,, \qquad (12.22)$$

where v_i are the velocities of the two electrons, and $P(\theta_{12})$ denotes the probability that the angle between them is θ_{12}.

For the no-neutrino decay mode, the angular distribution is sensitive to the dynamical mechanism giving rise to the decay. In anticipation of the next section, we call two important mechanisms, neutrino mass mechanism and left-right mixing mechanism. The angular distributions in these two cases are given by:

$$P_{0\nu,\text{mass}} \quad \propto \quad 1 - v_1 v_2 \cos\theta_{12}$$
$$P_{0\nu,\text{LR}} \quad \propto \quad 1 + v_1 v_2 \cos\theta_{12}\,. \qquad (12.23)$$

Thus, once $\beta\beta_{0\nu}$ is observed, studying its angular dependence one can pin-point the nature of interaction responsible for it.

12.3   Neutrinoless double beta decay in $SU(2)_L \times U(1)_Y$ models

Since neutrinoless double beta decay requires violation of lepton number by two units, it vanishes in the standard model. However, in models with the same gauge group as the standard model, the process can occur if lepton number is violated. We discuss various examples of the mechanism of lepton number violation and the induced rate of neutrinoless double beta decay.

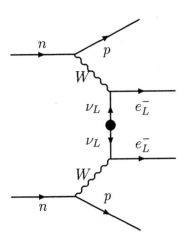

Figure 12.2: $\beta\beta_{0\nu}$ process induced by Majorana neutrino mass.

12.3.1   Light Majorana neutrino exchange

The simplest way to obtain $\beta\beta_{0\nu}$ decay is to have a Majorana neutrino i.e. a mass term of the form $m_\nu \nu_L^T C^{-1} \nu_L$, which violates lepton number by two units. Then, the diagram in Fig. 12.2 leads to $\beta\beta_{0\nu}$ decay. The effective interaction for this can be written as:

$$\mathcal{H}_{\beta\beta_{0\nu}} = 8G_F^2 \int d^4x \int d^4y \, T\left(J_{L\mu}^{(h)}(x) J_{L\lambda}^{(h)}(y)\right)$$
$$\overline{\psi}_e(x)\gamma^\mu L S_F(x-y)\gamma^\lambda R \widehat{\psi}_e(y) , \qquad (12.24)$$

where $S_F(x-y)$ is the Feynman propagator of the neutrino. Going to the momentum space, we get

$$\mathcal{H}_{\beta\beta_{0\nu}} = \frac{8G_F^2}{(2\pi)^4} \int d^4x \, d^4y \, d^4q \, d^4p_1 \, d^4p_2$$
$$T\left(J_{L\mu}^{(h)}(x) J_{L\lambda}^{(h)}(y)\right) e^{i(q-p_1)\cdot x} e^{i(q+p_2)\cdot y} M_{\mu\lambda}^{(\ell)} , \qquad (12.25)$$

where $M_{\mu\lambda}^{(\ell)}$ is the leptonic part, given by

$$M_{\mu\lambda}^{(\ell)} = \overline{u}_e(p_1)\gamma_\mu L \frac{\slashed{q} + m_\nu}{q^2 - m_\nu^2} \gamma_\lambda R v_e(p_2) - (1 \leftrightarrow 2)$$
$$= \frac{m_\nu}{q^2 - m_\nu^2} \left[\overline{u}_e(p_1)\gamma_\mu L \gamma_\lambda v_e(p_2) - (1 \leftrightarrow 2)\right] , \qquad (12.26)$$

q being the momentum of the virtual neutrino. We see that the double beta decay amplitude in this case is proportional to the neutrino mass. Thus, under the assumption that neutrino mass dominates the double beta decay amplitude, the measurement of the $\beta\beta_{0\nu}$ decay rate is a measurement of neutrino mass [17]. Turning this argument around, the present lower bounds on $\beta\beta_{0\nu}$ can be converted into upper limits on neutrino mass.

Of course, to extract information about neutrino mass from observations, we need knowledge of the nuclear matrix elements of the hadronic parts. Extensive discussion of this exists in the literature [13, 18]. An important aspect of the neutrino contribution is that, since $m_\nu^2 \ll q^2$, the neutrino pole in co-ordinate space leads to a $|\boldsymbol{x} - \boldsymbol{y}|^{-1}$ type potential. This implies that, nuclei not close by in the nucleus can undergo $\beta\beta_{0\nu}$-decay. This has the effect of enhancing the nuclear matrix element. In terms of an effective coupling coupling constant, we can write

$$G_{0\nu} \simeq G_F^2 m_\nu \left\langle \frac{1}{q^2} \right\rangle , \qquad (12.27)$$

where the matrix element of $1/q^2$ has to be taken within the initial and the final nuclear states.

To obtain a bound on the neutrino mass from the non-observation of $\beta\beta_{0\nu}$ so far, we note that lepton number violation is proportional to the neutrino mass. Thus, the dimensionless strength λ is given roughly by

$$\lambda \simeq m_\nu / \omega . \qquad (12.28)$$

Using $\omega \sim 200\,\mathrm{MeV}$ as in the last section, Eq. (12.21) gives the bound $m_\nu < 2\,\mathrm{eV}$. A more careful analysis of the $^{76}\mathrm{Ge}$ experiment [7] gives the upper limit on m_ν to be $0.65\,\mathrm{eV}$.

In presence of neutrino mixing, the above analysis is modified. For this, let us write

$$M_{\mu\lambda}^{(\ell)} = \frac{\langle m \rangle}{q^2} \left[\overline{u}_e(p_1)\gamma_\mu \mathsf{L}\gamma_\lambda v_e(p_2) - (1 \leftrightarrow 2) \right] , \qquad (12.29)$$

where

$$\langle m \rangle = \sum_\alpha U_{e\alpha}^2 m_\alpha , \qquad (12.30)$$

m_α being the mass of the eigenstate ν_α, and $U_{e\alpha}$ its mixing angle with the electron. We see that it is actually an *average* neutrino mass which contributes to $\beta\beta_{0\nu}$ decay. Since in general the elements of the mixing matrix are complex, it is possible in principle that the individual neutrino generations have large mass and yet conspire to cancel leading to a vanishing $\beta\beta_{0\nu}$-decay amplitude.

It is important to realize that such cancellation can happen even in the CP-conserving case [19]. For example, using Eq. (11.46), we can write $\langle m \rangle$ in terms of the CP eigenvalues of the neutrino states:

$$\langle m \rangle = \sum_\alpha |U_{e\alpha}|^2 \, m_\alpha \eta_{\Xi(\alpha)} \,. \tag{12.31}$$

Thus, if two Majorana neutrinos have opposite CP eigenvalues, their contributions would interfere destructively in the $\beta\beta_{0\nu}$ amplitude.

To see how this might happen, note that $\langle m \rangle$ actually represents the *ee*-element of a general Majorana neutrino mass matrix. Therefore it is possible to have a theory where the $\beta\beta_{0\nu}$ decay amplitude vanishes and yet the physical neutrino mass (e.g. the one which could be observed in tritium decay experiment) is non-zero. An example of such a case is the model of Zee described in §5.2.2, where the *ee*-element is zero because of the antisymmetry of the coupling. Huge cancellations also occur in a large class of $SO(10)$ grand-unified models [20].

However, there are other mechanisms that can lead to neutrinoless double beta decay without the intervention of neutrinos. We give examples of such mechanisms later in this chapter.

12.3.2   Heavy Majorana neutrino exchange

Neutrinoless double beta decay can also be mediated by heavy neutrinos [21]. The graph is the same as Fig. 12.2. In this case, since the mass satisfies $m_\nu^2 \gg q^2$, the neutrino propagator in Eq. (12.26) effectively reduces to $1/m_\nu$. The effective coupling is now given by

$$G_{0\nu} \simeq G_F^2 \frac{1}{m_\nu} \,, \tag{12.32}$$

and the lepton number violating strength is given by

$$\lambda \sim \omega/m_\nu \,. \tag{12.33}$$

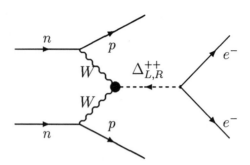

Figure 12.3: Contribution to $\beta\beta_{0\nu}$ process induced by doubly charged Higgs boson.

Since $\omega \approx 200\,\text{MeV}$, Eq. (12.21) would imply $m_\nu \gtrsim 10^7\,\text{GeV}$. The point to appreciate here is that, unlike the light neutrino case, here one gets a δ-function type potential between nucleons. Therefore, only nucleons close-by can contribute. This generally leads to a suppression of the nuclear matrix element. Thus, the bound on the effective coupling in this case is likely to be weaker than in Eq. (12.21).

12.3.3 Exchange of doubly charged Higgs boson

In many gauge models [22, 23] there exist doubly charged Higgs bosons which couple to two W_L bosons as well as to two electrons. As shown in Fig. 12.3, exchange of such doubly charged Higgs bosons can lead to $\beta\beta_{0\nu}$ decay without requiring a neutrino intermediate state [24]. While in some models this effect may be suppressed either due to small coupling [25] or due to a small nuclear matrix element [26], its effect can be important [27] in other models such as left-right models.

12.4 Neutrinoless double beta decay in Left-Right models

We saw in Ch. 6 that the left-right symmetric $SU(2)_L \times SU(2)_R \times U(1)_{B-L}$ models provide a very natural setting for small neutrino masses and violation of lepton number by two units. This model would therefore have built-in mechanisms for $\beta\beta_{0\nu}$ decay. To see this, let us remind the reader about the left and right leptonic doublets for mass eigenstate Majorana neutrinos:

$$\begin{pmatrix} \nu\cos\theta - N\sin\theta \\ e \end{pmatrix}_L \ , \quad \begin{pmatrix} \nu\sin\theta + N\cos\theta \\ e \end{pmatrix}_R \qquad (12.34)$$

where the heavy Majorana neutrino is N with $M_N = f v_R$ and the light one is ν, with $m_\nu \simeq m_D^2/M_N$. In the presence of three generations, there will be mixings between different generations. Furthermore, we recall that $M_{W_R} \gg M_{W_L}$. The general form of weak charged current Lagrangian is:

$$\begin{aligned} \mathcal{L}_{\text{cc}} = \ & \frac{g}{\sqrt{2}} \Big[(J^{(h)}_{L\mu} + J^{(\ell)}_{L\mu})(W_1^\mu \cos\zeta - W_2^\mu \sin\zeta) \\ & + (J^{(h)}_{R\mu} + J^{(\ell)}_{R\mu})(W_1^\mu \sin\zeta + W_2^\mu \cos\zeta) \Big] + \text{h.c.} . \end{aligned} \qquad (12.35)$$

Let us now turn to the discussion of $\beta\beta_{0\nu}$ process in this model. The strength of this process depends on the mass of the W_R bosons as well as mass of the light and heavy neutrinos. This process therefore provides a direct handle on the masses of these three particles. There are four independent mechanisms for double beta decay in left-right models. We discuss them one by one.

12.4.1   Light neutrino exchange

The Feynman diagram responsible in this case is shown in Fig. 12.2. This amplitude is given by

$$G_{0\nu}^{(m_\nu)} \simeq G_F^2 \sum_\alpha U_{e\alpha}^2 m_\alpha \left\langle \frac{1}{q^2} \right\rangle_{\text{Nuc}} . \qquad (12.36)$$

Here $U_{e\alpha}$ are mixing angles of various neutrinos to electron and we have taken the nuclear matrix element. In the limit of small intergenerational mixings, $\sum_\alpha U_{e\alpha}^2 m_\alpha$ is proportional to the Majorana mass of the neutrino. This process causes transitions between 0^+ states of the initial and final nucleus. Searches for this transition have been carried out in various nuclei such as ^{76}Ge, ^{82}Se, ^{100}Mo, ^{136}Xe, ^{150}Nd, leading to an upper bound $\langle m_\nu \rangle \leq 0.65\,\text{eV}$ where $\langle m_\nu \rangle$ stands for the nuclear average expression in Eq. (12.31).

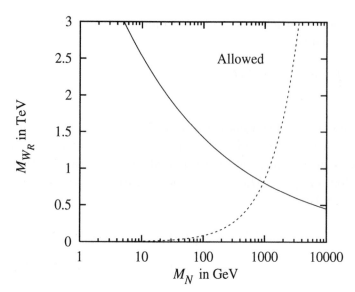

Figure 12.4: The regions below the solid and the dashed lines are ruled out by $\beta\beta_{0\nu}$ experiments and by vacuum stability arguments. The region above both the lines is allowed.

12.4.2 Heavy Majorana neutrino exchange

The effect of a heavy Majorana neutrino exchange on $\beta\beta_{0\nu}$ decay was discussed [21] in §12.3.2. However, this discussion is inapplicable to left-right models [28] since here, the heavy neutrinos couple predominantly to a heavier gauge boson. The Feynman diagram responsible is shown in Fig. 12.3, leading to the following amplitude:

$$G_{0\nu}^{(M_N)} \simeq G_F^2 \left(\frac{M_{W_L}}{M_{W_R}}\right)^4 \left\langle \frac{1}{M_N} \right\rangle_{\text{Nuc}}. \qquad (12.37)$$

This is also a $0^+ \to 0^+$ transition. Using the ^{76}Ge upper limit, one obtains the two dimensional plot in the M_{W_R}-M_N plane, as shown in Fig. 12.4. We have also shown a theoretical upper limit on M_N from considerations of vacuum stability. Together, they lead to a lower bound of 800 GeV for the W_R-mass.

12.4.3 Left-right mixing contribution

This contribution is shown in Fig. 12.5 and is a direct consequence of heavy and light neutrino mixing that arises in the see-saw mechanism.

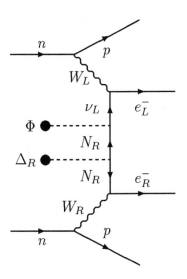

Figure 12.5: Left-right mixing contribution to $\beta\beta_{0\nu}$ decay.

The analog of Eq. (12.25) in this case is:

$$\mathcal{H}^{LR}_{\beta\beta_{0\nu}} = 8G_F^2 \frac{1}{(2\pi)^4} \sin\zeta \left(\frac{M_{W_1}}{M_{W_2}}\right)^2 \int d^4x\, d^4y\, d^4q\, d^4p_1\, d^4p_2$$
$$T\left(J^{(h)}_{L\mu}(x)J^{(h)}_{R\lambda}(y)\right) e^{i(q-p_1)\cdot x} e^{i(q+p_2)\cdot y} M^{(\ell)}_{\mu\lambda} \qquad (12.38)$$

where

$$M^{(\ell)}_{\mu\lambda} = \bar{u}_e(p_1)\gamma_\mu \mathsf{L}\frac{\slashed{q}+m_\nu}{q^2-m_\nu^2}\gamma_\lambda \mathsf{L}v_e(p_2) - (1 \leftrightarrow 2)$$

$$= \left[\bar{u}_e(p_1)\gamma_\mu \mathsf{L}\frac{\slashed{q}}{q^2-m_\nu^2}\gamma_\lambda v_e(p_2) - (1 \leftrightarrow 2)\right], \qquad (12.39)$$

This contribution has the distinctive feature that it is independent of the neutrino mass. Also, unlike the neutrino mass diagram, it leads to a $0^+ \rightarrow 2^+$ transition. The effective strength of this diagram can be written as:

$$G^{LR}_{0\nu} \approx G_F^2 \left(\frac{M_{W_1}}{M_{W_2}}\right)^2 \sin\zeta \left\langle\frac{1}{\slashed{q}}\right\rangle. \qquad (12.40)$$

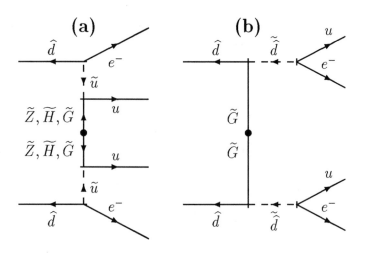

Figure 12.6: Contribution of R-parity breaking couplings to $\beta\beta_{0\nu}$ decay.

In the left-right model with the see-saw picture, one expects $\zeta \lesssim 10^{-5}$, as argued in Ch. 6. Since $M_{W_2} \geq 1.6\,\text{TeV}$, we thus obtain

$$G_{0\nu}^{LR} \ll G_{0\nu}^{(m_\nu)}, \tag{12.41}$$

provided the light Majorana neutrino masses are in the range of electron volts. Thus, we expect this to be negligible, pushing it beyond the reach of experiments, although $0^+ \to 2^+$ nature of the transition has a distinct signature of its own.

12.4.4 Higgs exchange contribution

In the left-right symmetric model, there are two doubly charged Higgs bosons: Δ_L^{++} and Δ_R^{++}, which can give potential contributions to $\beta\beta_{0\nu}$ decay. The contribution of Δ_L^{++} is negligible due to the small vev of Δ_L^0. However, Δ_R^{++} contribution need not be small [24, 27]. The diagram contributing to this process is shown in Fig. 12.3 where we keep the subscript R in all internal lines. Its effective strength is of order

$$G_{0\nu}^{(\Delta)} \simeq G_F^2 \left(\frac{M_{W_L}}{M_{W_R}}\right)^4 \left(\frac{M_{W_R}}{g M_{\Delta_R^{++}}^2}\right). \tag{12.42}$$

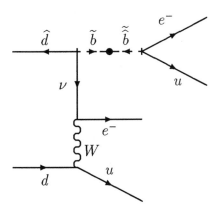

Figure 12.7: Joint vector-scalar exchange contribution to $\beta\beta_{0\nu}$ decay.

For $M_{W_R} \simeq 2\,\mathrm{TeV}$, $M_{\Delta_R^{++}} \simeq 30 \to 100\,\mathrm{GeV}$, $G_{0\nu}^{(\Delta)}$ may not be beyond the reach of experiments.

12.5 Neutrinoless double beta decay in supersymmetric models

In Ch. 8, we pointed out that R-parity violating couplings in the supersymmetric standard model lead to lepton number violating processes. This can lead to observable amplitudes for neutrinoless double beta decay [29] even though the neutrino mass can be very small. There are several diagrams that can lead to this processes, e.g. Fig. 12.6a and Fig. 12.6b. It was pointed out [29] that the dominant contribution arises from the graphs with gluino intermediate states. The amplitude for this process can be crudely estimated to be:

$$G_{0\nu} \simeq \frac{4\pi\alpha_s \lambda_{111}'^2}{M_{\tilde{G}} M_{\tilde{u}}^2} \qquad (12.43)$$

The matrix element for this process was carefully evaluated [30] and it was found that the present lower limit on the lifetime for neutrinoless double beta decay from the enriched ^{76}Ge by the Heidelberg-Moscow group [1] gives an upper limit on the allowed values for $\lambda_{111}' \leq 4 \times 10^{-4}$. This is a very stringent limit.

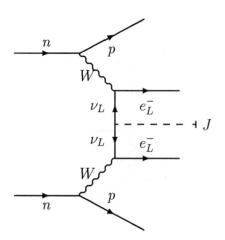

Figure 12.8: Majoron emission in $\beta\beta_{0\nu}$ decay.

A second class of contributions to neutrinoless double beta decay in the MSSM was pointed out recently [31], which involves a joint vector-scalar exchange as shown in Fig. 12.7. It involves a blob on the intermediate squark line, whose magnitude was explained in connection with Fig. 8.2. Using the discussion from there, we find that this contribution to the double beta amplitude is given by

$$G_{0\nu} \simeq \frac{G_F}{\sqrt{2}} \left(\frac{\lambda'_{113}\lambda'_{131} m_b (A + \mu \tan \beta)}{M_{\tilde{b}}^4} \right) \left\langle \frac{1}{\not{p}} \right\rangle_{\text{Nuc}} . \tag{12.44}$$

Again, for reasonable estimates of the nuclear matrix elements for this process [32], this implies the following bound on the R-parity violating couplings: $\lambda'_{113}\lambda'_{131} \leq 3 \times 10^{-8}$. The same graph also puts a limit on the product $\lambda'_{121}\lambda'_{112} \leq 10^{-6}$. These limits are also quite stringent.

12.6 Majoron emission in $\beta\beta_{0\nu}$ decay

In Ch. 5, we discussed the concept of the Majoron, which is a Goldstone boson which appears in a theory when the global $B - L$ symmetry is spontaneously broken [6]. There are three kinds of Majoron models: a) singlet Majoron [6], b) Triplet Majoron [23, 14], c) doublet Majoron [33, 34]. Majoron emission can occur in neutrinoless double beta decay [14,

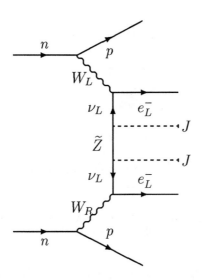

Figure 12.9: Double Majoron emission in $\beta\beta_{0\nu}$ decay.

35] via the diagram in Fig. 12.8. This process is most likely to happen in the triplet Majoron model, where the Majoron couples most strongly to the neutrinos. Denoting by $f_{\nu\nu J}$ the neutrino Majoron coupling, present experiments imply

$$f_{\nu\nu J} \leq 4 \times 10^{-4}. \tag{12.45}$$

It, however, appears that measurements of Z-width rule out the triplet Majoron model since the Majoron contributes equivalent of two extra neutrino species. The Z-width however still allows for the existence of singlet and perhaps even the doublet Majoron with some mixing.

In the supersymmetric doublet Majoron model [15, 36] there can be neutrinoless double beta decay with double Majoron emission via the diagram in Fig. 12.9. This leads to a $(T - y)^3$ type of spectrum [15], as discussed in Eq. (12.16). The only unknown particle physics parameter here is the Zino mass. Comparison with experiments, however, leads to only a very weak low bound of 1 to 2 GeV on the Zino mass, $M_{\tilde{Z}}$. However, a better analysis of the electron spectrum in the various $\beta\beta_{0\nu}$ process can lead to better bounds on $M_{\tilde{Z}}$ which will be an independent source of constraints on the supersymmetric theories.

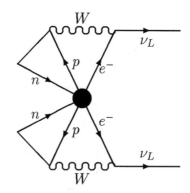

Figure 12.10: Diagram which leads to a Majorana neutrino mass from the $\beta\beta_{0\nu}$ amplitude. The blob in the middle represents the operator for the $\beta\beta_{0\nu}$ process.

12.7 Neutrino mass and $\beta\beta_{0\nu}$ decay

There appears to be an intimate connection between the neutrino mass and $\beta\beta_{0\nu}$ decay which has led to the speculation [17] that observation of $\beta\beta_{0\nu}$ decay will provide a measurement of m_ν. However, this conclusion relies on the assumption that the neutrino mass diagram (Fig. 12.2) dominates $\beta\beta_{0\nu}$ decay [17]. But as is clear from the discussion of §12.3, while neutrino mass diagram is perhaps the most plausible one, there exist many additional non-neutrino diagrams (Higgs, supersymmetry, right-handed neutrinos) which can play an equally important role in generating $\beta\beta_{0\nu}$-decay. Thus, observation of $\beta\beta_{0\nu}$ decay cannot immediately be translated into a value for the neutrino mass unless we have some independent way to make sure that the contribution of the other diagrams to $\beta\beta_{0\nu}$ decay is negligible.

It is however important to realize that while observation of $\beta\beta_{0\nu}$ decay cannot be directly translated to a value for the neutrino mass, it can certainly be used to infer the existence of a non-vanishing, Majorana neutrino mass regardless of whatever mechanism causes $\beta\beta_{0\nu}$ decay to occur.

The basic point is that once there is a lepton number violating interaction in a model, even though the neutrino mass may vanish at the tree level, it is bound to arise in some higher loop level. This can be seen as follows. One can think of neutrinoless double beta decay as being an effective scattering amplitude of the form $nn \rightarrow ppee$. One can then use the diagram in Fig. 12.10 to generate a Majorana mass for neutrino mass in higher orders [37]. The blob may contain for instance a doubly charged Higgs boson or whatever mechanism causes $\beta\beta_{0\nu}$ decay. This proves that the existence of $\beta\beta_{0\nu}$ decay implies the existence of a non-vanishing Majorana mass for the neutrino.

Chapter 13

Related processes

The obvious place to look for a signature of a massive neutrino is in processes involving neutrinos in the initial state, or the final state, or both. But that is not a necessity. One can look for processes involving other particles as well which cannot occur without the help of a massive neutrino as a virtual intermediate state. Such was the case of the neutrinoless double beta decay discussed in Ch. 12, where precisely the absence of neutrinos in the final state can be interpreted as a signature for neutrino mass. In this chapter, we discuss some more processes not involving neutrinos which nevertheless tell us about neutrino mass.

13.1 Lepton flavor changing processes

In the models presented in Part II of this book, we have repeatedly noticed that nonzero neutrino mass is almost invariably accompanied with neutrino mixing. Neutrino mixing, which is meaningless if the neutrinos are all massless, implies that the generational lepton numbers like the electron number, muon number and tau number are not conserved. This gives rise to flavor changing processes involving the charged leptons, a few of which we now discuss.

13.1.1 Radiative decays of muon and tau

In absence of generational quantum numbers, processes like

$$\ell \to \ell' + \gamma \tag{13.1}$$

can take place. Here, ℓ can denote the muon, in which case, ℓ' is the electron and we obtain the precess $\mu \to e + \gamma$. Also ℓ can be τ, in which case ℓ' can be either μ or e.

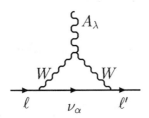

Figure 13.1: Diagram for $\ell \to \ell' + \gamma$. In renormalizable gauges, one must add other diagrams where one or both of the W lines are replaced by the unphysical Higgs.

The calculation of the rate of the process in Eq. (13.1) is similar to the analysis of the electromagnetic vertex of neutrinos made in Ch. 11. Following the arguments of that chapter, we can show that the amplitude of the process will in general be given as

$$\overline{u}'(\boldsymbol{p}') \left[\widetilde{F}(q^2) + \widetilde{F}_5(q^2)\gamma_5 \right] \sigma_{\lambda\rho} q^\rho u(\boldsymbol{p}) \epsilon^{*\lambda} \,, \tag{13.2}$$

where the form factors $\widetilde{F}$ and $\widetilde{F}_5$ depend on the model. The decay rate is given by a formula similar to Eq. (11.29):

$$\Gamma = \frac{(m_\ell^2 - m_{\ell'}^2)^3}{8\pi m_\ell^3} \left(\left|\widetilde{F}\right|^2 + \left|\widetilde{F}_5\right|^2 \right) \,. \tag{13.3}$$

One can also argue as before that in $SU(2)_L \times U(1)_Y$ models, absence of right-handed charged currents imply that the individual terms in the expression for the form factors $\widetilde{F}$ and $\widetilde{F}_5$ must be proportional either to m_ℓ or to $m_{\ell'}$. Since in nature $m_{\ell'} \ll m_\ell$ always, we might neglect the terms proportional to $m_{\ell'}$. In that case, the amplitude is proportional to m_ℓ. From Eq. (13.3), we thus see that the rate is proportional to m_ℓ^5. Despite the fact that this dependence is the same as that for the standard decay mode:

$$\Gamma(\ell \to \ell' \widehat{\nu}_{\ell'} \nu_\ell) = \frac{G_F^2 m_\ell^5}{192\pi^3} \,, \tag{13.4}$$

the radiative decays are suppressed, as explained below.

The one-loop diagram contributing to $\ell \to \ell' + \gamma$ is given in Fig. 13.1. This is similar to the radiative neutrino decay diagram in Fig. 11.2a.

The diagram corresponding to Fig. 11.2b is absent in this case since the internal fermion line is uncharged. Using the Feynman rules used in Fig. 11.3, one obtains

$$\widetilde{F} = \widetilde{K}(m_\ell + m_{\ell'}) \quad , \quad \widetilde{F_5} = \widetilde{K}(m_\ell - m_{\ell'}), \qquad (13.5)$$

where

$$\widetilde{K} = -\frac{eG_F}{8\sqrt{2}\pi^2} \sum_\alpha U^*_{\ell\alpha} U_{\ell'\alpha} \widetilde{f}(r_\alpha). \qquad (13.6)$$

where $\widetilde{f}$ is a function of

$$r_\alpha \equiv (m_\alpha/M_W)^2. \qquad (13.7)$$

Just as in the case of neutrino decay, we notice that the constant term in $\widetilde{f}(r_\alpha)$ cancels due to unitarity of the mixing matrix, which is the analog of the GIM suppression in the leptonic sector. So we get [1, 2, 3] the leading contribution

$$\Gamma = \frac{1}{2}\alpha G_F^2 \left(\frac{1}{32\pi^2}\right)^2 m_\ell^5 \left|\sum_\alpha U^*_{\ell\alpha} U_{\ell'\alpha} r_\alpha\right|^2, \qquad (13.8)$$

neglecting $m_{\ell'}$. Comparing with Eq. (13.4), we thus obtain the branching ratio

$$B(\ell \to \ell'\gamma) \equiv \frac{\Gamma(\ell \to \ell'\gamma)}{\Gamma(\ell \to \ell'\widehat{\nu}_{\ell'}\nu_\ell)} \simeq \frac{3\alpha}{32\pi} \left|\sum_\alpha U^*_{\ell\alpha} U_{\ell'\alpha} r_\alpha\right|^2. \qquad (13.9)$$

Since the elements of the mixing matrix are smaller than unity,

$$\sum_\alpha U^*_{\ell\alpha} U_{\ell'\alpha} r_\alpha < (m_{max}/M_W)^2, \qquad (13.10)$$

m_{max} being the largest neutrino mass. Thus, from Eq. (13.9), we obtain

$$B(\ell \to \ell'\gamma) < 10^{-4}(m_{max}/M_W)^4. \qquad (13.11)$$

Even if m_{max} denotes a mass close to the present upper limit of the mass of ν_τ of order 10 MeV, we obtain the upper limit on the branching ratio to be $O(10^{-17})$. A more conservative estimate would yield smaller numbers since the elements of the mixing matrix are presumably small. At any rate, the estimates of the branching ratios are orders of magnitude below the present experimental limits [4]

$$B(\mu \to e\gamma) \quad < \quad 4.9 \times 10^{-11}$$
$$B(\tau \to e\gamma) \quad < \quad 1.1 \times 10^{-4}$$
$$B(\tau \to \mu\gamma) \quad < \quad 4.2 \times 10^{-6}. \qquad (13.12)$$

One might ask whether the expected branching ratios can be larger in the model of §5.1.4, where Majorana masses were obtained after introducing right-handed neutrinos. Since the right handed neutrinos have large masses, as shown in Eq. (5.26), one might naively expect that the large values of r_α associated with these eigenstates can give a large contribution to the radiative decay rate.

But that argument is deceptive because the heavy neutrinos are also mainly $SU(2)_L$ singlets, so they do not interact with the $W^\pm$ to a first approximation. Of course the mass eigenstates do have a small mixture of doublet neutrino states, but that mixture is so small in see-saw type models that, despite large masses, the contribution of these neutrino states are negligible [5].

13.1.2 Decays of μ and τ into charged leptons

Apart from the radiative decays, another type of flavor-changing decays has been extensively searched for in experiments. These are decays of the μ and the τ which might contain only charged leptons and antileptons in the final state. For example, μ^- decays to $e^-e^-e^+$ (this final state is sometimes called $3e$ for the sake of brevity) is energetically possible. Experimental data indicate that [4]

$$B(\mu^- \to e^-e^-e^+) \quad < \quad 1.0 \times 10^{-12}. \qquad (13.13)$$

Similarly, we can think of some decay channels of the τ, quoting the presently known upper limits on the branching ratio in each case [4]:

$$B(\tau^- \to e^-e^-e^+) \quad < \quad 3.3 \times 10^{-6}$$
$$B(\tau^- \to \mu^-e^-e^+) \quad < \quad 3.4 \times 10^{-6}$$
$$B(\tau^- \to \mu^-\mu^-\mu^+) \quad < \quad 1.9 \times 10^{-6}. \qquad (13.14)$$

The theoretical estimates of these branching ratios are obviously model dependent. In the $SU(2)_L \times U(1)_Y$ model with right-handed

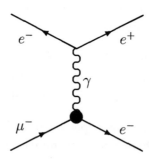

Figure 13.2: The decay $\mu \to 3e$ driven by the process $\mu \to e + \gamma$.

neutrinos which yields Dirac neutrinos and was discussed in §5.1.1, the process $\mu \to 3e$ for example goes through the effective $\mu \to e\gamma$ transition vertex, as shown in Fig. 13.2. Since the electron-photon vertex gives one factor of the electronic charge in the amplitude, in the rate we will naively expect $\Gamma(\mu \to 3e)/\Gamma(\mu \to e\gamma) \simeq \alpha/\pi$. However, the naive estimate is not accurate since in the calculation of the decay rate $\mu \to e\gamma$, the photon has to be taken as an on-shell physical particle, whereas in Fig. 13.2 the photon is virtual. Detailed calculation can again be done using the Feynman rules of Fig. 11.3, which yields [1, 6]

$$B(\mu \to 3e) \equiv \frac{\Gamma(\mu \to 3e)}{\Gamma(\mu \to e\widehat{\nu}_e\nu_\mu)} \simeq \frac{3\alpha^2}{16\pi^2} \left| \sum_\alpha U_{\ell\alpha}^* U_{\ell'\alpha} r_\alpha \ln r_\alpha \right|^2 . \quad (13.15)$$

The logarithmic factor can be quite large, but at any rate

$$B(\mu \to 3e) < \frac{3\alpha^2}{16\pi^2} \left(\frac{m_{\max}^2}{M_W^2} \ln \frac{m_{\max}^2}{M_W^2} \right)^2 \lesssim 10^{-17} \quad (13.16)$$

for $m_{\max} \simeq \mathcal{O}(10\,\text{MeV})$. In fact if the heaviest neutrino has a mass near $10\,\text{MeV}$, it will be mainly a ν_τ, which means $U_{\alpha\mu}$ and $U_{\alpha e}$ would be small. This will lower the expected branching ratio, which in any case is much smaller than the experimental bound quoted in Eq. (13.13). Similar conclusion can be drawn about the τ decays as well.

It must be emphasized that in any model where $\ell \to \ell'\gamma$ is allowed, there is a contribution to $\ell \to \ell'\ell''\widehat{\ell}''$ (where ℓ' and ℓ'' can be same or different charged leptons) via a diagram like the one in Fig. 13.2.

However, in specific models, there can be extra contribution to these non-radiative decays. For example, consider the $SU(2)_L \times U(1)_Y$ model presented in §5.2.1 where we added a triplet in the Higgs sector. The triplet has a doubly charged component Δ_{++}, which can have Yukawa couplings like

$$\sum_{\ell,\ell'} f^*_{\ell\ell'} \ell^T_L C^{-1} \ell'_L \Delta_{++} + \text{h.c.}, \qquad (13.17)$$

in the notation of Eq. (5.30). Thus, the exchange of a Δ_{++} can induce $\mu \to 3e$ decay [7] as shown in Fig. 13.3. This would give a decay rate similar in expression to the standard decay rate:

$$\Gamma(\mu \to 3e) \simeq \frac{G^2_{\text{eff}} m^5_\mu}{192\pi^3}, \qquad (13.18)$$

where now the effective strength is given by

$$G_{\text{eff}} \simeq \frac{f_{ee} f_{\mu e}}{M^2_{\Delta_{++}}}. \qquad (13.19)$$

Experimental constraints in Eq. (13.13) thus imply that the parameters of the model must satisfy

$$G_{\text{eff}} \leq 10^{-6} G_F \qquad (13.20)$$

in order to be acceptable. Similarly one can derive bounds from various τ-decay modes.

In the left-right symmetric electroweak models, it was shown in Ch. 6 that triplets of Higgs bosons enter very naturally in order to explain the smallness of neutrino masses. In order to maintain left-right symmetry in the Lagrangian, one needs a triplet Δ_L of the $SU(2)_L$ group, as well as a triplet Δ_R of the $SU(2)_R$ group. In that case, the doubly charged components of both Δ_L and Δ_R can mediate processes like $\mu \to 3e$ [8]. The implication of this fact on the right handed scale was discussed in Ch. 6.

13.1.3 Muonium-antimuonium transition

The conversion of a muonium $M \equiv \mu^+ e^-$ into an antimuonium involves a change of $|\Delta L_e| = 2$ and $|\Delta L_\mu| = 2$ and is, therefore, forbidden in the

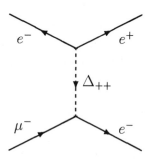

Figure 13.3: Tree diagram for $\mu \to 3e$ mediated by doubly charged scalars.

standard model. This process was suggested long before the advent of gauge theories [9] as a test of the multiplicative lepton number hypothesis. The phenomenology of this process is similar to that of K^0-$\widehat{K}^0$ mixing. Therefore to calculate the probability $P(M)$ of a free muonium to convert to an antimuonium, we have to consider the following evolution equation:

$$i\frac{d}{dt}\left(\begin{array}{c} M \\ \widehat{M} \end{array}\right) = \left(\begin{array}{cc} m_0 - i\Gamma & \delta \\ \delta & m_0 - i\Gamma \end{array}\right)\left(\begin{array}{c} M \\ \widehat{M} \end{array}\right) \qquad (13.21)$$

where m_0 is the mass of the muonium and Γ is its decay rate which is the same as the muon decay rate, $\Gamma = 0.5 \times 10^6\,\mathrm{s}^{-1}$. In Eq. (13.21) we have assumed that there is no energy splitting between the muonium and antimuonium such as could be caused by external electric field. From Eq. (13.21), we find

$$P(M) \simeq \frac{\delta^2}{2\Gamma}\,. \qquad (13.22)$$

If we assume that the transition is caused by an effective Hamiltonian of the form

$$\mathcal{H}_{M\widehat{M}} = \frac{4G_{M\widehat{M}}}{\sqrt{2}}[\overline{\mu}\gamma^\alpha Le][\overline{\mu}\gamma_\alpha Le] + \mathrm{h.c.}\,. \qquad (13.23)$$

then δ is given by [9]

$$\delta \equiv 2\left\langle M\left|\mathcal{H}_{M\widehat{M}}\right|\widehat{M}\right\rangle = \frac{16G_{M\widehat{M}}}{\sqrt{2}\pi a^3}\,, \qquad (13.24)$$

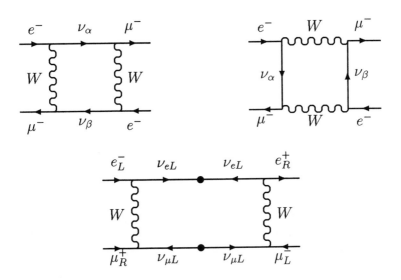

Figure 13.4: One loop diagrams for muonium-antimuonium transition in $SU(2)_L \times U(1)_Y$ models.

where a is the Bohr radius of the muonium. Thus, measurement of $P(M)$ will provide information on the coupling strength $G_{M\widehat{M}}$. There have been a number of experimental searches [10, 11], of which the most sensitive one provides the bound [11]

$$G_{M\widehat{M}} \leq 3 \times 10^{-3} G_F . \qquad (13.25)$$

Let us now turn to theoretical implications of a possible observation of M-$\widehat{M}$ transition. As mentioned before, it signals physics beyond the standard model. In $SU(2)_L \times U(1)_Y$ models with Dirac neutrino masses, it is mediated through neutrino mixing by the one-loop diagrams of Fig. 13.4a and Fig. 13.4b. If the neutrinos are Majorana particles, the diagram in Fig. 13.4c gives an additional contribution. This diagram does not even involve intergenerational mixing. Thus, it can dominate the amplitude. Using neutrino mass constraints from neutrinoless double beta decay, it was estimated [12] that the amplitude of Fig. 13.4c is less than about 3×10^{-5}.

A more dominant graph can arise in the left-right symmetric models

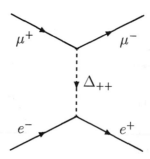

Figure 13.5: Muonium-antimuonium transition mediated by doubly charged scalars.

from the exchange [12, 13, 14] of doubly charged Higgs bosons Δ^{++} or in extensions of the standard model that contain $SU(2)_L$-singlet doubly charged Higgs bosons [15]. The relevant diagram is shown in Fig. 13.5. It gives an effective strength

$$G_{M\widehat{M}} = \frac{f_{ee}f_{\mu\mu}}{4\sqrt{2}M_{\Delta^{++}}^2}.$$ (13.26)

In the context of left-right models with an eV-keV-MeV type spectrum of neutrinos, if the cosmological mass density problem is obviated by $\nu_\mu \to 3\nu_e$ decay, one has to satisfy Eq. (6.39). Then, assuming natural values of Higgs coupling parameters, one obtains

$$G_{M\widehat{M}} \geq 4 \times 10^{-4} G_F.$$ (13.27)

This lower bound is above the contribution of the box graphs to the process. Therefore, observation of $G_{M\widehat{M}}$ above this bound would be an evidence in support of a doubly charged Higgs boson coupling to electrons and muons.

There are also supersymmetric contributions [16] to $M - \widehat{M}$ transition in the minimal supersymmetric standard model if the the R-parity breaking terms in the superpotential are not set to zero. The terms responsible are $\lambda_{132}L_eL_\tau\widehat{\mu} + \lambda_{231}L_\mu L_\tau\widehat{e}$. The tree level graph involving the exchange of $\tilde{L}_\tau$ leads to M-$\widehat{M}$ transition with a strength

$$G_{M\widehat{M}} \simeq \frac{\lambda_{132}\lambda_{231}}{M_{\tilde{L}_\tau}^2}$$ (13.28)

Given the present bounds on the λ's, we expect $G_{M\widehat{M}} \leq 10^{-2}G_F$, which is in the range accessible to the ongoing experiments.

13.1.4 Semi-leptonic processes

Lepton flavor violation has also been looked for in processes involving hadrons. Some examples on branching ratios of such processes are given below:

$$
\begin{aligned}
B(\pi^0 \to \mu^{\pm}e^{\mp}) &< 1.7 \times 10^{-8} \\
B(K_L \to \mu^{\pm}e^{\mp}) &< 3.3 \times 10^{-11} \\
B(\pi^+ \to \mu^+\nu_e) &< 8.0 \times 10^{-3} \\
B(K^+ \to \mu^+\nu_e) &< 4 \times 10^{-3},
\end{aligned}
\tag{13.29}
$$

where in each case, the rates are normalized to the total decay rate of the initial particle.

In the presence of neutrino mixing, the last two processes can be mediated by W-exchange at the tree level. Their suppression then puts direct bounds on the mixing of the first two generation of leptons. For example, comparing with the decay $K^+ \to \mu^+\nu_\mu$ which has a branching ratio of 63.5%, the last process gives $\tan^2\theta < 6 \times 10^{-3}$ or $\theta < 0.08$ for the ν_e-ν_μ mixing angle.

The process $K_L \to \mu^+e^-$ and $K_L \to \mu^-e^+$, on the other hand, is expected to be very suppressed even with strong neutrino mixing. The reason is that the hadronic current involves a transition between a d and an s quark. In $SU(2)_L \times U(1)_Y$ or even left-right models, flavor changing neutral current between quarks is absent at the tree level and suppressed at the one loop level due to the GIM mechanism. Thus, even the lepton flavor conserving process $K_L \to \mu^+\mu^-$ has a branching ratio of 9.5×10^{-9}. In processes like $K_L \to \mu^{\pm}e^{\mp}$, additional suppression factors are expected to come from lepton flavor changing.

Things can be different in grand-unified models. For example, the gauge group of $SO(10)$ has a subgroup $SU(4)_C$, as discussed in Ch. 7. Under this subgroup, the three colors of d quark appear in a quartet representation with the electron. Similarly, the muon forms a quartet with three colors of the strange quark. Thus, gauge bosons of $SU(4)_C$ can mediate the process $K_L \to \mu^+e^-$ through the tree level diagram of

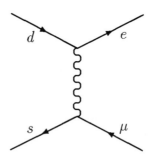

Figure 13.6: Gauge bosons of $SU(4)_C$ mediating the decay $K_L \to \mu^+ e^-$ at the tree level.

Fig. 13.6. The low branching ratio of this process then puts constraint on the $SU(4)_C$ breaking scale M_C, viz, [17] $M_C \gtrsim 10^5\,\text{GeV}$.

13.2 CP-violation in the leptonic sector

In the quark sector, the charged current involves a unitary mixing matrix, as shown in Eq. (2.20). In general, the elements of this matrix are complex, and this fact gives rise to CP violation in the quark sector [18]. In models with massive neutrinos, we have a mixing matrix in the leptonic sector as well. Complex numbers in this matrix would imply CP violation in the leptonic sector. As we will see in this section, the details of the consequences is sensitive to the Dirac or Majorana nature of neutrinos.

13.2.1 CP-violating phases in the fermion mass matrix

Let us first consider the case of Dirac neutrinos and examine how many CP-violating phases can appear in the mixing matrix. The charged current interaction in $SU(2)_L \times U(1)_Y$ models was written in Eq. (5.8), which was:

$$\frac{g}{\sqrt{2}} \sum_\ell \sum_\alpha U_{\ell\alpha} \bar{\ell}_L \gamma^\mu \nu_{\alpha L} W^-_\mu + \text{h.c.}, \qquad (13.30)$$

U being the mixing matrix. In order to keep the discussion general, let us consider that there are $\mathcal{N}$ generations of leptons. Then U is an $\mathcal{N} \times \mathcal{N}$

unitary matrix, which has $\mathcal{N}^2$ real parameters. If all elements of U were real, U would have become an orthogonal matrix, which has $\frac{1}{2}\mathcal{N}(\mathcal{N}-1)$ independent parameters which are the different mixing angles. Thus, the remaining $\frac{1}{2}\mathcal{N}(\mathcal{N}+1)$ parameters of U must be phases, which make U complex.

Exercise 13.1 *A general $\mathcal{N} \times \mathcal{N}$ matrix has $\mathcal{N}^2$ complex parameters, i.e., $2\mathcal{N}^2$ real parameters. Show that, if this matrix is unitary, the number of independent real parameters reduce to $\mathcal{N}^2$.*

However, not all of these phases are physically meaningful. For example, the interaction in Eq. (13.30) is invariant under the phase redefinitions

$$\nu_{\alpha L} \to e^{i\theta_\alpha} \nu_{\alpha L} \quad , \quad U_{\ell\alpha} \to e^{-i\theta_\alpha} U_{\ell\alpha} \,. \tag{13.31}$$

Since the overall phases of a Dirac field is physically irrelevant, by adjusting the phases of the complex neutrino fields we can absorb $\mathcal{N}$ phases corresponding to overall phases of each row of the matrix U. Similarly, we can adjust the phases of $\mathcal{N}$ charged lepton fields and absorb one phase in each column of U. It seems that we can thus get rid of $2\mathcal{N}$ phases, but this is not quite true. If we change all ν_α's by the same phase θ and change all the charged lepton fields by that same phase θ as well, it is clear from Eq. (13.30) that the mixing matrix U is unaffected. Thus the $2\mathcal{N}$ phases of redefining the fields must satisfy one constraint that they cannot be all equal. Out of the $\frac{1}{2}\mathcal{N}(\mathcal{N}+1)$ total phases in U, we thus see that

$$\frac{1}{2}\mathcal{N}(\mathcal{N}+1) - (2\mathcal{N}-1) = \frac{1}{2}(\mathcal{N}-1)(\mathcal{N}-2) \tag{13.32}$$

are physically meaningful. These phases are responsible for CP violation.

Needless to say, the above argument is exactly the same as that used in the quark sector. Indeed, the count for the CP-violating phases in Eq. (13.32) prompted Kobayashi and Maskawa [18] to postulate the third generation in days when it was experimentally unknown. In the leptonic sector also, the same conclusion holds, viz., CP violation can occur only if the number of fermion generations is three or more.

The situation changes dramatically if the neutrinos are Majorana particles. The point is that Majorana neutrinos are self-conjugate fields. Since the phase change of the conjugate field must be opposite to that of the field itself, it means that one cannot redefine the phase of a Majorana field. In Eq. (13.30), the only way to absorb phases from U is thus redefining the phases of the charged lepton fields. This way we can get rid of $\mathcal{N}$ phases and are left with

$$\frac{1}{2}\mathcal{N}(\mathcal{N}+1) - \mathcal{N} = \frac{1}{2}\mathcal{N}(\mathcal{N}-1) \tag{13.33}$$

phases which can inflict CP violation [19, 20, 21]. The dramatic change from the Dirac case is that now one can have CP violating effects even if there are two generations of fermions. For $\mathcal{N}$ generations, there are thus $\mathcal{N}-1$ additional CP violating phases if the neutrinos are Majorana particles.

Since Majorana masses are intimately related to $B-L$ violation, it is expected that these extra phases would appear only in $B-L$ violating processes. For example, if one looks at the process

$$e^+ + n + n \rightarrow p + p + e \,, \tag{13.34}$$

one can find that the rate is affected by these extra phases [22]. The phases can also play an important role in the $\beta\beta_{0\nu}$ process, which can be obtained from the process in Eq. (13.34) by crossing symmetry.

13.2.2 Electric dipole moment of the electron

It is well-known that if CP is conserved, an elementary fermion cannot have an electric dipole moment (EDM). This was alluded to in §11.1.1. Here we want to examine the magnitude of the EDM of charged leptons introduced by CP violation in the leptonic sector. Since the best experimental limits which exist concern the electron, viz.:

$$|d_e| \lesssim 10^{-26}\, e\text{-cm}\,, \tag{13.35}$$

our focus will mainly be on the EDM of the electron [23]. Our discussion of Ch. 11 means that the EDM is the coefficient of the $\sigma_{\lambda\rho}q^\rho\gamma_5$ term in the effective electromagnetic vertex. Such terms come necessarily from quantum corrections.

In $SU(2)_L \times U(1)_Y$ models, the EDM does not arise at one loop. This can be easily seen from our results on the $\ell \rightarrow \ell' + \gamma$ transitions, discussed in §13.1.1. In the present case, we are considering the electromagnetic vertex of a single charged lepton, i.e., $\ell = \ell'$. The EDM will then be given by the form factor $\tilde{F}_5$ defined in Eq. (13.2). However, the one-loop result for $\tilde{F}_5$, shown in Eq. (13.5), vanishes for $\ell = \ell'$. The result was first noted in the context of calculations in the quark sector [24], but is obviously valid in the leptonic sector as well. Of course, higher loop contributions give non-zero values of $\tilde{F}_5$ but they are suppressed by extra powers of coupling constants and loop integration factors.

Without going into all the calculations that go into finding Eq. (13.5), there is a very easy way to see the vanishing of EDM at one-loop level. The operator giving rise to the EDM is $\overline{\psi}\sigma_{\lambda\rho}q^\rho\gamma_5\psi$ for a fermion field ψ, and hermiticity of the Lagrangian implies that its coefficient must be purely imaginary since

$$(\overline{\psi}\sigma_{\lambda\rho}q^\rho\gamma_5\psi)^\dagger = -\overline{\psi}\sigma_{\lambda\rho}q^\rho\gamma_5\psi. \tag{13.36}$$

On the other hand, the coefficient of the $\overline{\psi}\sigma_{\lambda\rho}q^\rho\psi$ term should be real. Thus, looking for the EDM is essentially looking for imaginary parts of the vertex function.

Complex numbers can appear in the amplitude of Fig. 13.1 only through the elements of the mixing matrix U. However, for the same fermion on both outer legs, one of the vertices will contain a factor $U_{\ell\alpha}$ whereas the other will contain $(U^\dagger)_{\alpha\ell} = U_{\ell\alpha}^*$. Thus, the amplitude will involve only $|U_{\ell\alpha}|^2$, which is real. Consequently, the imaginary part vanishes at one-loop level, which signifies the vanishing of the EDM at this level.

The situation is different for left-right symmetric models. To discuss this, the first thing to realize is that in these models, complex numbers can appear not only in the fermionic mass matrices, but also in the mass matrix for charged gauge bosons. In Ch. 6, we assumed all VEV's to be real. If instead we consider the more general case when any number of them can be complex, the general structure of the mass terms in the charged gauge boson sector would be as follows:

$$\begin{pmatrix} W_L^+ & W_R^+ \end{pmatrix} \begin{pmatrix} M_{W_L}^2 & \Delta^2 \\ \Delta^{*2} & M_{W_R}^2 \end{pmatrix} \begin{pmatrix} W_L^- \\ W_R^- \end{pmatrix}. \tag{13.37}$$

Hermiticity of the mass matrix forces $M_{W_L}^2$ and $M_{W_R}^2$ to be real, but Δ can be complex. The mass eigenstates W_1 and W_2 will then be given, instead of Eq. (6.15), by

$$\begin{aligned} W_1 &= W_L \cos\zeta + W_R e^{i\alpha} \sin\zeta \\ W_2 &= -W_L \sin\zeta + W_R e^{i\alpha} \cos\zeta, \end{aligned} \qquad (13.38)$$

for suitably defined ζ and α. However, the charged current interaction involving leptons is given by

$$\mathcal{L}_{cc} = \frac{g}{\sqrt{2}} \sum_{\ell} \left(W_L^\mu \bar{\ell}_L \gamma_\mu \nu_{\ell L} + W_R^\mu \bar{\ell}_R \gamma_\mu N_{\ell R} \right) + \text{h.c..} \qquad (13.39)$$

We see that this interaction is invariant if we redefine $W_R \to e^{i\alpha} W_R$ and at the same time $N_R \to e^{-i\alpha} N_R$. By performing this operation, we can remove the phase α from Eq. (13.38), dumping it into the neutrino mass matrix. This means that the phases in the W mass matrix and those in the neutrino mass matrix are not independent. Without loss of generality, we can take the mixing matrix in the W sector to be real and write [25]

$$W_L = \sum_{i=1}^{2} O_{Li} W_i \quad , \quad W_R = \sum_{i=1}^{2} O_{Ri} W_i, \qquad (13.40)$$

in terms of the components of a 2×2 orthogonal matrix O, as mentioned in §11.3.3. The neutrino mass matrix can be diagonalized by using the matrices U and V introduced in Eq. (11.3). Using them, one obtains in a straightforward manner [2, 25, 26]

$$d_\ell = \sum_{i=1}^{2} \sum_{\alpha=1}^{2N} \frac{eg^2}{64\pi^2 M_{W_i}^2} O_{Li} O_{Ri} \text{Im}\, (U_{\ell\alpha} V_{\ell\alpha}) m_\alpha f(r_{\alpha i}), \qquad (13.41)$$

where

$$r_{\alpha i} \equiv m_\alpha^2 / M_{W_i}^2, \qquad (13.42)$$

m_α being the mass of the αth neutrino eigenstate. The function $f(r_{\alpha i})$ is given by

$$f(r) = \frac{4 - 11r + r^2}{(1-r)^2} - \frac{6r^2 \ln r}{(1-r)^3}. \qquad (13.43)$$

If all $r_{\alpha i} \ll 1$, we get $f(r) \simeq 4$ as the leading contribution, which gives

$$d_\ell = \frac{eg^2}{16\pi^2} \sum_{i=1}^{2} \frac{O_{Li} O_{Ri}}{M_{W_i}^2} \sum_{\alpha=1}^{2N} m_\alpha \mathrm{Im}\,(U_{\ell\alpha} V_{\ell\alpha})\,. \qquad (13.44)$$

However, following the diagonalization procedure, one can identify that

$$\sum_{\alpha=1}^{2N} m_\alpha U_{\ell\alpha} V_{\ell\alpha} = (m_D)_{\ell\ell}\,, \qquad (13.45)$$

a diagonal element in the Dirac mass terms for the neutrinos. Neglecting the term proportional to $1/M_{W_2}^2$, we thus obtain

$$d_e = (4 \times 10^{-24}\,e\text{-cm}) \times \sin 2\zeta\, \frac{\mathrm{Im}\,(m_D)_{ee}}{1\,\mathrm{MeV}}\,, \qquad (13.46)$$

assuming $r_{\alpha i} \ll 1$ as before. Since $\zeta \leq 10^{-2}$, as was mentioned in Eq. (6.27), we see that the calculated dipole moment might be close to the experimental bound of Eq. (13.35) if $\mathrm{Im}\,(m_D)_{ee} \simeq 100\,\mathrm{MeV}$.

Chapter 14

Neutrino properties in material media

In Ch. 10, we discussed neutrino oscillation in a material medium, which will again be used in the context of solar neutrinos in Ch. 15. The basic point is that, when a neutrino is traveling through a medium, its dispersion relation — i.e., the relation between its energy and the 3-momentum — is not the same as the corresponding relation in the vacuum. This was shown explicitly in Eq. (10.50).

This brings up an interesting question, however. The properties of neutrinos in material media seems, at least in this one instance, to be different from its properties in the vacuum. This should not be surprising, since we know that this is the case with photons, for example, which gives rise to the rich subject of electrodynamics in a medium. But we can ask whether for neutrinos, there are other physical properties which change when the neutrino travels through a medium. In this chapter, we try to give a glimpse of the basic techniques and some interesting results in this field. For the sake of simplicity, we will restrict ourselves to a medium which is in thermal equilibrium, and is homogeneous and isotropic. Also, the constituents of the medium are assumed to be only electrons and nucleons, along-with possibly their antiparticles. We start with re-deriving the dispersion relation of neutrinos in such a medium. This will extend the results obtained in Ch. 10 at the same time introducing some of the basic formalism involved.

14.1 The dispersion relation of neutrinos in a medium

14.1.1 The general structure

As with photons, the basic reason for the change of neutrino properties in a medium involves the interaction of neutrinos with the particles that constitute the medium. In Ch. 10, we showed how the interactions with the electrons and nucleons of the medium affect the dispersion relation of a neutrino.

There is, however, a more direct way of obtaining the dispersion relation. In quantum field theory, the dispersion relation of any particle is obtained by the pole of the propagator of that particle. Thus, the basic problem reduces to finding the neutrino propagator in a medium.

In the vacuum, the momentum-space Lagrangian of a free chiral (i.e., massless) neutrino is given by

$$\mathcal{L}_0 = \overline{\psi}_L(p) \not{p} \psi_L(p) \,, \tag{14.1}$$

so that the action is just obtained by integrating it over all of the momentum space. When interactions are introduced, a self-energy function appears:

$$\mathcal{L} = \overline{\psi}_L(p) \left(\not{p} - \Sigma \right) \psi_L(p) \,, \tag{14.2}$$

In the vacuum, the most general form for Σ is given by

$$\Sigma = a\not{p} \,, \tag{14.3}$$

where a is a Lorentz-invariant factor which can depend only on p^2. Thus, the self-energy has the same form as the free Lagrangian, and therefore it can be done away with after a renormalization of the field is performed.

In a medium, however, this is not the case. The reason is that the effects of the medium will show up in Σ in the form of terms dependent on the medium parameters. Since we assumed a homogeneous and isotropic medium, the only new 4-vector which appears in the problem is u^μ, the velocity of the center of mass of the medium from the frame of the observer. Thus, the most general form [1] for the self-energy is:

$$\Sigma = a\not{p} + b\not{u} \,, \tag{14.4}$$

where now a and b can in general depend on all Lorentz invariant quantities present in the problem. Since $u^2 = 1$, there are really only two such quantities. Let us denote them by

$$E = p \cdot u, \qquad P = \sqrt{E^2 - p^2}. \tag{14.5}$$

In the rest frame of the medium, E is the energy and P the magnitude of the 3-momentum of the neutrino.

So we can now rewrite Eq. (14.2) in the form

$$\mathcal{L} = \overline{\psi}(p)\mathsf{R}\slashed{V}\mathsf{L}\psi(p), \tag{14.6}$$

where

$$V_\mu = (1 - a)p_\mu - bu_\mu. \tag{14.7}$$

The propagator can then be identified as

$$S_F = \mathsf{L}\frac{\slashed{V}}{V^2}\mathsf{R}. \tag{14.8}$$

It is now clear that the dispersion relation of the chiral neutrino will be given by the solutions of $V^2 = 0$, i.e., by

$$(1 - a)E - b = \pm(1 - a)P. \tag{14.9}$$

It has to be borne in mind that in this equation a and b are functions of E and P, whose explicit forms must be known before one can express E as a function of P. However, even without the explicit forms, some characteristics of the solution can be seen from here. Let us denote the solutions of Eq. (14.9) with the positive and the negative signs on the right side by E_P and E'_P respectively. Then, in analogy with the situation in the vacuum, E_P is the dispersion relation for the neutrino, and $\overline{E}_P = -E'_P$ is for the antineutrino. Thus, putting all the dependences of a and b, we observe that $\overline{E}_P$ satisfies the equation

$$\left[1 - a(P, -\overline{E}_P)\right]\overline{E}_P + b(P, -\overline{E}_P) = \left[1 - a(P, -\overline{E}_P)\right]P, \tag{14.10}$$

whereas E_P satisfies

$$[1 - a(P, E_P)]E_P - b(P, E_P) = [1 - a(P, E_P)]P. \tag{14.11}$$

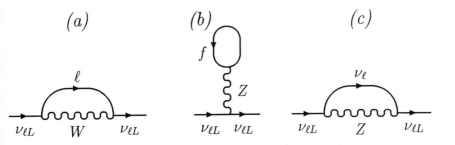

Figure 14.1: One-loop self-energy diagrams for neutrinos in the standard model. In diagram (a), the internal fermion line can only be the charged lepton which appears in the same doublet as the neutrino. In diagram (b), the loop can contain all fermions which couple to the Z. Diagram (c) is present if there are neutrinos in the background medium.

These are not the same equations, and hence in general

$$\overline{E}_P \neq E_P. \tag{14.12}$$

This is to be expected since the interactions of neutrinos and antineutrinos with, e.g., the electrons in the medium are not the same. In the special case where the background medium is CP symmetric and the interactions are also CP invariant, the neutrinos and the antineutrinos obey the same dispersion relation.

> **Exercise 14.1** *If the Lagrangian is CP-symmetric and so is the medium, show that a is an even function of E whereas b is an odd one [2]. Hence, show that $\overline{E}_P = E_P$ in this case.*

14.1.2 Propagators in the thermal medium

In order to proceed beyond the general analysis given above, we should calculate the self-energy of neutrinos in a specific model. Let us stick to the standard electroweak model, where the 1-loop self-energy diagrams for neutrinos are given in Fig. 14.1. Here, the fermions in the loop can be the fermions which constitute the medium, which will give a non-trivial contribution to the self-energy. Thus, to calculate these diagrams, one needs the propagators of these thermal fermions. These can be derived in many equivalent ways [3, 4, 5]. Here, we follow the derivation by Nieves [6], which brings out the physical ideas most directly.

For quantum field theoretic calculations in the vacuum, one defines the propagator by the relation

$$iS_F(x - y) = \left\langle 0 \left| \mathcal{T}\psi(x)\overline{\psi}(y) \right| 0 \right\rangle , \qquad (14.13)$$

where $|0\rangle$ is the vacuum state, and $\mathcal{T}$ denotes the time-ordering operator. In a thermal medium, the obvious generalization of this definition is

$$iS_F(x - y, \mu, \beta) = \left\langle \mathcal{T}\psi(x)\overline{\psi}(y) \right\rangle , \qquad (14.14)$$

where now the angular brackets indicate the expectation value in the thermal state with temperature $T = 1/\beta$ and chemical potential μ.

To see where it makes the crucial difference, consider the momentum expansion of a Dirac field operator given in Eq. (4.1), which contains two terms. When we put it in Eq. (14.13), four terms result for each time-ordering, but only one of them is non-zero since $f_s(k)|0\rangle = \widehat{f}_s(k)|0\rangle = 0$, and $\langle 0| \widehat{f}_s^\dagger(k) = \langle 0| f_s^\dagger(k) = 0$. This, however, is not the case when we talk about the thermal state instead of the vacuum state. In fact, the average number of particles at a certain energy in a thermal state is given by the Fermi-Dirac distribution, which means

$$\left\langle f_s^\dagger(k')f_s(k) \right\rangle = \delta^3(k - k')f_F(k, \mu, \beta) , \qquad (14.15)$$

$$\left\langle \widehat{f}_s^\dagger(k')\widehat{f}_s(k) \right\rangle = \delta^3(k - k')f_F(k, -\mu, \beta) , \qquad (14.16)$$

where

$$f_F(k, \mu, \beta) = \frac{1}{e^{\beta(k \cdot u - \mu)} + 1} . \qquad (14.17)$$

Since the fermionic creation and the annihilation operators satisfy the anti-commutation relation

$$\left\{ f_s^\dagger(k'), f_s(k) \right\} = \delta^3(k - k') , \qquad (14.18)$$

and the antifermionic operators satisfy a similar equation, Eqs. (14.15) and (14.16) imply the following expectation values:

$$\left\langle f_s(k)f_s^\dagger(k') \right\rangle = \delta^3(k - k')\left[1 - f_F(k, \mu, \beta) \right] , \qquad (14.19)$$

$$\left\langle \widehat{f}_s(k)\widehat{f}_s^\dagger(k') \right\rangle = \delta^3(k - k')\left[1 - f_F(k, -\mu, \beta) \right] . \qquad (14.20)$$

Thus, when we insert the momentum expansion of the fermion field from Eq. (4.1) into the expression for the propagator of Eq. (14.14), all four terms would contribute in this case. The final result for the propagator is:

$$
iS_F(k,\mu\,\beta) = (\not{k} + m)\left[\frac{i}{k^2 - m^2 + i0} - 2\pi\delta(k^2 - m^2)\eta_F(k,\mu,\beta)\right],
$$
$$(14.21)$$

where η_F contains the distribution functions for both fermions and antifermions:

$$
\eta_F(k,\mu,\beta) = \Theta(k\cdot u)f_F(k,\mu,\beta) + \Theta(-k\cdot u)f_F(-k,-\mu,\beta)\,, \quad (14.22)
$$

Θ being the step function which is zero if its argument is negative, and unity if the argument is positive.

Exercise 14.2 *Follow similar lines of reasoning to obtain the following thermal propagator of a scalar field of mass M:*

$$
i\Delta_F(k,\mu,\beta) = \frac{i}{k^2 - M^2 + i0} + 2\pi\delta(k^2 - M^2)\eta_B(k,\mu,\beta)\,, \quad (14.23)
$$

where η_B is defined in a manner similar to η_F, but with the Bose-Einstein distribution function.

14.1.3 Calculation of the dispersion relation of neutrinos

We can now easily calculate the contributions to the self-energy arising out of the various diagrams in Fig. 14.1. For a medium consisting of electrons, nucleons and their antiparticles only, Fig. 14.1c gives no contribution from the medium. As for Fig. 14.1a, the contribution exists only if the external lines represent ν_e. This contribution is

$$
-i\Sigma_a(p) = \left(\frac{ig}{\sqrt{2}}\right)^2 \int \frac{d^4k}{(2\pi)^4}\gamma^\rho\mathsf{L}\,iS_e(k,\mu,\beta)\,\gamma^\lambda\mathsf{L}\,i\Delta_{\rho\lambda}(q) \quad (14.24)
$$

where S_e denotes the thermal electron propagator and $\Delta_{\rho\lambda}$ is the W-propagator. The W-momentum is taken to be $q = k - p$, and we assume that the temperature is much lower than the W-mass, so that the thermal parts of the W-propagator can be neglected. Also, in order to avoid dealing with the diagrams with unphysical Higgs bosons, we will consider the W-propagator in the unitary gauge. This corresponds to the choice $\xi \to \infty$ in the Feynman rules of Fig. 11.3.

For the electron propagator, we use the expression derived in Eq. (14.21). Putting this expression into Eq. (14.24), we see that the first term in the square brackets appearing in the propagator gives just the self-energy of the ν_e in the vacuum. This part is not of interest to us. The matter-induced self-energy, which we will denote with a prime, is then given by

$$\Sigma'_a(p) = \frac{g^2}{2} \int \frac{d^4k}{(2\pi)^3} \delta(k^2 - m_e^2) \eta_F(k,\mu,\beta) \, \gamma^\rho \slashed{k} \gamma^\lambda \mathsf{L} \Delta_{\rho\lambda}(q) \,. \quad (14.25)$$

We can now evaluate this correction order-by-order in the Fermi constant G_F. For this, we write the W-propagator as a power series in $1/M_W^2$:

$$\Delta_{\rho\lambda} = \frac{-g_{\rho\lambda} + \frac{q_\rho q_\lambda}{M_W^2}}{q^2 - M_W^2} = \frac{g_{\rho\lambda}}{M_W^2} + \frac{q^2 g_{\rho\lambda} - q_\rho q_\lambda}{M_W^4} + \cdots \quad (14.26)$$

For the corrections to the leading order in $1/M_W^2$, we note that in the rest frame of the medium where $u^\mu = (1,0,0,0)$, we can rewrite the δ-function as

$$\delta(k^2 - m_e^2) = \frac{1}{2\Omega_K} \left[\delta(k_0 - \Omega_K) + \delta(k_0 + \Omega_K) \right], \quad (14.27)$$

where $\Omega_K = \sqrt{\mathbf{k}^2 + m_e^2}$. Using now the identity $\gamma_\rho \slashed{k} \gamma^\rho = -2\slashed{k}$ as well as the expression for the number densities of the electrons and positrons in terms of the distribution function,

$$n_{e\mp} = 2 \int \frac{d^3k}{(2\pi)^3} \frac{1}{e^{\beta(\Omega_K \mp \mu)} + 1}, \quad (14.28)$$

we obtain

$$\Sigma'_a(p) = \frac{g^2}{4M_W^2} \left(n_{e^-} - n_{e^+} \right) \gamma^0 \mathsf{L} \,. \quad (14.29)$$

In a general frame, the factor of γ^0 should be replaced by $\slashed{u}$. Thus, in the notation introduced in §14.1.1, we obtain [7, 8, 2]

$$\begin{aligned} a &= 0 \,, \\ b &= \sqrt{2} G_F \left(n_{e^-} - n_{e^+} \right) \,. \end{aligned} \quad (14.30)$$

From Eq. (14.9), we can now write down the dispersion relation for the neutrinos:

$$E_P = P + \sqrt{2} G_F \left(n_{e^-} - n_{e^+} \right) , \qquad (14.31)$$

whereas for the antineutrinos it is

$$\overline{E}_P = P - \sqrt{2} G_F \left(n_{e^-} - n_{e^+} \right) , \qquad (14.32)$$

In normal matter where no positrons are present, Eq. (14.31) is the same as the relation obtained in Eq. (10.50). Similarly, contribution of the diagram Fig. 14.1b is equivalent to the potential given in Eq. (10.53).

Exercise 14.3 *Find the contribution of Fig. 14.1c to the neutrino self-energy in a medium containing neutrinos and antineutrinos.*

Although this correction to the dispersion relation is sufficient for the purpose of discussing neutrino propagation in terrestrial or stellar media, it may prove insignificant in the early universe where $n_{e^-} \approx n_{e^+}$, and so the correction term is small. For such cases, it is useful to go to the next order in the expansion of Eq. (14.26). In this case, unlike the lowest order case, both a and b will be non-zero, and will also depend on E and P [7, 9, 10]. In addition, there is an extra complication coming from the fact that if we work in the renormalizable R_ξ-gauges introduced in §11.3.1, the contributions to a and b turn out to be gauge dependent at this order of calculation. However, the dispersion relations are independent of the gauge parameter [10] ξ.

Exercise 14.4 *Consider a model in which the Majorana neutrinos interact with a massless real scalar field S through the interaction*

$$\mathcal{L}_{\text{int}} = h \nu_L^{\mathsf{T}} C^{-1} \nu_L S + \text{h.c.} . \qquad (14.33)$$

In a background consisting of the neutrinos and the S particle only, calculate a and b at the 1-loop level. (For the sake of simplicity, assume $T \gg m_\nu$.) From these results, show that [2] $E_P = \overline{E}_P$.

14.2 Electromagnetic properties of neutrinos in a medium

14.2.1 General considerations

Electromagnetic form factors of a neutrino were discussed in §11.1. The vertex function Γ_λ was introduced in Eq. (11.3), and subsequently it was shown that the electromagnetic gauge invariance imposes the following constraint on it:

$$q^\lambda \Gamma_\lambda = 0 . \tag{14.34}$$

Subject to this constraint, the most general form for Γ_λ applicable in the vacuum was written down in Eq. (11.8).

This form involves the photon momentum q only. In a medium, as we discussed earlier in this chapter, a new 4-vector u^μ appears in the problem. As a result, additional terms will be allowed. We denote these terms with a prime [11, 12]:

$$\begin{aligned}
\Gamma'_\lambda &= (q \cdot u \gamma_\lambda - \slashed{q} u_\lambda)(C_E + i D'_E \gamma_5) \\
&\quad + \varepsilon_{\lambda \rho \alpha \beta} \gamma^\rho q^\alpha u^\beta (C_M + i D'_M \gamma_5) .
\end{aligned} \tag{14.35}$$

Using the explicit form of the plane wave solutions given in §4.3, it is easy to verify that in the non-relativistic limit, the terms C_E and C_M vanish, whereas the D'_E term gives a new contribution to the electric dipole moment. Similarly, the D'_M term corresponds to a new contribution to the magnetic dipole moment in that limit. These can be called the matter-induced electromagnetic form factors of the neutrino. Hermiticity of the effective Lagrangian implies the condition given in Eq. (11.4), which dictates that the form factors appearing in Eq. (14.35) are all real. In addition, the form factors may satisfy some extra conditions depending on the symmetry under some discrete operations like P, C, T or their combinations.

However, regarding the question of symmetry under discrete operations, one thing should be understood clearly. Suppose we are talking about CP. The constraints imposed by the CP symmetry will be valid if the particle interactions are CP invariant and the background medium is CP symmetric as well. If any one of these conditions is not valid, the constraints imposed by the CP symmetry will not be realized. With this in mind, one can check the consequences of the discrete symmetries. For this purpose, let us denote the vertex function by $\Gamma_\lambda(p, p', u)$, where the incoming and the outgoing fermion momenta were defined in Fig. 11.1, and the dependence on the medium appears through the center of mass velocity u. Then, the effect of P, C and T can be written as

$$\Gamma_\lambda(p, p', u) \xrightarrow{P} \Gamma^P_\lambda(p, p', u) \tag{14.36}$$

Table 14.1: The transformation properties of various quantities under the discrete operations P, C and T.

	δ_P	δ_C	δ_T
1	$+$	$+$	$+$
i	$+$	$+$	$-$
γ_λ	$+$	$-$	$-$
$\gamma_\lambda \gamma_5$	$-$	$+$	$-$
$\varepsilon_{\lambda\rho\alpha\beta}$	$-$	$+$	$-$

$$\Gamma_\lambda(p, p', u) \xrightarrow{C} -\Gamma_\lambda^C(-p', -p, u) \tag{14.37}$$

$$\Gamma_\lambda(p, p', u) \xrightarrow{T} -\Gamma_\lambda^T(-p, -p', -u), \tag{14.38}$$

where Γ^P, e.g., can be obtained from Γ by multiplying each quantity appearing in Γ by its parity phase δ_P given in Table 14.1. The CP and CPT phases of any quantity can be obtained simply by multiplying the individual phases, i.e., $\delta_{CP} = \delta_C \delta_P$ etc.

If the Lagrangian as well as the medium are symmetric with respect to any of these operations, the arrow appearing in the transformation equations should be replaced by an equality sign. Proceeding this way, we obtain the following conditions [11] on the matter-induced form factors in the static limit (i.e., $q \to 0$), when they depend only on E and P as defined in Eq. (14.5):

$$
\begin{aligned}
&\text{P symmetry :} &&D'_E = 0, \; C_M = 0 , \\
&\text{C symmetry :} &&D'_{E,M} \text{ odd functions of } E, \\
& &&C_{E,M} \text{ even functions of } E, , \\
&\text{T symmetry :} &&C_M = 0, \; D'_E = 0 \\
&\text{CP symmetry :} &&C_E \text{ and } D'_M \text{ odd in } E, \\
& &&C_M \text{ and } D'_E \text{ even in } E, , \\
&\text{CPT symmetry :} &&D'_{E,M} \text{ odd functions of } E, \\
& &&C_{E,M} \text{ even functions of } E, ,
\end{aligned}
\tag{14.39}
$$

Exercise 14.5 *Show that, without imposing the consequence of hermiticity, T symmetry gives C_M, D'_E imaginary and C_E, D'_M real.*

For Majorana neutrinos, there is the extra condition

$$\Gamma_\lambda(p, p', u) = \Gamma_\lambda^C(-p', -p, u), \tag{14.40}$$

which implies

$$C_{E,M}(-E,P) = -C_{E,M}(E,P) \,,$$
$$D'_{E,M}(-E,P) = D'_{E,M}(E,P) \,. \tag{14.41}$$

Thus, if the background medium is CPT-symmetric, all the matter in-duced form-factors must vanish for Majorana neutrinos. Normal matter, however, is not CPT-symmetric, so these form factors can be non-zero even for Majorana neutrinos. In particular, this implies that they can have matter-induced electric and magnetic dipole moments.

14.2.2 Calculation of the vertex in a background of electrons

In the standard model, the 1-loop diagrams for the electromagnetic ver-tex is obtained from the diagrams of Fig. 14.1 by attaching a photon line to every possible internal line. Thus, Fig. 14.1a will give rise to two diagrams since the photon line can attach to either of the inter-nal lines, while Fig. 14.1b will give one diagram corresponding to each charged fermion in the loop. As an illustration, we will perform the calculation for a medium whose effects come predominantly from elec-trons [13, 14, 15].

Restricting ourselves to the temperature range $T \ll M_W$, we can write down the amplitudes for the diagrams easily, using the thermal electron propagator from Eq. (14.21). This will produce three kinds of terms. First, there will be terms with no occurrence of the factor η_F. These give the vacuum contributions to the vertex, which we do not discuss here. Second, in Fig. 14.1b, there will be a term with two factors of η_F. This will give an absorptive part in the amplitude, which also will not be discussed. We will discuss only the third kind of term, which contains only one factor of the distribution functions. We will denote this by Γ'_λ.

The contribution from the Z-mediated diagram of Fig. 14.1b can be written as [14]

$$\Gamma'^{(Z)}_\lambda = \mathcal{T}^{(Z)}_{\lambda\rho}\gamma^\rho \mathsf{L} \,, \tag{14.42}$$

where, to the leading order in the Fermi constant,

$$T^{(Z)}_{\lambda\rho} = -\frac{eg_Z^2}{M_Z^2} \int \frac{d^4k}{(2\pi)^3} \text{Tr} \left[(\not{k} - \not{q} + m_e)\gamma_\lambda(\not{k} + m_e)\gamma_\rho(g_V^{(e)} + g_A^{(e)}\gamma_5) \right]$$
$$\times \left[\frac{\delta[(k-q)^2 - m_e^2]\eta_F(k-q,\mu,\beta)}{k^2 - m_e^2} + \frac{\delta[k^2 - m_e^2]\eta_F(k,\mu,\beta)}{(k-q)^2 - m_e^2} \right].$$

$$(14.43)$$

Here, we have used the shorthand

$$g_Z = g/(2\cos\theta_W), \qquad (14.44)$$

and $g_V^{(e)}$ and $g_A^{(e)}$ were defined in Eq. (2.28). Carrying out the traces in Eq. (14.43) and making a change of variable in one of the terms, we obtain

$$T^{(Z)}_{\lambda\rho} = -\frac{4eg_Z^2}{M_Z^2} \left\{ g_V^{(e)} \left\langle \frac{2k_\lambda k_\rho + (k_\lambda q_\rho + q_\lambda k_\rho) - g_{\lambda\rho}k\cdot q}{q^2 + 2k\cdot q} + (q \to -q) \right\rangle_+ \right.$$
$$\left. - ig_A^{(e)}\varepsilon_{\lambda\rho\alpha\beta}q^\alpha k^\beta \left\langle \frac{1}{q^2 + 2k\cdot q} + (q \to -q) \right\rangle_- \right\}, \quad (14.45)$$

where in this equation

$$k^\mu = (\Omega_K, \boldsymbol{k}), \qquad \Omega_K = \sqrt{k^2 + m_e^2} \qquad (14.46)$$

and the angular brackets stand for

$$\langle x \rangle_\pm = \int \frac{d^3k}{(2\pi)^3 2\Omega_K} \, x \, (f_F(k,\mu,\beta) \pm f_F(k,-\mu,\beta)) . \qquad (14.47)$$

Exercise 14.6 *Write down the contribution of the W-mediated diagrams to* Γ'_λ. *Use the identity*

$$\gamma_\lambda L A \gamma^\lambda L = -\gamma_\lambda L \text{Tr}\,(A\gamma^\lambda L) \qquad (14.48)$$

which is valid for any 4×4 matrix A to show that, in the leading order in G_F, $\Gamma'^{(W)}_\lambda$ can be obtained from $\Gamma'^{(Z)}_\lambda$ by making the substitutions

$$\frac{g_Z^2}{M_Z^2} \to \frac{g^2}{2M_W^2}, \quad g_V^{(e)} \to \frac{1}{2}, \quad g_A^{(e)} \to -\frac{1}{2}. \qquad (14.49)$$

Of course, $\mathcal{T}_{\lambda\rho} = \mathcal{T}_{\lambda\rho}^{(Z)} + \mathcal{T}_{\lambda\rho}^{(W)}$. Using the expression given in Eq. (14.45), we see that it satisfies the condition

$$q^\lambda \mathcal{T}_{\lambda\rho} = 0 \,, \tag{14.50}$$

which is a consequence of the electromagnetic gauge invariance. However, the 1-loop expression given in Eq. (14.45) also satisfies [14]

$$q^\rho \mathcal{T}_{\lambda\rho} = 0 \,, \tag{14.51}$$

which is an accidental symmetry, not expected to be valid at higher orders. The most general form of $\mathcal{T}_{\lambda\rho}$ consistent with these two conditions can be written as [16]

$$\mathcal{T}_{\lambda\rho} = \mathcal{T}_T R_{\lambda\rho} + \mathcal{T}_L Q_{\lambda\rho} + \mathcal{T}_P P_{\lambda\rho} \,, \tag{14.52}$$

where $\mathcal{T}_{T,L,P}$ are form factors associated with the tensors

$$
\begin{aligned}
R_{\lambda\rho} &= g_{\lambda\rho} - \frac{q_\lambda q_\rho}{q^2} - Q_{\lambda\rho} \\
Q_{\lambda\rho} &= \frac{\tilde{u}_\lambda \tilde{u}_\rho}{\tilde{u}^2} \\
P_{\lambda\rho} &= \frac{i}{Q} \varepsilon_{\lambda\rho\alpha\beta} q^\alpha u^\beta \,,
\end{aligned}
\tag{14.53}
$$

in which we have introduced the notations

$$\tilde{u}_\lambda = u_\lambda - \frac{\omega q_\lambda}{q^2} \tag{14.54}$$

and

$$\omega = q \cdot u \,, \qquad Q = \sqrt{\omega^2 - q^2} \,. \tag{14.55}$$

Exercise 14.7 *The most general tensor with two independent 4-vectors was given in Eq. (2.72). Use this for the vectors q and u, and impose the conditions of Eqs. (14.50) and (14.51), to arrive at the form for $\mathcal{T}_{\lambda\rho}$ given in Eq. (14.52).*

It is now straight forward to find the expressions for the form factors [14]:

$$
\begin{aligned}
\mathcal{T}_T &= -2eG_F(g_V^{(e)} + 1)\left(A - \frac{B}{\tilde{u}^2}\right) \,, \\
\mathcal{T}_L &= -4eG_F(g_V^{(e)} + 1)\frac{B}{\tilde{u}^2} \,, \\
\mathcal{T}_P &= 4eG_F(g_A^{(e)} - 1)QC \,,
\end{aligned}
\tag{14.56}
$$

where

$$A = \left\langle \frac{2m_e^2 - 2k \cdot q}{q^2 + 2k \cdot q} \right\rangle_+ + (q \to -q) \,,$$

$$B = \left\langle \frac{2\Omega_K^2 + 2\Omega_K \omega - k \cdot q}{q^2 + 2k \cdot q} \right\rangle_+ + (q \to -q) \,,$$

$$C = \left\langle \frac{k \cdot \tilde{u}/\tilde{u}^2}{q^2 + 2k \cdot q} \right\rangle_- + (q \to -q) \,. \tag{14.57}$$

The expressions are complicated, and analytic expressions [14] can be obtained only in the limits of either ω or Q small. Rather than giving those expressions, let us try to see some of the possible physical consequences of the induced electromagnetic vertex.

14.2.3   Induced electric charge of neutrinos

The effective charge of a fermion is defined in terms of the electromagnetic vertex function in the static limit $(\omega = 0)$, when the 3-momentum transfer is vanishing, i.e.:

$$e_{\text{eff}} = \frac{1}{2E} \overline{u}(p) \Gamma_0(\omega = 0, Q \to 0) u(p) \,. \tag{14.58}$$

It is easy to see from Eq. (11.8) that the vertex function obtained in the vacuum vanishes in the specified limit, so the neutrino charge is zero, as it should be. In a medium, however, this need not be the case.

To check this, we use the expression of Γ'_λ obtained in §14.2.2 and try to interpret it in the rest frame of the medium. For Γ'_0, we need only the components $\mathcal{T}_{0\rho}$ due to the definition in Eq. (14.42). For $\omega = 0$, Eq. (14.53) tells us that

$$R_{0\rho} = P_{0\rho} = 0 \,, \quad Q_{\lambda\rho} = u_\lambda u_\rho \tag{14.59}$$

in this frame. Thus, only the form factor $\mathcal{T}_L$ is relevant for our purpose here, and so we need only to evaluate the quantity B defined in Eq. (14.57). Notice, however, that B does not vanish at $\omega = 0$ when $Q \to 0$.

Rather, in this frame,

$$
\begin{aligned}
B\big|_{\omega=0} &= \left\langle \frac{-4\Omega_K^2 q^2 + 4(\mathbf{k}\cdot\mathbf{q})^2}{q^4 - 4(\mathbf{k}\cdot\mathbf{q})^2} \right\rangle_+ \\
&\xrightarrow{q\to 0} \left\langle \frac{\Omega_K^2 - K^2\cos^2\theta}{K^2\cos^2\theta} \right\rangle_+ ,
\end{aligned} \tag{14.60}
$$

where θ is the angle between $\mathbf{k}$ and $\mathbf{q}$.

> **Exercise 14.8** *If the background consists of a non-relativistic electron gas, show that, for $\omega = 0$ and small Q,*
>
> $$ A = B = 2m_e C = -\frac{1}{4}n_e\beta + O(Q^2). \tag{14.61} $$

For a non-relativistic background of electrons, using Eq. (14.61) into the definition of Eq. (14.58), we obtain [17, 18]

$$
e_{\text{eff}} = \frac{1}{4} e G_F n_e \beta (g_V^{(e)} + 1) \tag{14.62}
$$

for the effective charge of ν_e's. For ν_μ or ν_τ, since the charged current diagram is absent, the factor $(g_V^{(e)} + 1)$ should be replaced by $g_V^{(e)}$.

14.2.4 Radiative neutrino decay in a medium

The results described so far in this chapter are valid for neutrinos in the standard model, which are massless in the vacuum. However, if the neutrinos are massive in the vacuum, then other effects can arise. For example, we can use the results derived in §14.2.2 to calculate the rate of radiative decays of neutrinos:

$$
\nu_\alpha \to \nu_{\alpha'} + \gamma. \tag{14.63}
$$

Assuming that the mechanism given rise to neutrino masses does not give rise to flavor-changing neutral currents, only the charged current mediated vertex, $\Gamma_\lambda^{(W)}$, is relevant for our purpose. The vertex will now involve elements of the mixing matrix U, and has to be evaluated at $q^2 = 0$, since the matter effects on the dispersion relations of photons and neutrinos will produce higher order corrections to the decay amplitude. From Eq. (14.56), it is easy to see that the form factors $\mathcal{T}_{L,P}$ vanish for

$q^2 = 0$ since both of them contain a factor of $1/\tilde{u}^2$, and $\tilde{u}^2 = -Q^2/q^2$ from the definition in Eq. (14.54). Thus we are left with

$$\Gamma'_\lambda = U_{e\alpha} U^*_{e\alpha'} \mathcal{T}_T R_{\lambda\rho} \gamma^\rho \mathsf{L} \,, \tag{14.64}$$

with $\mathcal{T}_T$ obtained from Eq. (14.57) with $q^2 = 0$:

$$\mathcal{T}_T = -4\sqrt{2} e G_F \int \frac{d^3k}{(2\pi)^3 2\Omega_K} \left(f_F(k, \mu, \beta) + f_F(k, -\mu, \beta) \right) \,. \tag{14.65}$$

Using the photon polarization sum

$$\sum_{\text{transverse modes}} \epsilon^*_\lambda(q) \epsilon_\rho(q) = -R_{\lambda\rho} \,, \tag{14.66}$$

the matter-induced decay rate can be easily obtained in the rest frame of the medium [19]:

$$\Gamma' = \frac{m_\alpha^2 - m_{\alpha'}^2}{16\pi m_\alpha} |U_{e\alpha} U^*_{e\alpha'} \mathcal{T}_T|^2 F(\mathcal{V}) \,, \tag{14.67}$$

where $\mathcal{V}$ is the magnitude of the 3-velocity of the incoming neutrinos in this frame, and [20]

$$F(\mathcal{V}) = \sqrt{1 - \mathcal{V}^2} \left[\frac{2}{\mathcal{V}} \ln \frac{1 + \mathcal{V}}{1 - \mathcal{V}} - 3 + \frac{m_{\alpha'}^2}{m_\alpha^2} \right] \,. \tag{14.68}$$

For a non-relativistic electron background, Eq. (14.65) implies

$$\mathcal{T}_T = -\sqrt{2} e G_F n_e / m_e \,, \tag{14.69}$$

so that the decay rate calculated from Eq. (14.67), assuming $m_{\alpha'} \ll m_\alpha$, is

$$\Gamma' = \frac{1}{2} \alpha G_F^2 |U_{e\alpha} U^*_{e\alpha'}|^2 F(\mathcal{V}) \times \frac{m_\alpha n_e^2}{m_e^2} \,. \tag{14.70}$$

For normal matter densities, this can be many orders of magnitude larger than the corresponding decay rate in the vacuum. For example, if we disregard the third generation of fermions and assume $m_2 \gg m_1$, the rate for $\nu_2 \to \nu_1 + \gamma$ in the vacuum, derived from Eq. (11.39), is given by

$$\Gamma = \frac{\alpha}{2} \left(\frac{3 G_F}{32\pi^2} \right)^2 m_2^5 \sin^2\theta \cos^2\theta \left(\frac{m_\mu}{M_W} \right)^4 \,, \tag{14.71}$$

whereas the corresponding rate in the medium is

$$\Gamma' = \frac{\alpha}{2}G_F^2 \sin^2\theta \cos^2\theta F(\mathcal{V}) \times \frac{m_2 n_e^2}{m_e^2}. \tag{14.72}$$

Thus,

$$\frac{\Gamma'}{\Gamma} = 9 \times 10^{23} F(\mathcal{V}) \left(\frac{n_e}{10^{24}\,\mathrm{cm}^{-3}}\right)^2 \left(\frac{m_2}{1\,\mathrm{eV}}\right)^{-4}. \tag{14.73}$$

The main reason for the increase in the decay rate is that the leptonic GIM suppression, which reduces the decay rate in the vacuum, is not applicable in this case, since the background is not flavor symmetric in the sense that it contains electrons but not muons or taus. Further enhancements are obtained if one considers the effect of the photons in the background medium [21].

14.3   Other effects

As promised at the beginning of this chapter, we have only tried to give a glimpse of how physics can be modified when the neutrinos are traveling through a medium. Here, we cannot possibly discuss many other such effects in detail. Therefore, we are giving a brief summary of the type of applications of the general ideas discussed in this chapter.

The electromagnetic vertex can be used to calculate the decay of plasmons into $\nu\widehat{\nu}$ pairs [14], which is an important source of energy loss in stars. The coupling to an external magnetic field can be derived [14] from the formulation of §14.2.2. This modifies the dispersion relations of neutrinos in the presence of an external magnetic field. The resulting dispersion relations are anisotropic due to the directionality of the magnetic field [14]. This can have important consequences for [22] in resonant neutrino oscillations, and can explain some interesting properties of pulsars [23].

On the other hand, if the modification of photon dispersion relations are taken into account, it is possible to show that even a single neutrino can emit a photon in a medium, giving rise to the Čerenkov effect [24, 25]. The effect is present even for neutrinos in the standard model, but it can also be induced if the neutrino has a magnetic moment [24, 26].

In Ch. 5, we discussed various Majoron models, where we commented that the decay rate of a neutrino to a lighter neutrino with the emission of a Majoron is very small. This is because the transformation that diagonalizes the neutrino mass matrix also diagonalizes their coupling to the Majoron, at least to a very high order of accuracy. However, in a medium, the mass matrix changes, so this conclusion is not valid any more. The majoronic decay is therefore highly enhanced [27]. In addition, the dispersion relation of a neutrino is not the same as that of its antineutrino within a medium, as shown in §14.1.1. This brings in the extra possibility that a neutrino can decay to its own antineutrino by emitting a Majoron [28].

Part IV

Astrophysics and Cosmology

Chapter 15

Solar neutrinos

Neutrinos are an essential part of the process of stellar evolution. The sun shines because of the energy production in the following nuclear fusion reaction in its deep interior core:

$$4p \rightarrow \alpha + 2e^+ + 2\nu_e + 28\,\text{MeV}\,. \tag{15.1}$$

This reaction represents in a compact form a series of successive reactions and is at the core of the whole new field of neutrino astronomy. The released 28 MeV binding energy of the α-particle diffuses out of the core, getting degraded in frequency in the process to appear as sunlight. Our concern here is with the two neutrinos which have energies roughly in the range of a few MeVs, depending on the details of the nuclear reaction. Recalling that the density of a typical stellar core is roughly $100\,\text{g}/\text{cm}^3$ and typical $\nu_e e$ scattering cross section is about $10^{-43}\,\text{cm}^2$, the mean free path of a neutrino is of order $10^{17}\,\text{cm}$, which is much larger than the radius of the sun and of typical stars. Thus, neutrinos escape the sun and stars. From the observed solar luminosity, one can determine that the flux of neutrinos from the sun is about $6 \times 10^{10}\,\text{cm}^{-2}\,\text{s}^{-1}$ on the earth. These neutrinos carry away about 2-3% of the total energy emitted by the sun. Thus neutrinos from stars are another messenger of physics information in addition to the usual electromagnetic radiation. An advantage of neutrinos over electromagnetic radiation is that they carry information about the core and therefore detailed study of stellar neutrinos is bound to provide useful information on stellar interior as well as validity of existing theoretical models for the structure and evolution of the sun and other stars. With these motivations, in the early sixties Bahcall and his collaborators and Davis and his co-workers pioneered a detailed program of studying the solar neutrinos: Bahcall its theoretical aspects and Davis, the experimental detection [1].

15.1 Production and detection of solar neutrinos

15.1.1 Source of neutrinos in the sun

As mentioned earlier, neutrinos are an essential byproduct in the hydrogen burning process responsible for energy generation in the sun. The main process responsible for helium production is the so-called pp-chain described below [2]. 86% of the neutrinos are produced in the ppI chain:

$$
\begin{aligned}
\underline{\text{pp I}}: \qquad p + p &\rightarrow d + e^+ + \nu_e \,(\leq .42\,\text{MeV}) \\
p + e^- + p &\rightarrow d + \nu_e \,(1.44\,\text{MeV}) \\
d + p &\rightarrow \gamma + {}^3\text{He} \\
{}^3\text{He} + {}^3\text{He} &\rightarrow {}^4\text{He} + p + p \,.
\end{aligned} \tag{15.2}
$$

Here and elsewhere in this chapter, the number in the parentheses after each ν_e denotes the energies of the neutrinos emitted.

The ${}^3\text{He}$ produced in the above chain leads to the following chains, called hep chain which contributes only a small fraction ($\sim 10^{-7}$) of the neutrino flux and ppII chain which contributes about 14% of the neutrino flux.

$$
\begin{aligned}
\underline{\text{hep}}: \qquad {}^3\text{He} + p &\rightarrow {}^4\text{He} + e^+ + \nu_e \,(\leq 18.77\,\text{MeV}) \\
\underline{\text{pp II}}: \qquad {}^3\text{He} + {}^4\text{He} &\rightarrow {}^7\text{Be} + \gamma \\
{}^7\text{Be} + e^- &\rightarrow {}^7\text{Li} + \nu_e \,(0.861\,\text{MeV}) \\
{}^7\text{Li} + p &\rightarrow {}^4\text{He} + {}^4\text{He} \,.
\end{aligned} \tag{15.3}
$$

Finally, there is the ppIII chain which is responsible for a fraction 1.5×10^{-4} of the total neutrino flux. Despite the smallness of this fraction, it is of direct interest to us since these are the neutrinos detected in the Kamiokande experiment, and also constitute the major source in the experiment of Davis to be described below. It uses the ${}^7\text{Be}$ produced in the ppII chain:

$$
\begin{aligned}
\underline{\text{ppIII}}: \qquad {}^7\text{Be} + p &\rightarrow {}^8\text{B} + \gamma \\
{}^8\text{B} &\rightarrow {}^8\text{Be}^* + e^+ + \nu_e \,(\leq 14.06\,\text{MeV}) \\
{}^8\text{Be}^* &\rightarrow {}^4\text{He} + {}^4\text{He} \,.
\end{aligned} \tag{15.4}
$$

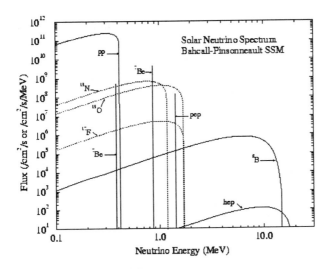

Figure 15.1: Fluxes of neutrinos emitted from the sun in various reactions. For the continuous spectra, the unit is $cm^{-2} s^{-1} MeV^{-1}$. For the discrete lines, the plot represents the number of neutrinos received per cm^2 per second. (Courtesy J N Bahcall.)

There is also the CNO cycle, originally proposed by Bethe [3], which is responsible for the production of neutrinos.

$$
\begin{aligned}
\text{CNO cycle}: \quad {}^{12}\text{C} + p &\rightarrow {}^{13}\text{N} + \gamma \\
{}^{13}\text{N} &\rightarrow {}^{13}\text{C} + e^+ + \nu_e \ (\leq 1.2\,\text{MeV}) \\
{}^{13}\text{C} + p &\rightarrow {}^{14}\text{N} + \gamma \\
{}^{14}\text{N} + p &\rightarrow {}^{15}\text{O} + \gamma \\
{}^{15}\text{O} &\rightarrow {}^{15}\text{N} + e^+ + \nu_e \ (\leq 1.73\,\text{MeV}) \\
{}^{15}\text{N} + p &\rightarrow {}^{12}\text{C} + \alpha \\
{}^{15}\text{N} + p &\rightarrow {}^{16}\text{O} + \gamma \\
{}^{16}\text{O} + p &\rightarrow {}^{17}\text{F} + \gamma \\
{}^{17}\text{F} &\rightarrow {}^{17}\text{O} + e^+ + \nu_e \ (\leq 1.74\,\text{MeV}) \\
p + {}^{17}\text{O} &\rightarrow \alpha + {}^{14}\text{N} .
\end{aligned}
\tag{15.5}
$$

This cycle dominates over the pp chain only if the temperature exceeds 1.8×10^7 K. For the sun, this condition is not met and therefore, this

Table 15.1: Flux of solar neutrinos from various reactions [4].

Chain	Reaction	Flux in $\mathrm{cm^{-2}\,s^{-1}}$
pp	pp	5.91×10^{10}
	pep	1.40×10^8
	hep	1.21×10^3
	^{7}Be	5.15×10^9
	^{8}B	6.62×10^6
CNO cycle	^{13}N decay	6.18×10^8
	^{15}O decay	5.45×10^8
	^{17}F decay	6.48×10^6

contributes only 1.5% to neutrino production.

The calculation of solar neutrino flux for these various chains depends on many factors such as the solar temperature, relative abundance of elements, nuclear reaction rates as well as the hydrodynamics of the solar interior. This calculation was pioneered by Bahcall and co-workers and subsequently done by other groups as well. We summarize the most recent results [4] of the calculations in Table 15.1. As is seen from the table, the main flux comes from the pp reaction. Since two neutrinos share the total ^{4}He binding energy of 28 MeV, a back of the envelope calculation gives a rough order of magnitude of the neutrino flux as

$$
\begin{aligned}
\Phi &= \frac{\text{(Luminosity)}}{4\pi d^2 \times \frac{1}{2}(\text{Binding energy of } ^4\text{He})} \\
&\approx \frac{4 \times 10^{33}\,\text{erg}/\,\text{s}}{4\pi \times \left(1.5 \times 10^{13}\,\text{cm}\right)^2 \times 14\,\text{MeV}} \\
&\simeq 6 \times 10^{10}\,\text{cm}^{-2}\,\text{s}^{-1}.
\end{aligned}
\tag{15.6}
$$

While most solar model calculations are in reasonable agreement with each other, there are however models with different input assumptions that argue in favor of noticeably different predictions for the two crucial neutrino fluxes i.e. of ^{8}B and ^{7}Be [5]. The discussion of this issue will intensify in the coming years as various experiments keep adding to the evidence for the solar neurtrino deficits in the different energy components.

15.1.2   Detection of solar neutrinos

The detection techniques for the solar neutrinos can be broadly classified into three groups: (*i*) radiochemical, (*ii*) geochemical and (*iii*) Water Čerenkov techniques.

Radiochemical method :   In the radiochemical method, the detector chemical, on interaction with neutrinos, converts into a radioactive isotope of another element with half-life of a few weeks. By using chemical techniques, the few atoms of the radioactive end product are extracted and counted. Nuclei which have been used for such experiments are listed in Table 15.2, where we have also listed the capture rates, the half-life of the daughter nucleus, the threshold energy of each experiment. In addition, we have shown how some of the key reactions for neutrino production contribute to the flux detected in any given experiment.

The pioneering experiment for the detection of solar neutrinos, carried out by Davis and collaborators [6] in the Homestake mines in South Dakota, USA, uses the radiochemical method with ^{37}Cl as the detecting material. This experiment started taking data around 1970 and is still ongoing. We will discuss the results in §15.1.3. The main source of neutrinos in this detector is the ones produced in the decay of ^{8}B. Because of high threshold energy, this experiment cannot detect the pp neutrinos, as shown in Table 15.2.

The calculation of the capture rate for this reaction has gone through some dramatic turns. Bahcall and his collaborators gradually improved their calculations since the mid-1960s, and by 1988, they arrived at a capture rate of [7] 7.9 ± 0.9 SNU, where the uncertainty quoted is at the 1σ level. (SNU, or the *solar neutrino unit*, is a unit suitable for these experiments, which represents the number of captures occurring per day in 10^{36} atoms in the detector.) However, in the same year, another calculation from a group in Saclay [8] indicated a much lower capture rate of 5.8 ± 1.3 SNU. It was later pointed out that the Saclay group used a cross-section for the ^{7}Be $+ p$ reaction which was much too low, and over-corrected some of the opacity calculations existing in the literature. Once these are taken into account, their results became quite close to those of Bahcall's group. Meanwhile, Bahcall and Pinsonneault [9, 4] improved on the earlier calculations of Ref. [7]. In their latest version [4],

Table 15.2: Reactions suitable for radiochemical detection of solar ν_e's. All reactions are of the form $\nu_e + X \to e^- + Y$ for suitable nuclei X and Y which are listed. The rates are given from Ref. [4]. The total capture rate also includes contributions from neutrinos from the CNO cycle, which are not shown separately here.

Nuclei		E_{th}	τ	Capture rates in SNU				
X	Y	(MeV)	(days)	pp	pep	^{7}Be	^{8}B	Total
^{37}Cl	^{37}Ar	0.814	35	0.00	0.22	1.24	7.36	$9.3^{+1.2}_{-1.4}$
^{71}Ga	^{71}Ge	0.233	11.4	69.7	3.0	37.7	16.1	137^{+8}_{-7}

they have included the effects of diffusion of different elements produced in the sun. Since no other group has taken these effects into account yet, we will accept the results of this version [4] to be the standards, which we show in Table 15.2.

Another element suitable for this purpose is Gallium [10]. In the early 1990s, two radiochemical experiments started producing results using Gallium as the active element in the detector. The experiment carried out in Italy's Gran Sasso Laboratory by the GALLEX collaboration uses 30 tons of liquid gallium chloride. The other one, conducted by the SAGE group and located in the Baksan mountains in the Caucasus region of Russia, uses 60 tons of metallic gallium. The main advantage of using Gallium is its low threshold energy, which enables the detection of the pp neutrinos. Because of this, the capture rate is quite high.

Several other possible detector materials are: ^{7}Li, ^{81}Br [11], ^{127}I [12].

Geochemical method : The various geochemical experiments use abundance of naturally occurring elements and therefore require end products with long half-lives ($\sim 10^6$) yrs. Various isotopes suitable for this method have been discussed, but so far there has not been any experiment employing this technique. One such suggestion uses ^{205}Tl, which turns to ^{205}Pb for a threshold neutrino energy of 0.062 MeV. Thus, it can detect the pp neutrinos and the capture rate is expected to be high, about 263 SNU. Other suggestion include ^{97}Mo, ^{98}Mo, ^{81}Br etc.

Table 15.3: Results of solar neutrino experiments. In the second column, the first error is statistical, and the second one is systematic. The units for the Kamiokande result is 10^6 CGS units of flux, i.e., $10^6 \, \mathrm{cm}^{-2} \mathrm{s}^{-1}$. For the other experiments, the units is SNU, which is explained in the text. In the third column, comparison has been made with the flux calculated in the model by Bahcall and Pinsonneault [4].

Experiment	Ref.	Result (1σ)	$\dfrac{\text{Result}}{\text{BP SSM}}$
Homestake	[14]	$2.55 \pm 0.17 \pm 0.18$ SNU	0.27 ± 0.03
Kamiokande	[15]	$3.0 \pm 0.41 \pm 0.35$ CGS	0.44 ± 0.06
SAGE	[16]	$69 \pm 10 \pm 6$ SNU	0.50 ± 0.09
Gallex	[17]	$79 \pm 10 \pm 6$ SNU	0.56 ± 0.07

Čerenkov detectors : This technique has been used in the Kamiokande collaboration [13], and is going to be used in various upcoming detectors. Basically, it uses the νe elastic scattering reaction, which has a forward peaked angular distribution within an angle

$$\theta \sim \sqrt{\frac{m_e}{E_\nu}}. \tag{15.7}$$

Using the directionality of the neutrino, one can identify the solar neutrinos. Notice that this method can detect even ν_μ's and ν_τ's. However, the efficiency of detection is not the same as that of ν_e's since the cross-sections are different, as shown in §2.3.

15.1.3 The solar neutrino puzzle

The experimental situation in the detection of solar neutrinos on earth is currently in a state of continuous improvement. So far there have been four experiments which have measured the neutrino flux from the sun. The first is the pioneering experiment of Davis, which started in the mid-1960s and the measurements have been going on except for a brief interruption for the breakdown of a pump. The second experiment is by the Kamiokande collaboration, using water Čerenkov detector, which started in the mid-1980s. The other two experiments, carried out by the SAGE and the GALLEX collaborations, use Gallium detectors, as mentioned earlier. The cumulative results of these four experiments are shown in Table 15.3.

Table 15.4: Properties of the sun.

Luminosity	$3.86 \times 10^{33}\,\mathrm{erg/s}$
Baryon number	$\sim 10^{57}$
Radius	$6.96 \times 10^{10}\,\mathrm{cm}$
Approximate thickness of the convective zone	$1.8 \times 10^{10}\,\mathrm{cm}$
Age	$4.5 \times 10^{9}\,\mathrm{yr}$
Surface temperature	$5772\,\mathrm{K}$
Core temperature	$1.56 \times 10^{7}\,\mathrm{K}$
Approximate core density	$148\,\mathrm{g/cm^3}$

The results of all experiments are lower than the prediction of the standard solar model. Moreover, different detectors see different suppression ratios compared to the standard solar model. This is known as the *solar neutrino puzzle*. Since the different detectors are sensitive to different energy regions of the solar neutrino spectrum, the proper understanding of this puzzle must incorporate this feature in it.

From an astrophysical point of view, it might be tempting to think that this puzzle may be an indication of some hitherto unknown property of the sun. After all, in calculating the neutrino flux from the sun, many properties about the sun are needed as input. Table 15.4 summarizes some of these properties. The principal assumptions that go into the calculation of these parameters are the following [1, 2]:

- Hydrostatic equilibrium of the sun, which is known to be an excellent approximation.

- Energy transport by photons, which requires knowledge of sun's opacity.

- Energy generation by nuclear fusion, which should also be an excellent approximation.

- Relative abundance of different elements is determined solely by nuclear reactions.

Of course, any of these assumptions could be faulty. For instance, if there are some new kinds of particles which participate in energy transport (such as weakly interacting massive particles or WIMPs), this would effect the flux predictions. This is because the assumption about energy transport by photons only would not be true any more. The temperature of the solar core would be less. To see how this can effect the neutrino flux prediction, we note that the ^{8}B neutrino flux depends on core temperature as T_c^{18}. As a result, only about 10% lowering in temperature of the core would explain the deficit in the chlorine and Kamiokande experiments. But the pp neutrinos have an inverse dependence on core temperature : $\Phi(pp) \sim T_c^{-1.5}$. Although the dependence is weak, this makes the Gallium results look more puzzling than before.

The calculation of neutrino fluxes from the sun has also been criticized [18] because the standard solar model does not take many things into consideration, e.g., solar rotation, magnetic fields, turbulent diffusion, and also does not satisfactorily account for the abundances of various nuclei, most notably ^{7}Li.

In spite of these criticisms and skepticisms, a strong argument has been advanced [19] to the effect that modifications of the solar model alone can never solve the solar neutrino puzzle. The argument goes roughly like this. No matter what is the spectrum and flux of neutrinos emitted from the sun, if they all arrive at the detectors, the Chlorine detector should detect more than the Kamioka one, since the former has a lower threshold. However, the results are the opposite, so it suggests that there is an energy-dependent suppression of the neutrino flux on its way from the solar core to the detectors.

A more quantitative solar model independent analysis along this line has been carried out by various groups [20] who have shown that if the neutrino absorption cross-sections in the various detectors are not different from what is assumed in the various analyses, then the constraint of solar luminosity along with the Chlorine, Gallium and Kamiokande data imply that the ^{7}Be neutrino flux must be negative, which is clearly unacceptable. In fact the same conclusion also emerges if we take any two of the three experiments.

Thus, it is clear [21] that we will have to consider another alternative for the resolution of the solar neutrino puzzle, viz, new properties of neutrinos beyond those given by the standard model of electroweak interactions. We remind the reader that, in the standard model, neutrino is a massless particle. A common denominator of all the new properties of the neutrino to be discussed below is that neutrino must be massive.

15.2 Solution using neutrino oscillations

15.2.1 Vacuum oscillations

As discussed in Ch. 10, massive neutrinos, in analogy with massive quarks, can undergo oscillations from one flavor to another. The general formula for oscillation is determined by the form of the mass matrix. The nature of the mass matrix is in turn determined by whether the neutrino propagates in vacuum or in matter, also discussed in Ch. 10. Again, in each case, the oscillation can be either to another flavor of left-handed neutrino which is active with respect to normal weak interactions, or to a sterile neutrino which has none or very feeble interaction with matter.

In order to resolve the solar neutrino puzzle by the vacuum oscillation, we can use the results of §10.1 where the fraction of ν_e's that survives as ν_e was given in Eq. (10.20) for a mono-energetic beam of neutrinos. In an experiment which detects a range of energies, the survival probability is thus

$$P_{\nu_e \nu_e}(x) \;=\; 1 - \sin^2 2\theta \int dE\, R(E) \sin^2\left(\frac{\Delta}{4E}\, x\right), \qquad (15.8)$$

where $R(E)$ is the fraction of neutrinos with energies between E among the captured neutrinos, with $\int dE\, R(E) = 1$. For x, we can use the earth-sun distance, which gives

$$\frac{\Delta}{4E}\, x \;=\; \left(\frac{\Delta}{1\,\mathrm{eV}^2}\right)\left(\frac{1\,\mathrm{MeV}}{E}\right) \cdot 1.9 \times 10^{11}. \qquad (15.9)$$

Since the energies of neutrinos detected are roughly in the 1 to 10 MeV range, we see that if $\Delta \gtrsim 10^{-11}\,\mathrm{eV}^2$, the integral gives a value $\frac{1}{2}$, so that the survival probability is always greater than $\frac{1}{2}$. This cannot explain the results of the Chlorine experiment. Careful analysis shows [22, 23] that fit to all the experiments can be obtained for a range of values with $\Delta \simeq 10^{-11}\,\mathrm{eV}^2$, and $\sin^2 2\theta$ almost equal to unity. These ranges are shown in Fig. 15.2.

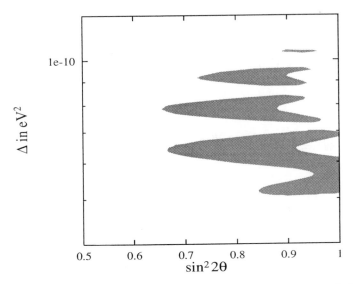

Figure 15.2: Allowed regions in the parameter space of Δ and $\sin^2 2\theta$ for the vacuum oscillation solution, using combined data from Davis, Kamiokande, and Gallium experiments. Adapted from Ref. [22].

If one considers all three generations, solutions can be obtained [24] even for much larger values of Δ. In fact, if the values of $\Delta_{\alpha\beta}$'s in Eq. (10.10) are all large compared to $10^{-11}\,\mathrm{eV}^2$, one obtains after energy averaging

$$P_{\nu_e \nu_e} = \sum_\alpha |U_{e\alpha}|^4 , \qquad (15.10)$$

using Eq. (10.9). The mixing matrix elements $U_{e\alpha}$ can be expressed in terms of two mixing angles. The allowed probabilities will then put constraints on these two angles. Note that the survival probability here is energy-independent, so that one should expect the same suppression in flux for all experiments. This is not possible at the 2σ level from the present experimental data, but at 3σ level it is marginally possible.

15.2.2 Resonant oscillation in solar matter

As pointed out in §10.4, the situation changes drastically once matter effects of the solar interior are included. Recall that neutrino propagation in matter is governed by the equation

$$i\frac{d}{dx}\left(\begin{array}{c} \nu_e \\ \nu_\mu \end{array}\right) = \frac{1}{2E}M^2\left(\begin{array}{c} \nu_e \\ \nu_\mu \end{array}\right),\qquad(15.11)$$

where [25, 26]

$$M^2 = \left(\begin{array}{cc} -\frac{1}{2}\Delta\cos 2\theta + A & \frac{1}{2}\Delta\sin 2\theta \\ \frac{1}{2}\Delta\sin 2\theta & \frac{1}{2}\Delta\cos 2\theta \end{array}\right).\qquad(15.12)$$

In writing this equation, we have put $x \approx t$ since the neutrinos are relativistic, and also ignored terms proportional to the unit matrix which are unimportant for the discussion about mixing. As in Ch. 10, we use the notations $A = 2\sqrt{2}G_F n_e E$ and $\Delta = m_2^2 - m_1^2$.

The effective mixing angle, which controls the oscillation probability, is given by

$$\sin^2 2\tilde{\theta} = \frac{\Delta^2\sin^2 2\theta}{(\Delta\cos 2\theta - A)^2 + \Delta^2\sin^2 2\theta}.\qquad(15.13)$$

We see that when $A = \Delta\cos 2\theta$, the mixing angle is maximal and transition from ν_e to ν_μ occurs. This relation is equivalent to Eq. (10.55), but shows more readily that it is a resonance condition, with width $\Gamma \approx \Delta\sin 2\theta$.

To discuss this phenomenon in detail, let us write down the eigenvalues of the matrix M^2 in Eq. (15.12):

$$\tilde{m}_{1,2}^2 = \frac{1}{2}\left[A \mp \sqrt{(\Delta\cos 2\theta - A)^2 + \Delta^2\sin^2 2\theta}\right].\qquad(15.14)$$

Propagation through matter is determined by these two eigenvalues, with the corresponding eigenstates $\tilde{\nu}_\alpha$ related to the weak eigenstates as follows:

$$\left(\begin{array}{c} \nu_e \\ \nu_\mu \end{array}\right) = \left(\begin{array}{cc} \cos\tilde{\theta} & \sin\tilde{\theta} \\ -\sin\tilde{\theta} & \cos\tilde{\theta} \end{array}\right)\left(\begin{array}{c} \tilde{\nu}_1 \\ \tilde{\nu}_2 \end{array}\right).\qquad(15.15)$$

The angle $\tilde{\theta}$ is given in Eq. (15.13). The equation of propagation for the $\tilde{\nu}_\alpha$'s can be obtained by using Eqs. (15.11), (15.12) and (15.15), remembering that $\tilde{\theta}$ depends on the electron density which varies with distance [27]:

$$i\frac{d}{dx}\begin{pmatrix} \tilde{\nu}_1 \\ \tilde{\nu}_2 \end{pmatrix} = \begin{pmatrix} \frac{\tilde{m}_1^2}{2E} & i\frac{d\tilde{\theta}}{dx} \\ -i\frac{d\tilde{\theta}}{dx} & \frac{\tilde{m}_2^2}{2E} \end{pmatrix}\begin{pmatrix} \tilde{\nu}_1 \\ \tilde{\nu}_2 \end{pmatrix}. \tag{15.16}$$

Notice that if the density is constant, the off-diagonal terms in this matrix vanish, so that the states denoted by $\tilde{\nu}_\alpha$ are indeed the stationary eigenstates. For varying density as in the case of the sun, one can define the stationary eigenstates only for the Hamiltonian at a given point. However, as long as the off-diagonal elements are small, any matter eigenstate $\tilde{\nu}_\alpha$ traverses unchanged with their relative admixture of ν_e and ν_μ changing adiabatically according to the value of electron density at a given point. However, we argued before that terms proportional to the unit matrix do not affect oscillation probabilities, so the only physical parameter in the diagonal elements must be their difference. Thus the *adiabaticity condition* is

$$\left|\frac{d\tilde{\theta}}{dx}\right| \ll \left|\frac{\tilde{m}_2^2 - \tilde{m}_1^2}{2E}\right|. \tag{15.17}$$

Using the resonant value of $\tilde{\theta}$ from Eq. (15.13) as well as the resonant density, we can rewrite it as

$$\gamma \gg 1, \tag{15.18}$$

where γ is called the *adiabaticity parameter*, defined to be

$$\gamma = \frac{\Delta}{E} \cdot \frac{\sin^2 2\theta}{\cos 2\theta} \left|\frac{d}{dx}\ln n_e\right|_{\text{res}}^{-1}, \tag{15.19}$$

The physical meaning of this condition is the following. The resonance condition $A = \Delta\cos 2\theta$ is approximately satisfied within the half-width Γ, i.e., for a range of electron density

$$\delta n_e \approx \frac{\Gamma}{2\sqrt{2}G_F E}. \tag{15.20}$$

Denoting $\delta n_e = |dn_e/dx|_{\text{res}}\,\delta x$ and using $\Gamma = \Delta\sin 2\theta$, we see that the resonance condition is approximately valid in a distance

$$\delta x \approx \frac{\Delta\sin 2\theta}{|dn_e/dx|_{\text{res}}\,2\sqrt{2}G_F E}. \tag{15.21}$$

The oscillation length at resonance, on the other hand, is given by

$$L_{\text{res}} = \frac{4\pi E}{\Delta \sin 2\theta}, \tag{15.22}$$

following the definition in Eq. (10.10). The ν_μ-fraction of the wave in ν_e has enough time to build if $\delta x \gg L_{\text{res}}$, which is precisely the resonance condition of Eq. (15.19).

To see the essence of resonant matter oscillation (or MSW effect, as it is often called after Mikheyev, Smirnov [26] and Wolfenstein [25] who played pivotal roles in discovering this effect) we remind the reader that [28] when the adiabaticity condition is satisfied, $\tilde{\nu}_\alpha$ remains unchanged. Let us assume that ν_e starts at a point where the electron density is infinite. Then $\tilde{\theta}_0 = \pi/2$, where the subscript 0 denotes the production point of the neutrino for the rest of this section. Eq. (15.15) then implies that $\tilde{\nu}_2$ corresponds to ν_e. On the way out of the solar core, the ν_e component in the beam, however, decreases continuously, since at a general point, $\tilde{\nu}_2 = \nu_e \sin\tilde{\theta} + \nu_\mu \cos\tilde{\theta}$, and $\tilde{\theta}$ decreases with decreasing density. When finally the neutrino beam comes out of the sun, $n_e \approx 0$ so that $\tilde{\theta}$ equals the vacuum mixing angle θ. Thus, the probability of finding the ν_e component of the beam decreases from almost unity to the final value of $\sin^2\theta$. This is the *resonant conversion* phenomenon, which is most dramatic if θ is very small, so that the ν_e beam is almost totally depleted.

Exercise 15.1 *Assuming that the propagation of neutrinos is adiabatic, show that the survival probability of ν_e's outside the sun is given by*

$$P_{\nu_e\nu_e}^{(\text{ad})} = \frac{1}{2}\left[1 + \cos 2\tilde{\theta}_0 \cos 2\theta + \sin 2\tilde{\theta}_0 \sin 2\theta \cos\left(\int dx' \frac{\tilde{m}_2^2 - \tilde{m}_1^2}{2E}\right)\right] \tag{15.23}$$

Argue that, when one integrates over different energies of neutrinos, the interference term is negligible if $\Delta \gg 10^{-10}$ eV^2 in case of real experiments where the neutrino energies span a range of at least a few MeV's.

In the more general case when the adiabaticity condition is not necessarily satisfied, the oscillation probability is given by [29]

$$P_{\nu_e\nu_e} = \frac{1}{2}\left[1 + (1 - 2P_{\text{LZS}})\cos 2\tilde{\theta}_0 \cos 2\theta\right], \tag{15.24}$$

where

$$P_{\text{LZS}} = \exp\left(-\frac{\pi}{4}\gamma\right) \tag{15.25}$$

is the Landau-Zenner-Stuckelberg probability of jumping from one eigen-state to another, assuming the density to be linearly varying near the resonance and that γ is not very much smaller than unity. More accurate expressions, for exponentially varying density [30] and extending to regions where $\gamma \ll 1$, [31] has been worked out in the literature.

Exercise 15.2 *Derive Eq. (15.24). The argument for the limiting case of $P_{\mathrm{LZS}} = 0$, $\tilde{\theta}_0 = \pi/2$ has been presented in the text. If a neutrino is produced in the far half of the sun so that it undergoes resonance twice, once on the way towards the core and once on its way out, show that*

$$P_{\nu_e \nu_e} = \frac{1}{2}\left[1 + (1 - 2P_{\mathrm{LZS}})^2 \cos 2\tilde{\theta}_0 \cos 2\theta\right] . \tag{15.26}$$

To see how it applies in the sun, we need to pay attention to two conditions. First, since the electron density decreases on the way out, the resonance is never reached unless the neutrino is produced at a region where the density is higher than the resonant density, i.e., unless $A_0 > \Delta \cos 2\theta$. Most neutrinos are produced at the solar core, where the electron density is roughly

$$n_0 \approx 6 \times 10^{25}\,\mathrm{cm}^{-3} . \tag{15.27}$$

This gives the following condition for the occurrence of the resonance:

$$\frac{\Delta}{E} \cos 2\theta < 1.5 \times 10^{-11}\,\mathrm{eV} \equiv m_{\mathrm{res}} . \tag{15.28}$$

The second condition is that of the onset of adiabaticity as given in Eq. (15.19). To use it in the present context, note that the electron density in the sun (except for the inner 10% of the radius), can be given by

$$n_e(x) = 2n_0\,e^{-10x/R_\odot} \tag{15.29}$$

where $R_\odot$ is the solar radius given in Table 15.4. Using Eq. (15.19), we thus find that the adiabatic condition is satisfied if

$$\frac{\Delta}{E} \cdot \frac{\sin^2 2\theta}{\cos 2\theta} \gg \frac{10}{R_\odot} = 3 \times 10^{-15}\,\mathrm{eV} \equiv m_{\mathrm{ad}} . \tag{15.30}$$

Let us now consider a beam of neutrinos of some average energy E. For the sake of simplicity, we consider the range of energies to be small enough so that we can treat the beam as monochromatic, but large enough so that the interference term of Eq. (15.23) can be ignored. As an illustration, suppose that one has obtained experimentally the survival probability of ν_e's to have some value $P_0 < 0.5$. We can seek for the values of Δ/E and θ which will give the required suppression in flux.

To find these solutions, let us first consider that region of the parameter space where Eq. (15.30) is satisfied, i.e., neutrino propagation is adiabatic. If $\theta \ll 1$, resonance occurs for $\Delta/E < m_{\text{res}}$. If $\Delta/E \gg m_{\text{res}}$, matter effects are negligible, so that the survival probability is given by the neutrino oscillation formulas in the vacuum, and is therefore always larger than $\frac{1}{2}$. If $\Delta/E = m_{\text{res}}$, Eq. (15.13) implies $\tilde{\theta}_0 = \pi/4$, so that the survival probability is exactly $\frac{1}{2}$. And if $\Delta/E \ll m_{\text{res}}$, $P_{\text{surv}} = \sin^2 \theta$, as argued earlier. Thus, $P_{\text{surv}} = P_0$ must be obtained for a value of Δ/E which is somewhat smaller than m_{res}. This solution is independent of the value of θ as long as $\theta \ll 1$, and is valid unless non-adiabatic effects are important, which implies, from Eq. (15.30), $4\theta^2 \gg m_{\text{ad}}/m_{\text{res}} = 2 \times 10^{-4}$. In Fig. 15.3, we show the contours of equal probability for several values of P_0, where this solution appears roughly as a horizontal line for each value of P_0.

There is a second way of satisfying $P_{\text{surv}} = 0.3$ in the adiabatic region. This occurs for $\sin^2 \theta = P_0$ if $\Delta/E \ll m_{\text{res}}$ so that we can use the argument given in the text for $\tilde{\theta}_0 = \pi/2$. Since this branch has a constant value of θ, it appears as a vertical line in the plot of Fig. 15.3.

Finally, if the non-adiabatic effects are important, putting $\tilde{\theta}_0 \approx \pi/2$, we see that Eq. (15.24) reduces to

$$P_{\nu_e \nu_e} = \sin^2 \theta + P_{\text{LZS}} \cos 2\theta \,, \tag{15.31}$$

which reduces to just P_{LZS} if $\theta \ll 1$. This leads to the non-adiabatic solution [32] given by the condition

$$\frac{\Delta}{E} \cdot \sin^2 2\theta = -\frac{40 \ln P_0}{\pi R_\odot} \,. \tag{15.32}$$

This is the diagonal solution in Fig. 15.3.

Thus, taking all the three branches together, we see that any equal probability line in the plot is roughly of the form of a triangle, with a horizontal, a vertical and a diagonal branch. Of course, near the vertices,

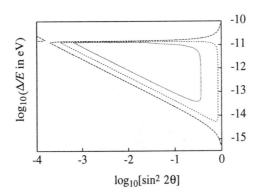

Figure 15.3: Contours of equal survival probability for ν_e's, using Eq. (15.24). Starting from the innermost one, the survival probabilities are 0.1, 0.3 and 0.5.

the triangle is a little rounded since the extreme assumptions that we took do not apply there.

The fit for experimental data can now easily be understood. First, one needs to integrate over the spectrum of energies relevant for any given experiment. Second, one can take into account the distribution of production points of the neutrinos in the sun, but this sophistication does not have much impact on the result. Third, one can use a better density profile for the sun which is valid even for the core. One such profile is given by [33]

$$n_e(x) = n_0 \exp\left(-\frac{az^2}{z+b}\right), \qquad (15.33)$$

where $z = x/R_\odot$, $a = 11.1$ and $b = 0.15$. Since any experiment gives a range of probabilities because of finite error bars, the allowed region in the plane of Δ vs $\sin^2 2\theta$ is in the form of a band whose limiting curves have roughly the shape of a rounded triangle, somewhat similar to those in Fig. 15.3.

All these regions presently allowed by the data of Davis, Kamiokande, and the two Gallium experiments can be marked out in a plot similar

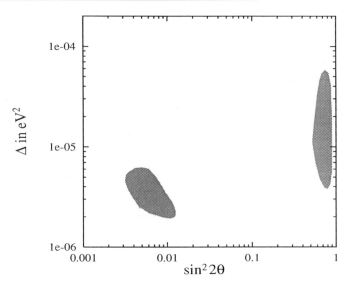

Figure 15.4: Allowed regions in the parameter space of Δ and $\sin^2 2\theta$ for the matter oscillation solution, using combined data from Davis, Kamiokande, and Gallium experiments. Adapted from Ref. [22].

to the one in Fig. 15.3. The allowed region for each experiment shows clearly three regions, a horizontal branch, a vertical one and a slanted one. Below, we give a qualitative understanding of the nature of these three branches of the solution. There are only two regions which are allowed by all the experiments [22, 34]. These are shown in Fig. 15.4.

We argued before that as long as the condition $\Delta \cos 2\theta < Em_{\rm res} = A_0$ is satisfied for some energy, the resonance condition will be reached somewhere in the sun. In the horizontal branch, for the most part, $\cos 2\theta \approx 1$. Thus, for a fixed value of Δ, virtually all neutrinos having an energy greater than a certain value are converted to ν_μ and only the less energetic neutrinos survive. The lower the value of Δ, the smaller the ν_e flux because lower and lower energies can satisfy the resonance condition and are converted to ν_μ.

In the vertical branch, $\cos 2\theta$ is small enough that even very low energy neutrinos satisfy Eq. (15.28) and undergo resonant conversion. The survival probability, $\sin^2 \theta$, is practically same for neutrinos with all energies. The range of values of θ in the vertical branch is such that $\sin^2 \theta$ gives the range of survival probabilities suggested by experiments.

Both these branches are bounded, for small Δ and/or small θ, by the validity of adiabaticity condition given in Eq. (15.19). Finally, the slanted branch denotes the range of values where non-adiabatic oscillation can produce the required depletion through Eq. (15.24). This can occur only if P_{LZS} is large, i.e., the energies of the neutrinos are large. Thus, in this branch, the higher energy neutrinos survive whereas the low energy ones are depleted, in contrast to the horizontal branch.

15.3 Solution using neutrino decay

If the electron neutrino is unstable and decays in flight from the sun to the earth, this would explain the deficit of solar neutrinos observed on earth [35]. At the time this idea was proposed there was no plausible particle physics scenario available for $\nu_e \to \nu' + \Phi$. With the introduction of the massless pseudo-scalar boson, Majoron [36], this proposal was revived [37]. This required a lifetime of the neutrino $E_{\nu_e} \tau_{\nu_e}/m_{\nu_e} \leq 500\,\mathrm{s}$. This scenario is, however, already ruled by observation of neutrinos from SN1987A, which implies a neutrino (ν_e) lifetime of at least 160,000 years, since it took that long for neutrinos to reach from SN1987A on earth. A combination of decay and oscillation, discussed in §10.3, might be able to explain the puzzle [38]. Since the survival probability in the case of fast decay is $\cos^4 \theta$, as shown in Eq. (10.38), we need $\theta \approx 45°$ in order to obtain a survival probability of about 0.25.

15.4 Solution using neutrino magnetic moment

If the neutrinos have a non-zero magnetic moment μ_ν, they can undergo spin flip $\nu_L \to \nu_R$ when they pass through the magnetic field in the sun. Since the right-handed neutrinos have extremely weak interactions with matter, they will escape detection, thereby explaining the depletion of the observed neutrino flux on earth [39, 40].

To discuss this phenomenon quantitatively, let us assume that the mass of the ν_L produced in the weak interactions is m_L, whereas the mass of the ν_R is m_R. Then the propagation of these states are governed by the equation

$$i\frac{d}{dx}\begin{pmatrix} \nu_L \\ \nu_R \end{pmatrix} = \begin{pmatrix} m_L^2/2E & \mu_\nu B \\ \mu_\nu B & m_R^2/2E \end{pmatrix}\begin{pmatrix} \nu_L \\ \nu_R \end{pmatrix}. \tag{15.34}$$

Thus, in an original beam of left-handed neutrinos, the survival probability of ν_L's after a distance x is given by

$$\begin{aligned} P_{\nu_L \nu_L}(x) &= |\langle \nu_L(x) \,|\, \nu_L(0) \rangle|^2 \\ &= 1 - \sin^2 \beta \, \sin^2 \Omega x, \end{aligned} \tag{15.35}$$

where

$$\begin{aligned} \Delta_{LR} \equiv m_L^2 - m_R^2 &= 4E\Omega \cos\beta, \\ \mu_\nu B &= \Omega \sin\beta. \end{aligned} \tag{15.36}$$

Exercise 15.3 *Show that the solution of Eq. (15.34), apart from a possible overall phase, is given by*

$$\begin{pmatrix} \nu_L(x) \\ \nu_R(x) \end{pmatrix} = [\cos\Omega x - i(\sigma_3 \cos\beta + \sigma_1 \sin\beta)\sin\Omega x]\begin{pmatrix} \nu_L(0) \\ \nu_R(0) \end{pmatrix}, \tag{15.37}$$

where σ_3 and σ_1 are the usual Pauli matrices. From this, deduce Eq. (15.35).

Eq. (15.35) immediately shows two important points. First, for the survival probability to be noticeably smaller than unity, one must have $\beta \gtrsim 1$, i.e.,

$$\Delta_{LR} \lesssim 4E\mu_\nu B. \tag{15.38}$$

Otherwise, the magnetic field will be too weak to cause spin flip. The implication of this constraint will be discussed later.

Second, we should have $\Omega x \gtrsim 1$, i.e., the neutrino should traverse a non-trivial distance before the flip becomes noticeable. In view of Eq. (15.38), this condition can be written as

$$\mu_\nu B x \gtrsim 1. \tag{15.39}$$

Since the path length x cannot be larger than the solar radius, and educated guesses about the magnetic fields in the sun are $B \simeq 10^3$ to 10^4 G, one needs $\mu_\nu \simeq (0.1 \text{ to } 1) \times 10^{-10}\mu_B$ to satisfy Eq. (15.39), where

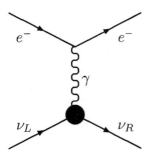

Figure 15.5: $\nu_e e$ scattering via photon exchange, which arises if the neutrinos have a magnetic moment.

$\mu_B \equiv e/2m_e$ is the Bohr magneton. Before proceeding further, it is wise to check whether such a large value of μ_ν allowed by present information in physics and astrophysics.

From $\nu_e e$ scattering experiments, one can get information on μ_ν as follows. As discussed in Ch. 2, $\nu_e e$ scattering proceeds via W and Z exchange in the standard model. In the presence of a non-vanishing magnetic moment, a new contribution to $\nu_e e$-scattering arises via the diagram of Fig. 15.5. Since the final state in this diagram has a ν_R, it does not interfere with the other diagrams and one has

$$\sigma_{\nu_e e}^{\text{tot}} = \sigma_{W,Z} + \sigma_\gamma, \tag{15.40}$$

where [41], for energies above a threshold energy E_{th}, the cross-section for the magnetic part is given by

$$\sigma_\gamma = \alpha \mu_\nu^2 \left[\frac{E_{\text{th}}}{E_\nu} - \ln \left(\frac{E_{\text{th}}}{E_\nu} \right) - 1 \right]. \tag{15.41}$$

In Table 15.5, we give the presently known upper limits on κ from laboratory a well as astrophysical sources [42]. The constraint from nucleosynthesis will be discussed in Ch. 17. The constraint from stellar energy loss arise from the fact that for $\mu_\nu \neq 0$, the plasmons in the stellar plasma can decay to $\nu\hat{\nu}$, adding a new source of energy loss for stars. It is interesting that the required value for $\mu_\nu \simeq (0.1 \text{ to } 1) \times 10^{-10} \mu_B$ is consistent with all laboratory and astrophysical observations except SN1987A bounds given in the last row of Table 15.5, which will be

Table 15.5: Upper limits on neutrino magnetic moment.

Experiment	Ref.	Limit on μ applies to	Upper limit in $10^{-10}\mu_B$
Laboratory experiments			
CHARM II	[45]	ν_μ & $\hat{\nu}_\mu$	20
BNL	[46]	ν_μ & $\hat{\nu}_\mu$	8.5
E225	[47]	ν_μ & $\hat{\nu}_\mu$	8.2
		ν_e	10.4
UCI	[48]	$\hat{\nu}_e$	2.8
Astrophysical constraints			
Nucleosynthesis	[49]	at least two species	0.1
Stellar energy loss	[44]	all	0.7
SN 1987A	[50]	all	10^{-2}-10^{-3}

discussed more fully in Ch. 16. In this case also, one must take into account matter effects but it has been shown that they do not change the conclusions [43]. In any case, ignoring the supernova bound for the moment (see discussion later), and using $|\mu_\nu| \lesssim 10^{-10}\mu_B$ together with the estimates of B discussed earlier, we obtain from Eq. (15.38) the constraint [44]

$$|\Delta_{LR}| \lesssim 10^{-7}\,\mathrm{eV}^2\,. \tag{15.42}$$

For a single Dirac neutrino, it seems that this condition is trivially satisfied since $\Delta_{LR} = 0$. But it is not so, since one must use the effective masses in the medium. Using the results of §10.4, we can write $m_L^2 = m^2 + 2E(V_{cc} + V_{nc})$ where m is the mass in the vacuum, and $m_R^2 = m^2$ since the ν_R does not have any weak interaction. This implies

$$n_e - \frac{1}{2}n_n \lesssim 10^{22}\,\mathrm{cm}^{-3}\,. \tag{15.43}$$

which can be satisfied only in the convective zone of the sun, i.e., in the outer part.

However, although this kind of direct magnetic moment provides a potential solution to the solar neutrino puzzle, it is not a favorable one since the magnetic moments are subject to the supernova constraint, and therefore must be much smaller. However, in Ch. 11, we also discussed the concept of a transition magnetic moment between two different neutrinos. This possibility is more relevant here since, as we will see in the following chapter, this kind of interaction is free from supernova as well as nucleosynthesis bounds. To work out this possibility in detail, one needs to consider a 4×4 matrix involving the left and right chiralities of both flavors of neutrinos involved and find out oscillation probabilities corresponding to that matrix. This process has been named *spin-flavor oscillation.* Ignoring matter effects, then, the condition for oscillation for this case is given by

$$|m_{\nu_1}^2 - m_{\nu_2}^2| \lesssim 10^{-7} \, \text{eV}^2 \,, \tag{15.44}$$

using Eq. (15.42), where the transition moment is between two neutrinos ν_1 and ν_2. Matter effects in this case have also been discussed. Once they are taken into account, one can allow [51] $\Delta \leq 10^{-4} \, \text{eV}^2$. Moreover, notice that the oscillation probabilities are energy dependent, as seen from the energy dependence of Ω and β in Eq. (15.35), so it is conceivable that different experiments would give different suppressions in flux. Detailed calculations, using some plausible magnetic field distribution in the convective zone of the sun, has been carried out [52].

Theoretical models [53, 54] for such scenarios generally imply that ν_e and ν_ℓ ($\ell = \mu$ or τ) pair up to form a Dirac neutrino with practically no Majorana mass entry to a high precision.

The possibilities of direct or transition moments can however be distinguished in a fairly straightforward manner in future solar neutrino detectors that can measure the neutral current effects due to ν_μ and ν_τ such as the Canadian heavy-water detector at Sudbery [55] (SNO) and BOREX [56].

15.5 Violation of the equivalence principle for neutrinos

The weak equivalence principle of general relativity has been tested to about 1 part in 10^{11} for ordinary matter. It has been shown [57] that a violation of this principle for neutrinos at the level of 1 part in 10^{14} can be responsible for a depletion of the solar neutrino flux.

In a gravitational field, the Dirac equation for the neutrinos give rise to the following Klein-Gordon equation:

$$\left[(g_{\lambda\rho} + f_\alpha h_{\lambda\rho})\partial^\lambda \partial^\rho + m_\alpha^2\right]\nu_\alpha(x) = 0\,, \tag{15.45}$$

where $g_{\lambda\rho}$ is the flat-space metric, $h_{\lambda\rho}$ is the part responsible for gravitation, and f_α is a coefficient with which the neutrino eigenstate ν_α interacts with the gravitational field. In the general theory of relativity, $f_\alpha = 1$ for any particle. Thus, differences between different f_α's measure the violation of the equivalence principle.

Now, assuming spherical symmetry and taking the weak field approximation to write $h_{\lambda\rho} = 2\varphi\delta_{\lambda\rho}$ where φ is the Newtonian gravitational potential, the time-independent Klein-Gordon equation can be written as

$$\left[-i\frac{d}{dr} - K(r)\right]r\xi = 0\,, \tag{15.46}$$

where ξ is the radial part of the wave function, and

$$K(r) = \sqrt{E^2[1 + 4f\varphi(r)] - m^2} \approx E - \frac{m^2}{2E} + 2fE\varphi(r)\,, \tag{15.47}$$

where E would have been the energy of the neutrino in absence of mass and gravitational field.

So far, we considered a single flavor of neutrino. For two flavors, discarding terms which are proportional to the unit matrix, the Hamiltonian can be written as

$$H = \frac{1}{2E}\begin{pmatrix} 0 & 0 \\ 0 & Q(r) \end{pmatrix}\,, \tag{15.48}$$

where

$$Q(r) = \Delta - 4\,\delta f\,E^2\varphi(r)\,, \tag{15.49}$$

where $\delta f \equiv f_2 - f_1$ gives the measure of violation of the equivalence principle.

Level crossing will now occur if there is a region of high gravitational field where $Q(r) < 0$. This can induce oscillation if there is already some mixing in the vacuum. Denoting the vacuum mixing angle by θ as before, we see that, in the flavor basis, the evolution equation is still of the form given in Eq. (15.11), with

$$M^2 = \begin{pmatrix} -\frac{1}{2}Q\cos 2\theta + A & \frac{1}{2}Q\sin 2\theta \\ \frac{1}{2}Q\sin 2\theta & \frac{1}{2}Q\cos 2\theta \end{pmatrix}. \qquad (15.50)$$

Most of the analysis of §15.2.2 now applies with Δ replaced by Q. However, notice that the energy dependence of the δf term is very different from that of the weak interaction term, so this effect can be distinguished from the effect described in §15.2.2.

The strength of the effects mentioned here can be parameterized by the relative importance of the two terms in $Q(r)$ in Eq. (15.49):

$$\eta \equiv \left| \frac{4\,\delta f\, E^2 \varphi(r)}{\Delta} \right|. \qquad (15.51)$$

For an order-of-magnitude estimate, one can take the density of the sun to be uniform, at the value of the mean solar density. Then, for $r \ll R_\odot$, one finds [57]

$$\eta \approx 10^{14} |\delta f| \times \left(\frac{E}{1\,\mathrm{MeV}} \right)^2, \qquad (15.52)$$

assuming $\Delta \approx 10^{-4}\,\mathrm{eV}^2$, which are values relevant for the solution of the solar neutrino puzzle. Thus, a violation of universality at a level of 1 part in 10^{14} will affect the conventional MSW effect.

15.6 Implications and outlook

15.6.1 Time variation of the solar neutrino flux

We have discussed various possible solutions of the solar neutrino puzzle. Future experiments, with more data, should be able to distinguish between them. One way to distinguish is by measuring the time variation of the solar neutrino flux.

Consider, first, that it is the magnetic moment of the neutrino which causes the suppression in flux. The magnetic field in the sun exhibits a 11 year cycle connected with the appearance of sun-spots. When more sun-spots appear the magnetic field in the solar convective zone increases. When that happens, more neutrinos flip helicity and become invisible.

Thus, the flux measured on earth should exhibit an anti-correlation with the number of sun-spots. It has been argued that the data of Davis does indeed indicate an 11-year variation. In view of the large errors, it is not conclusive yet but as more data from more experiments accumulate, one can reach a definite conclusion.

Another test of the magnetic moment hypothesis is the biannual variation of the observed flux [40]. The origin of the biannual variation can be understood by noting the two following facts. First, the axis of the sun is not perpendicular to the plane of earth's rotation; there is an angle of about 7° between the axis of the sun and the perpendicular to the orbital plane of the earth. Second, in the equatorial plane of the sun, the magnetic field in the solar interior is close to zero. Thus neutrinos coming through the equatorial plane do not undergo as much spin-flip. Because of the tilt of the solar axis, the earth passes through the sun's equatorial plane twice a year. Therefore during those times, the neutrino capture rate on earth should be higher than during the rest of the year.

If vacuum neutrino oscillations are responsible for the suppression in the solar neutrino flux, then one would expect an annual variation, because the distance between the sun and the earth changes throughout the year.

And if the matter-induced resonant neutrino oscillations hold the key to the solar neutrino puzzle, then we should expect a day-night effect, i.e., a diurnal variation [58]. This is because, at night the neutrinos have to come through the earth, and matter effects in the earth would affect the neutrino flux.

15.6.2   Outlook

It is clear that measurements of the solar neutrino flux are going to add to our knowledge in an important way. From a particle physicist's point of view, the worst scenario would be a resolution that uses some as yet unknown property of the sun. This will add to our understanding of what goes on in the sun. In fact, there may even be some benefit to particle physics in it if it turns out that new particles are required to transport energy from the core. A class of such particles suggested are the so-called WIMPs [59]. However, as we argued in §15.1.3, this seems to be the unlikely solution since the Chlorine detector detects a smaller flux compared to the Kamioka detector.

Equally unexciting from a particle physicist's point of view is the solution via the violation of the Equivalence principle, although this might also lead to the idea of forces exchanged by some unknown particle whose couplings depend on the properties of the particles in a non-standard way. However, as we argued in §15.5, the violation alone cannot explain the depletion of the flux. In addition, one needs some vacuum mixing anyway, which implies neutrino masses in the vacuum.

It thus seems that the solar neutrino problem is pointing at the possibility of new neutrino physics. Since in the standard model neutrino neither has any mass nor magnetic moment, the possibility implies very interesting new physics beyond the standard model. In particular, in all these cases, $m_\nu \neq 0$. Therefore, experimental search for neutrino mass and oscillations become interesting. In fact, if all the experimental results on solar neutrino flux are to be taken seriously, the inescapable conclusion is that neutrino has mass, which in itself is an exciting development.

Chapter 16

Neutrinos from supernovae

For massive stars, the end of stellar evolution process is a gigantic explosion known as the supernova, after which the stellar core becomes either a neutron star or a black hole whereas the mantle explodes in a brilliant display of fireworks throwing its debris of heavy elements into interstellar space that eventually end up in planets. The aspect of the supernova that concerns us again is the neutrino emission during supernova explosion. As we will discuss later, during the supernova, almost 99% of the energy released comes out in the form of neutrinos, with only 1-2% coming out as light. These neutrinos carry in their spectrum key information not only about the detailed nature of supernova collapse but also about properties of neutrino, not yet explorable in the laboratory. Indeed, SN1987A which was first observed on February 23, 1987 and has been quite well studied by many earth-based experiments, has not only confirmed the existing theories of supernova but has enabled us to push further the frontiers of our knowledge in neutrino physics. It is the purpose of this chapter to discuss these topics [1].

16.1 Qualitative picture of supernova collapse

As stars evolve, they first get their energy burning hydrogen to helium. The helium, being heavier, settles to the core of the star. The duration of this process depends on the mass of the star – the process lasting longer for less massive stars and shorter for more massive ones. Towards the end the hydrogen burning period, a period of gravitational contraction heats up the core and starts the phase of He burning to Carbon. The carbon being heavier will settle to the center with He and H floating above it. The process of contraction and heating repeats itself towards the end of the helium burning phase when carbon will start burning to neon. Similar process then leads from neon to oxygen, and from oxygen to silicon. If a star is more massive than 5-10 solar masses, silicon burning can start at $T \simeq 3.4 \times 10^9$ K giving rise to iron. As the iron core grows, the mass of the core will exceed the Chandrasekhar mass ($1.4M_\odot$), at which point the (degenerate) electron Fermi pressure will fail to support the gravitational pull and the core will begin to contract and get hotter. However, iron being the most stable nucleus, its burning cannot lead to any other heavier nucleus. Instead, iron will photo disintegrate as follows:

$$^{56}\text{Fe} \rightarrow 13\,^4\text{He} + 4n - 124.4\,\text{MeV}\,. \tag{16.1}$$

This absorbs energy and accelerates the collapse even further. At this time the temperature is high enough that electrons get absorbed by protons to give a neutron and a neutrino. The absorption of electron leads to further loss of electron pressure, again adding to the collapse rate. This increases nuclear density and temperature until the collapse is suddenly halted by the hard core nuclear repulsion leading to a bounce back. In this process of bounce back, through a mechanism not completely understood, the stellar envelope explodes causing the brilliant firework that is seen.

In the process of the collapse, the stellar core of about 1.4 to 2 solar masses forms a neutron star (or a black hole, if the mass is larger). The rest of the mass of the star gets ejected into the intergalactic space. This is called a Type II supernova. As the core collapses, it gets more tightly bound gravitationally, so it releases the extra energy. The energy release ΔE is given by

$$\Delta E = \left[-\frac{G_N M^2}{R} \right]_{\text{star}} - \left[-\frac{G_N M^2}{R} \right]_{\text{NS}} \tag{16.2}$$

Note that R_{star} is several times 10^{10} cm where a neutron star has a radius of only a few kilometers. Therefore, even though M is bigger than M_{NS}, ΔE is dominated by the second term and we get,

$$\Delta E = 5.2 \times 10^{53}\,\text{erg} \cdot \left(\frac{10\,\text{km}}{R_{\text{NS}}} \right) \left(\frac{M_{\text{NS}}}{1.4M_\odot} \right)^2\,. \tag{16.3}$$

Where does this energy go? The nuclear binding energy is about 3.2 MeV per nucleus. Since the number of nuclei per unit mass is 6×10^{23}, total energy of photo disintegration is $M_{NS} \times 6 \times 10^{23} \times 3.2\,\text{MeV} \approx$ $(6M_{NS}/1.4M_\odot) \times 10^{51}$ erg. The kinetic energy in the explosion is

$$\frac{1}{2}Mv^2 = 2.5 \times 10^{51}\,\text{erg} \cdot \left(\frac{M}{10M_\odot}\right)\left(\frac{v}{5000\,\text{km/s}}\right)^2, \qquad (16.4)$$

which is small even under very extreme assumptions about the mass and the velocity. The optical energy ($\lesssim 10^{49}$ erg) and the gravitational energy emitted are much smaller. It is therefore clear that bulk (about 99%) of the gravitational binding energy released must be carried away by the neutrinos. Below we study the characteristics of the neutrino flux [2] that is a result of this process.

16.2 Flux of supernova neutrinos

To study the characteristics of neutrinos emitted from the supernova, we have to discuss the mechanism for their production. There are basically two components to the neutrino flux. The first one occurs during the first few milliseconds of stellar collapse, when electrons get absorbed into protons to give neutrinos via the inverse β-decay process

$$e^- + p \to n + \nu_e. \qquad (16.5)$$

These are known as the *deleptonization* neutrinos, which are very energetic. This burst lasts for about a hundred milliseconds.

As the core collapse proceeds, a second stage of neutrino emission begins. The flux of these neutrinos consist of ν_e, $\widehat{\nu}_e$, ν_μ, $\widehat{\nu}_\mu$, ν_τ and $\widehat{\nu}_\tau$. They have energy in the range of 15 to 20 MeV. This corresponds to an emission temperature of about 5-6 MeV if one assumes a thermal distribution with zero chemical potential. These neutrinos originate from reactions such as $e^+e^- \to \nu_\ell \widehat{\nu}_\ell$, $n + p \to n + p + \nu_\ell + \widehat{\nu}_\ell$ etc. These processes thermalize the neutrinos which bounce back and forth before being finally emitted, within a sphere called the *neutrino sphere*, whose size is much larger than the collapsed core radius. The temperature in the neutrino sphere depends on the distance from the core. The stronger the interaction of a neutrino, the further from the core its sphere extends. Since the emission temperature corresponds to the surface temperature, the neutrinos having stronger interaction emerge with lower energy. Since $\nu_e e$ scattering gets a contribution from the charged current interactions unlike $\nu_\mu e$ and $\nu_\tau e$ scattering, it therefore follows that the ν_e and $\hat{\nu}_e$ will leave the neutrino sphere at lower energy than the ν_μ and ν_τ. This theoretical expectation of course has not been tested because a water Čerenkov detector can detect only fast electrons and positrons, whereas ν_μ and ν_τ can interact only with the detector via neutral current interactions such as $\nu_\mu N \to \nu_\mu N$.

There were four neutrino detectors in operation at the time the supernova 1987A signal reached the earth: i) Kamiokande II in Japan; ii) IMB in the United States; iii) Baksan in Soviet Union and iv) Mont Blanc in Europe. The signals at Mont blanc [3] occurred 4.7 hours before the signals were observed simultaneously at Kamiokande and IMB. For various reasons, it is not conclusively established whether these signals are from SN1987A. First, the energies of the observed recoil electron are too close to the threshold. Secondly, Kamiokande with a much larger detector did not observe any signal coinciding with them. Similarly, the Baksan detector reported [4] a burst of six events within about 25 seconds of the first IMB event. One out of the six events is attributed to the background.

The neutrinos observed by Kamiokande II [5] and IMB [6] are believed to be from SN1987A from their energy as well as event characteristic. Eleven events were observed by Kamiokande and eight by IMB. These have been demonstrated in Table 16.1. Since the angular distribution in both the experiments is isotropic, these events are supposed to arise from the reaction

$$\hat{\nu}_e + p \to e^+ + n. \tag{16.6}$$

The $\nu_e e$ scattering is forward peaked and also has a much lower cross section (see Ch. 2). For example, at $E_\nu = 10\,\text{MeV}$, $\sigma_{\nu_e e}/\sigma_{\hat{\nu}_e p} \simeq 10^{-2}$. By using these events, one can calculate the total neutrino luminosity $L_{\hat{\nu}}$ of SN1987A and compare it with the gravitational binding energy estimated in §16.1 to test the supernova models.

To estimate $L_{\hat{\nu}}$, we assume that the number of $\hat{\nu}_e$ emitted is $N_{\hat{\nu}_e}$, and these have an average energy $E_{\hat{\nu}_e}$. Let D be the distance from the earth to the supernova. In the case of the SN1987A, this as well as other parameters are summarized in Table 16.2. The Kamiokande detector contains 2.2 kTon (i.e., 2.2×10^9 g) of water, which has $(2.2 \times 10^9) \times (6 \times 10^{23})$ or 1.32×10^{33} nucleons. Of these, a fraction 2/18, i.e., $6.6 \cdot 10^{32}$

Table 16.1: The neutrino events of SN1987A. The angle θ_e gives the direction of the electron produced in the detector with respect to the direction of the Large Magellanic Cloud where the progenitor star of SN1987A was located.

Event	t in s	E_e in MeV	θ_e
Kamiokande data			
1	0	20.0 ± 2.9	18 ± 18
2	0.107	13.5 ± 3.2	40 ± 27
3	0.303	7.5 ± 2.0	108 ± 32
4	0.324	9.2 ± 2.7	70 ± 30
5	0.507	12.8 ± 2.9	135 ± 23
6	0.686	6.3 ± 1.7	68 ± 77
7	1.541	35.4 ± 8	32 ± 16
8	1.728	21.0 ± 4.2	30 ± 18
9	1.915	19.8 ± 3.2	38 ± 22
10	9.219	8.6 ± 2.7	122 ± 30
11	10.433	13.0 ± 2.6	49 ± 26
12	12.439	8.9 ± 1.9	91 ± 39
IMB data			
1	0	38 ± 7	80 ± 10
2	0.412	37 ± 7	44 ± 15
3	0.650	28 ± 6	56 ± 20
4	1.141	39 ± 7	65 ± 20
5	1.562	36 ± 9	33 ± 15
6	2.684	36 ± 6	52 ± 10
7	5.010	19 ± 5	42 ± 20
8	5.582	22 ± 5	104 ± 20

are Hydrogen nucleus or proton. The number of events observed in this detector is given by

$$
\begin{aligned}
\text{No. of events} &= \frac{N_{\hat{\nu}_e}}{4\pi D^2} \sigma_E(\hat{\nu}_e p) \times 6.6 \cdot 10^{32} \\
&= \frac{N_{\hat{\nu}_e} \times 9.75 \cdot 10^{-42}\,\text{cm}^2 \cdot (E_{\hat{\nu}_e}/10\,\text{MeV})^2 \times 6.6 \cdot 10^{32}}{4\pi(1.44 \cdot 10^{23}\,\text{cm})^2},
\end{aligned}
$$

(16.7)

using the cross section from Ch. 2. Since the number of events is about 20, we get

$$
N_{\hat{\nu}_e} = 4 \times 10^{56} \left(\frac{E_{\hat{\nu}_e}}{15\,\text{MeV}}\right)^{-2}.
$$

(16.8)

Assuming average neutrino energy to be $\langle E \rangle \simeq 15\,\text{MeV}$, we get the $\hat{\nu}$ luminosity to be about $N_{\hat{\nu}_e} \langle E \rangle \simeq 3 \times 10^{52}\,\text{erg}$. Multiplying this by six for three neutrinos and their antineutrinos, we get the total energy emitted to be roughly $2 \times 10^{53}\,\text{erg}$. Thus, this rough calculation leads almost to the gravitational binding energy release calculated in Eq. (16.3). From more detailed analysis, it indeed appears that almost all the energy released in the stellar collapse is emitted in neutrinos as is expected from theoretical models of supernovae.

16.3 Neutrino properties implied by SN1987A observations

There are three different ways to obtain constraints on new physics from supernova observations:

1. The fact that all observed neutrino events are clustered within a time of 10-12 seconds.

2. All observed neutrino energies are less than 60 MeV.

3. Observed neutrino events account for practically all the gravitational binding energy released in the process of core collapse.

If any property of the neutrino (such as its mass or charge) will cause time spreading of the signal, the first point above will constrain that

Table 16.2: [
]Average parameters of SN1987A for $t \leq 1\,\mathrm{s}$.

Distance to Earth	55 kpc
	($\simeq 1.4 \times 10^{23}\,\mathrm{cm}$)
Core density (ρ_{core})	$8 \times 10^{14}\,\mathrm{g/\,cm^3}$
Radius (R)	$10\,\mathrm{km}$
Mass	$(1.4 \text{ to } 2) \times M_\odot$
Binding Energy	$(2 \text{ to } 4) \times 10^{53}\,\mathrm{erg}$
Core temperature (T_c)	30 to 60 MeV
Magnetic field in the core (B_c)	$10^{12}\,\mathrm{G}$
Fermi energy of electrons ($E_F^{(e)}$)	300 MeV
Fermi energy of neutrinos ($E_F^{(\nu)}$)	200 MeV
Fractional number density of neutrinos (Y_ν)	0.04
Fractional number density of electrons (Y_e)	0.3
Density of neutrons	$4.8 \times 10^{38}/\,\mathrm{cm^3}$

property. Similarly, if the introduction of a new particle (such as right-handed neutrino, Majoron or axion) can add a new mechanism for energy loss from core collapse, properties of those particles (e.g. their mass and couplings) will be constrained. Let us apply these considerations to obtain constraints on the properties of the neutrino, the weak interactions of the ν_R etc.

16.3.1 Neutrino mass

The basic idea here is that, for a given energy, a heavier mass neutrino will travel slower and reach later [7]. To quantify, one needs the relation

$$v_\nu \simeq 1 - \frac{1}{2}\frac{m_\nu^2}{E_\nu^2}. \tag{16.9}$$

If t_e is the time when a particular neutrino is emitted from the supernova and t_a is the time when it arrived on the earth, then, using Eq. (16.9) we get

$$t_a - t_e = D\left[1 + \frac{1}{2}\frac{m_\nu^2}{E_\nu^2}\right]. \tag{16.10}$$

where D is the distance traveled. For two neutrino events of energies E_1 and E_2, this implies

$$|\Delta t_a - \Delta t_e| = \frac{1}{2} D m_\nu^2 \cdot \frac{|E_1^2 - E_2^2|}{E_1^2 E_2^2} . \tag{16.11}$$

Experimentally, $\Delta t_a \leq 12\,\text{s}$. At this point, one needs to do a statistical analysis to see which pair of events will provide the most stringent constraint on m_ν. For the sake of illustration, if we choose events number 6 and 11 of Kamiokande, the time separation and energy difference is large enough. We get,

$$|9.8\,\text{s} - \Delta t_e| = 0.09\,\text{s} \cdot (m_\nu / 1\,\text{eV})^2 . \tag{16.12}$$

Assuming $\Delta t_e \leq 10\,\text{s}$, we get $m_\nu \leq 15\,\text{eV}$. More careful analysis of this kind [7] comes up with upper limits on m_ν ranging from 19 eV to 30 eV.

If neutrinos are Dirac particles, neutrino scattering in the supernova core will produce the corresponding right handed neutrinos which have no weak interaction. Therefore, they will drain energy from the core very efficiently. Since the rate of this process is proportional to the mass of the neutrino, one can obtain upper limits on ν_τ and ν_μ mass [8] of about 14 keV.

16.3.2 Neutrino lifetime

The fact that electron antineutrinos were observed implies that they did not decay in flight. If $\tau_{\widehat{\nu}_e}$ is the lifetime in the rest frame, we get

$$(E_\nu / m_\nu) \cdot \tau_{\widehat{\nu}_e} \geq 5 \times 10^{12}\,\text{s} , \tag{16.13}$$

taking time dilation into account. This implies

$$\tau_{\widehat{\nu}_e} \cdot (E_\nu / 20\,\text{MeV}) \geq (m_\nu / 1\,\text{eV}) \cdot 2.5 \times 10^5\,\text{s} . \tag{16.14}$$

The result is interesting since it rules out neutrino decay as a way to resolve the solar neutrino puzzle.

16.3.3 Magnetic moment of the neutrino

As discussed in Ch. 11, one can talk about two kinds of magnetic moments in relation to the neutrinos: the direct magnetic moment that couples the left-handed neutrino with its right-handed counterpart, or transition magnetic moment connecting different species of neutrinos to one another. We will see now that supernova constraints on magnetic moments [9] applies to the direct magnetic moment.

In the presence of the magnetic moment interaction, the ν_{eL} in the supernova core will scatter against electrons and protons to produce right handed neutrinos. The cross section for this scattering is obtained from the following square of the scattering amplitude $\mathcal{M}_a$, where a denotes the target:

$$|\mathcal{M}_a|^2 = \frac{2e^4\kappa^2}{m_e^2} \cdot \frac{t(s-m_a^2)(u-m_a^2)}{(t-m_\gamma^2)^2}. \tag{16.15}$$

Here s, t, and u are the standard Mandelstam variables, κ is the magnetic moment in units of the Bohr magneton $e/(2m_e)$, and m_γ is the Debye mass of the photon in a relativistic plasma, given by $m_\gamma^2 = (4\alpha/3\pi)\tilde{\mu}_e^2$ where $\tilde{\mu}_e$ is the electron chemical potential. From this, one can calculate the total ν_R production via the magnetic moment interaction. Once the ν_R is produced, it can interact with matter only via the magnetic moment interaction or via any possible right-handed current interaction as in left-right symmetric theories (see Ch. 6). The cross section for magnetic moment interaction is

$$\sigma \simeq \frac{4\pi\alpha^2}{m_e^2}\kappa^2 \ln(q^2/m_\gamma^2) \simeq 2.5 \times 10^{-25}\kappa^2 \, \mathrm{cm}^2, \tag{16.16}$$

choosing $q^2 = (3.1T)^2$, T being the core temperature. For $\kappa \approx 10^{-10}$, we obtain a cross section of order $10^{-45} \, \mathrm{cm}^2$, leading to a ν_R mean free path

$$\lambda_{\nu_R} \simeq \frac{1}{\sigma n_e} \approx 2.6 \times 10^6 \, \mathrm{cm} \gg R_{\mathrm{NS}}, \tag{16.17}$$

using ρ_{core} from Table 16.2 to estimate n_e.

Similarly, for right-handed gauge interactions, we get

$$\sigma \simeq 6\pi B G_F^2 E_e^2, \tag{16.18}$$

where

$$B = \zeta^2 + \left(\frac{M_{W_1}}{M_{W_2}}\right)^4. \tag{16.19}$$

For $M_{W_R}, M_{Z'} \gg 1\,\text{TeV}$, we obtain

$$\lambda_{\nu_R} \simeq \frac{1}{\sigma n_e} > 10^7\,\text{cm} \gg R_{\text{NS}}. \qquad (16.20)$$

Thus, if ν_R has only these interactions, its mean free path is bigger than the radius of the neutron star. So it escapes, opening a new channel for energy loss from core collapse. Furthermore, since these ν_R come from the core, their average energy is in the range of 100–200 MeV. One can make a rough estimate of this energy loss as follows:

$$Q_{\nu_R} \approx V_{\text{core}} n_e n_\nu \langle E \rangle \sigma_{\nu_R}. \qquad (16.21)$$

If the cross section of ν_R is dominated by the magnetic moment term in Eq. (16.16), we get

$$Q_{\nu_R} \approx 5 \times 10^{55} (10^{10}\kappa)^2\,\text{erg/s}. \qquad (16.22)$$

We now require that the energy loss is less than about one-third of the total binding energy of the neutron star, and that the ν_R emission lasts for about 1 s. Thus, the luminosity is

$$Q_\nu \leq 10^{53}\,\text{erg/s}. \qquad (16.23)$$

Under these very generous constraints, we obtain $\kappa \leq 5 \times 10^{-12}$, which is more stringent than limits from other sources (see Table 15.5).

This limit is further improved once we realize that the ν_R produced in the magnetic moment interaction can flip its helicity back to ν_L, since for the galactic magnetic field $B_{\text{gal}} \approx 10^{-6}\,\text{G}$ and the distance given in Table 16.2,

$$\mu_\nu B_{\text{gal}} D \approx 2 \times 10^{13} \kappa, \qquad (16.24)$$

which can be large for $\kappa \gtrsim 10^{-13}$. In that case, these should have produced high energy ν_L in the underground detectors. No such signal is seen, which translates into a bound $\kappa \lesssim 10^{-13}$.

The bound derived above becomes meaningless if there are strong Higgs interactions of ν_R which would make the mean free path smaller than the core radius. Such interactions might anyway be needed theoretically [9] to generate large magnetic moment of neutrino.

At this point, one should consider whether the ν_R produced via the magnetic moment interaction could flip back to ν_L through spin rotation in the magnetic field of the core. As noted in our discussion of the solar neutrino puzzle, the condition for this rotation to be effective is $\mu_\nu BL \simeq 1$. In the case of the neutron star, $L \simeq 10^6$ cm, $B \simeq 10^{12}$ G. For $\kappa \simeq 10^{-10}$, we thus get $\mu_\nu BL \gtrsim 1$. Therefore, naively one would expect rapid flipping back of ν_R.

However, in discussing the flipping, one must also include the matter effect, since in the neutron star the density of n, p, e and ν_e are rather high. The matter effect on ν_L is characterized by the matter contribution to the diagonal term in the neutrino mass matrix [9]:

$$\Delta m = \sqrt{2} G_F \left(n_e - \frac{1}{2} n_n + 2 n_\nu \right) \tag{16.25}$$

where n_a represents the number density of the particle a. The first two terms in this formula represent the contributions of electron, proton and neutron. As shown in Ch. 10, the neutral current contribution of the electron and the proton cancel, so that the term n_e gives the charged current contribution from $\nu_e e$ scattering and the term n_n is the neutral current contribution of the neutron. In addition to the analysis of Ch. 10, here the background contains neutrinos as well, which gives the n_ν term. This term has a factor 2 because of identity of ν_e's in the initial and final states, which gives two diagrams.

In the core, the dominant component is neutrons. In fact, defining Y_a to be the fraction of particles of type a compared to nucleons, we have $Y_n + Y_p = 1$. Also, $Y_e = Y_p$ from charge neutrality. Thus, only Y_e and Y_ν are independent. The estimates of Y_a in the supernova core are theoretical and therefore model dependent. According to Burrows and Lattimer [2], $Y_e \simeq 0.3$, $Y_\nu \simeq 0.04$. Using ρ_{core} from Table 16.2 to estimate the number densities, one now obtains in the supernova core,

$$\Delta m \simeq \sqrt{2} G_F n_e \simeq 32 \, \text{eV} \,, \tag{16.26}$$

whereas $\mu_\nu B \simeq 10^{-9}$ eV. Thus, the matter effect dominates, suppressing the flip-back of ν_R to ν_L. If, however, the term within the parentheses in Eq. (16.25) goes through a zero, ν_R flipping back to ν_L can occur, thereby avoiding the bound on μ_ν.

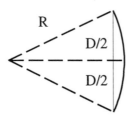

Figure 16.1: The curved line is the bent path of a charged particle in a magnetic field. The solid straight line denotes the path of an uncharged particle.

16.3.4 Electric charge of neutrino

If the neutrino had a tiny electric charge its path would bend in the galactic magnetic field B_{gal}. This would lead to a longer path traveled by the less energetic neutrinos than the more energetic ones [12]. To estimate the extra distance, let us look at Fig. 16.1. If neutrino has a charge Q_ν, it will travel the curved path with radius given by

$$R = \frac{m_\nu v}{Q_\nu e B_{\text{gal}}}. \qquad (16.27)$$

From Fig. 16.1 using simple geometry, we find the extra distance traveled by a neutrino of energy E_ν to be

$$2R\theta - D = \frac{D^3 Q_\nu^2 e^2 B_{\text{gal}}^2}{24 m_\nu^2} \left(1 + \frac{m_\nu^2}{E_\nu^2}\right). \qquad (16.28)$$

So the fractional time delay between two neutrinos with energies E_1 and E_2 is given by

$$\Delta t = \Delta t_e + \frac{D^3 Q_\nu^2 e^2 B_{\text{gal}}^2}{24} \left(\frac{1}{E_1^2} - \frac{1}{E_2^2}\right). \qquad (16.29)$$

Setting $\Delta t_e \leq 20\,\text{s}$ and considering as an example Kamioka events number 7 and 12, we get $Q_\nu \leq 10^{-18}$ using $B_{\text{gal}} \simeq 10^{-6}\,\text{G}$. A more careful statistical analysis [13], taking the non-uniformity of the galactic magnetic field into account, leads to $Q_\nu \leq 10^{-17}$.

16.3.5 Strength of right-handed weak interactions

Consider the left-right model of Ch. 6 where there are gauge bosons $W_R^\pm$ which couple to charged currents of right handed fermions. There is also a neutral gauge boson Z' which couples to right handed neutrinos. In such a case, if the ν_R has a mass less than 10-100 MeV, it can be produced in the supernova core via the following interactions:

$$
\begin{aligned}
e_R + p &\to n + \nu_R \\
n + p &\to p + p + e_R + \hat{\nu}_L \\
e^+ + e^- &\to \nu_R + \hat{\nu}_L \\
n + p &\to n + p + \nu_R + \hat{\nu}_L \,.
\end{aligned}
\tag{16.30}
$$

The first two interactions involve the exchange of $W_R^{\pm}$ whereas the second two involve the exchange of Z'. The argument now goes similar to the one in §16.3.3. Once the ν_R is produced, it escapes unless M_{W_R} or $M_{Z'}$ is less than about $4M_W$. This would drain the energy from core collapse without leaving enough energy for the left-handed neutrinos. Using this, one can show [14] that

$$
Q_{\nu_R} = \frac{6G_F^2}{\pi} B v n_p n_e \langle E_\nu \rangle \,,
\tag{16.31}
$$

where B is defined in Eq. (16.19). Using the constraint from Eq. (16.23), we then get

$$
B \leq 10^{-10} \,.
\tag{16.32}
$$

This implies that $\zeta \leq 10^{-5}$ and $M_{W_2} \geq 23\,\text{TeV}$, which are rather stringent bounds. Similarly, one obtains $M_{Z'} \geq (30 \to 85)M_Z$ from the last two processes of (16.30).

16.3.6   Radiative decay of neutrinos

At the time SN1987A was observed, the Solar Maximum Mission (SMM) was taking data with a gamma ray spectrometer. It did not find any γ-ray signal above the background [15]. For the period of about $10\,\text{s}$ in which all the neutrinos arrived, it set the following limits on the time-integrated flux (often called *fluence*) of γ-rays:

$$
F_\gamma \leq
\begin{cases}
0.9\,\text{cm}^{-2} & \text{for} \quad 4.1\,\text{MeV} \leq E_\gamma \leq 6.4\,\text{MeV} \\
0.4\,\text{cm}^{-2} & \text{for} \quad 10\,\text{MeV} \leq E_\gamma \leq 25\,\text{MeV} \\
0.6\,\text{cm}^{-2} & \text{for} \quad 25\,\text{MeV} \leq E_\gamma \leq 100\,\text{MeV} \,.
\end{cases}
\tag{16.33}
$$

If a massive neutrino decays radiatively, the photon from the decay would contribute to the fluence. Thus, one can use Eq. (16.33) to constrain radiative lifetimes of neutrinos.

To get the bound, we first note that there would be a delay in the arrival of the photon flux from the decay of a neutrino. This is because the parent neutrino, being massive, travels slower than light. The delay is characterized by $\Delta t \simeq \frac{1}{2} D m_\nu^2 / E_\nu^2 \simeq 0.02\,\mathrm{s}\,(m_\nu/1\,\mathrm{eV})^2$, taking the average neutrino energy to be $12\,\mathrm{MeV}$. The spreading in the arrival times of photon would also be of the same order of magnitude. If $m_\nu \lesssim 20\,\mathrm{eV}$, we get $\Delta t \lesssim 10\,\mathrm{s}$, so the decay photons should have reached the earth. For a certain species of neutrinos, the fluence from the supernova is given by

$$
f_0 = \frac{\text{Total energy released}}{\text{Average } \nu \text{ energy} \times \text{number of } \nu \text{ species} \times 4\pi D^2}
$$
$$
\simeq 1.4 \times 10^{10}\,\mathrm{cm}^{-2}. \tag{16.34}
$$

Thus, if B_γ is the branching ratio of the neutrinos decays into radiative modes, the fluence of photons would be

$$
f_\gamma = f_0 B_\gamma \left[1 - \exp\left(-\frac{D m_\nu}{E_\nu \tau_\nu} \right) \right] \tag{16.35}
$$
$$
= f_0 B_\gamma \cdot \frac{D m_\nu}{E_\nu \tau_\nu}, \tag{16.36}
$$

assuming $\tau_\nu \gg D m_\nu / E_\nu$ at the last step. If one assumes a blackbody distribution for the emitted neutrinos with a temperature of $4\,\mathrm{MeV}$ and that $E_\gamma = \frac{1}{2} E_\nu$, one obtains that about a quarter of the decay photons would be in the range of 4.1 to 6.4 MeV. Thus, using the bound in the first line of Eq. (16.33), one obtains [16]

$$
\frac{\tau_\nu}{B_\gamma} \geq 1.7 \times 10^{15}\,\mathrm{s} \cdot \left(\frac{m_\nu}{1\,\mathrm{eV}} \right) \quad \text{for } m_\nu \lesssim 20\,\mathrm{eV}. \tag{16.37}
$$

In other mass ranges, one can perform similar analysis. As long as $\tau_\nu \gg D m_\nu / E_\nu$, we get [16]

$$
\frac{\tau_\nu}{B_\gamma} \geq 8.4 \times 10^{17}\,\mathrm{s} \cdot \left(\frac{1\,\mathrm{eV}}{m_\nu} \right) \quad \text{for } 100\,\mathrm{eV} \lesssim m_\nu \lesssim \text{few MeV},
$$
$$
\frac{\tau_\nu}{B_\gamma} \geq 3.4 \times 10^{16} \gamma_{\mathrm{GRS}}^{-\frac{1}{2}}\,\mathrm{s} \quad \text{for } 20\,\mathrm{eV} \lesssim m_\nu \lesssim 100\,\mathrm{eV}. \tag{16.38}
$$

Here $\gamma_{\rm GRS}$ is the instrument background of the gamma ray spectrometer. In the other extreme, i.e., $\tau_\nu \ll D m_\nu / E_\nu$, Eq. (16.35) gives

$$f_\gamma \simeq f_0 B_\gamma \,. \tag{16.39}$$

Similar analysis then puts bounds on the photonic branching ratios B_γ:

$$B_\gamma \; \lesssim \; 1.4 \times 10^{-11} \gamma_{\rm GRS}^{-\frac{1}{2}} \left(\frac{m_\nu}{1\,{\rm eV}} \right) \quad \text{for } 20\,{\rm eV} \lesssim m_\nu \lesssim 100\,{\rm eV}\,,$$
$$\lesssim \; 2.8 \times 10^{-10} \quad \text{otherwise}\,. \tag{16.40}$$

A similar limit can also be obtained by considering the γ-ray background from all previous supernovae [17]:

$$\frac{\tau_\nu}{m_\nu B_\gamma} \geq 3 \times 10^{16}\,{\rm s/\,eV}\,. \tag{16.41}$$

16.3.7 Bounds on Majoronic decay modes

The time profile and energetic considerations for the observed electron neutrinos from the SN1987A implies upper and lower bounds on the lifetimes for the heavier neutrinos such as ν_μ and ν_τ decaying to ν_e (or $\hat{\nu}_e$) $+ J$, where J is an invisible, weakly interacting particle such as the Majoron or familon.

Let us first discuss the upper bound. Consider a heavier neutrino with mass less than 3–4 MeV so that as it emerges from the neutrino sphere, it is relativistic. If it is unstable and has a lifetime τ, then its travel time to the earth, t_0, can be divided into two parts if $\tau \ll t_0$. If we can make the instantaneous decay approximation, then for a time $E\tau/m_{\nu_H}$, it will travel with a velocity of p/E. After that time, it will decay to $\nu_e + J$, of which the ν_e will travel with the speed of light. The time T from the emission of the heavy neutrino to the arrival of the secondary ν_e on the earth is

$$T = \frac{E}{m_{\nu_H}}\tau + \left(L - \frac{p\tau}{m_{\nu_H}} \right)\,. \tag{16.42}$$

The time delay Δt of the secondary neutrinos with respect to the primary ones is [18]:

$$\Delta t \simeq \left(\frac{E - p}{m_{\nu_H}} \right)\tau \simeq \frac{m_{\nu_H}}{2E_{\nu_e}}\tau\,. \tag{16.43}$$

The delayed ν_e's would have roughly similar energy profile as the original. It was estimated that the delayed pulse would roughly contain 4-15 neutrinos [19]. Neither IMB nor Kamiokande experiments observed any delayed pulse between 15 s to 1 hour. This translates to an upper limit on $\Delta t \leq 15$ s leading to a constraint on heavy neutrino lifetime of about

$$\tau_{\nu_H} \leq \frac{300}{m_{\nu_H}/1 \text{ MeV}} \text{ s} . \tag{16.44}$$

In order to derive a lower bound note that if the lifetime of ν_H is τ, then in time $\Delta t \simeq E\tau/m_{\nu_H}$, half the energy in the $\nu_H + \bar{\nu}_H$ components would escape the core of the supernova in the form of J's, due to their decay. For low mass relativistic ν_H's, this energy can be estimated to be:

$$W_J \simeq \frac{1}{2}\left(W_{\nu_H} + W_{\bar{\nu}_H}\right) = \frac{1}{6}W_{\text{tot}} . \tag{16.45}$$

Since weak interaction processes continuously replenish the ν_H content of the supernova, it takes time $t^J \simeq 6\frac{\tau E_{\nu_H}}{m_{\nu_H}}$ for the total energy to decrease by a e-fold. For this cooling mechanism to be ineffective, we require that $t^J \geq 15$ s. (which is the typical SN ν-emission time). This implies the lower bound

$$\tau_{\nu_H} \geq \left(\frac{m_{\nu_H}}{8 \text{ MeV}}\right) \text{ s} . \tag{16.46}$$

For a 1 MeV ν_H, one gets $0.1 \text{ s} \lesssim \tau \lesssim 300 \text{ s}$.

16.3.8 Bounds on $\nu_\tau \to \nu_e e^+ e^-$ decay mode

Non-observation of MeV-range γ-rays by SMM satellite detector during the SN1987A can also be used to set an upper limit on the branching ratio $B(\nu_\tau \to \nu_e e^+ e^-)$ for electron-positron decay mode of the ν_τ which can be as heavy as 32 MeV without contradicting laboratory bounds. Let us remind the reader that the present laboratory upper limits on ν_μ mass (170 keV) does not allow it to decay into the $e^+ e^-$ mode.

There are many ways to derive this upper limit [19] but we present here the simplest of them which uses the photon bremsstrahlung of the $e^+ e^-$ in the final state, i.e., $\nu_\tau \to e^+ e^- \nu_e \gamma$. Denote the width

$\Gamma(\nu_\tau \to \nu_e e^+ e^-)$ by $B_{e^+ e^-}/\tau$ and let us assume $\tau \geq 10\,\mathrm{s}$ so that the ν_τ is out of the supernova before it decays. The branching ratio for the above bremsstrahlung process can be estimated to be roughly $\frac{\alpha}{\pi} B_{e^+ e^-} \simeq 10^{-3} B_{e^+ e^-}$. Note that the total number of ν_τ's emitted in the supernova is 10^{58} so that the photon flux at the SMM, Φ_γ, will be given by:

$$\Phi_\gamma = \frac{10^{55} B_{e^+ e^-} 2E}{4\pi L^2 \tau m_{\nu_\tau}} = \frac{1\,\mathrm{MeV}}{m_{\nu_\tau}} \times 10^9\,\mathrm{cm}^{-2}\Gamma. \qquad (16.47)$$

The SMM restriction of $\Phi_\gamma \leq .1\,\mathrm{cm}^{-2}\,\mathrm{s}^{-1}$ then implies that

$$\Gamma(\nu_\tau \to \nu_e e^+ e^-) \leq 10^{-10} \left(\frac{m_{\nu_\tau}}{1\,\mathrm{MeV}}\right) \mathrm{s}^{-1}. \qquad (16.48)$$

Similar bounds follow from $e^+ e^-$ annihilation photons subsequent to decay.

16.3.9   Bound on neutrino mixings

It appears that one way to understand the abundance of heavy elements in the universe is to assume that the hot bubble surrounding a Type II supernova provides a neutron-rich environment, where heavy elements are synthesized by nuclear reaction among the nucleons in the ejector from the supernova. This is the so-called r-process nucleosynthesis. Crucial to the success of this mechanism is the presence of an abundance of neutrinos. The neutron abundance of course depends on the weak interaction rates in that environment; for instance if there are too many high energy ν_e's, then the reaction $\nu_e + n \to p + e^-$ will deplete the neutron abundance making the r-process nucleosynthesis less effective. It has recently been argued [20] that for a certain range of $\Delta_{\nu_e \nu_x}$ (where $x = e$ or μ) and mixing angles, the MSW mechanism makes the conversion of higher energy ν_μ's and ν_τ's $(E_{\nu_{\mu,\tau}} \simeq 2E_{\nu_e})$ so efficient that the hot bubble surrounding the supernova contains an excess of energetic ν_e's, which deplete the neutron content, thereby retarding severely the r-process nucleosynthesis. Therefore, to the extent that r-process nucleosynthesis is accepted as an explanation for the heavier element abundance in the universe, a certain range of values of $(\Delta, \sin^2 2\theta)$ must be forbidden. Roughly, it is found [20] that for $\Delta \geq 2\,\mathrm{eV}^2$, one must have $\sin^2 2\theta \leq 10^{-4}$. This is a very severe constraint in a very interesting mass range for the neutrinos. This is, of course, independent of whether the neutrinos are Dirac or Majorana type.

16.3.10 Test of weak equivalence principle for neutrinos

Observation of the neutrino burst within a few hours of the optical burst from SN1987A implies that neutrinos and photons obey roughly the same equivalence principle. More quantitatively, if there were a new force coupling to neutrinos whose range is bigger than galactic distance scales, then its strength must be less than $5 \times 10^{-3} G_N$.

It is clear from the above discussion that the SN1987A has provided a great deal of extremely useful constraints on the properties of neutrino beyond the standard model, thereby strongly constraining the kind of new physics beyond the standard model one can envisage. It is also hoped that more information both about particle physics as well as stellar physics will be gained from future supernovae. At least some of the particle physics information such as the limit on neutrino magnetic moment or the strength of right-handed currents could not have been gained from laboratory experiments. New detectors are being planned to keep the supernova watch. It is believed that roughly one to two supernova collapse occurs in our own galaxy in every thirty years and the observed extra-galactic rate [22] is about of order one per hundred years per $10^{10} L_b$ where L_b is the measured blue luminosity density of the universe. This translates to a present rate for type II supernova of about $10^{-85} \, \text{cm}^{-3} \, \text{s}^{-1}$.

Chapter 17

Neutrino cosmology

In the standard big-bang cosmology, neutrinos are the most abundant form of matter in the universe next to radiation. Therefore, not only do the neutrinos play an important role in the evolution of the universe from its beginnings to the present state but cosmological observations also imply important restrictions on the possible properties of the neutrino beyond the standard model. In this chapter, we will present an overview of neutrino cosmology.

17.1   The Big Bang model

One of the basic assumptions of the standard big bang model of the universe [1] is the homogeneity and isotropy of the universe. The universe on large scales is observed to be homogeneous. The microwave background radiation, which is believed to be a relic of the early universe, is observed to be isotropic to a very high degree of accuracy on very large scales. Friedman, Robertson and Walker proposed the following metric which describes the homogeneous and isotropic space that our universe is supposed to be:

$$ds^2 = dt^2 - a^2(t) \left(\frac{dr^2}{1 - kr^2} + r^2 d\theta^2 + r^2 \sin^2 \theta \, d\phi^2 \right) . \qquad (17.1)$$

Here (t, r, θ, ϕ) are the coordinates, $a(t)$ is the cosmic scale factor, k takes values $+1, 0, -1$ for spaces with positive, zero and negative curvature. In order to describe the evolution of the universe from the moment after the big bang (or Planck time $\simeq 10^{-43}$ s) to now, we have to study the variation of $a(t)$ with time by solving the Einstein equation

$$R^{\mu\nu} - \frac{1}{2} R g^{\mu\nu} = -8\pi G T^{\mu\nu} , \qquad (17.2)$$

318

where $T^{\mu\nu}$ is the energy-momentum tensor, $R^{\mu\nu}$ is the Ricci tensor and $R \equiv g_{\mu\nu}R^{\mu\nu}$. Assuming the universe to be a perfect fluid, we can write

$$T^{\mu}{}_{\nu} = \text{diag}\,(\rho, -p, -p, -p) \qquad (17.3)$$

where ρ is the energy density and p is the pressure. Conservation of energy, $\partial_{\mu}T^{\mu}{}_{\nu} = 0$, then implies

$$d(\rho a^3) + p\,da^3 = 0\,. \qquad (17.4)$$

For radiation (i.e., any collection of relativistic particles)

$$p = \frac{1}{3}\rho \qquad (17.5)$$

so we get

$$\rho \propto a^{-4}\,, \qquad (17.6)$$

by solving Eq. (17.4). On the other hand, pressure is negligible compared to the mass density of non-relativistic particles (called *matter* in cosmological literature). Thus, putting $p = 0$ for matter, we get

$$\rho \propto a^{-3}\,. \qquad (17.7)$$

Substituting the metric of Eq. (17.1) and using Eq. (17.3), we can rewrite the Einstein equation as:

$$\left(\frac{\dot{a}}{a}\right)^2 + \frac{k}{a^2} = \frac{8\pi G}{3}\rho(t)\,. \qquad (17.8)$$

If we ignore the k-term which is negligible compared to the density term in the early universe, we get

$$H^2(t) = \frac{8\pi G}{3}\rho(t)\,, \qquad (17.9)$$

where

$$H(t) \equiv \dot{a}/a \qquad (17.10)$$

is called the *Hubble parameter* or the expansion rate of the universe. Eq. (17.9) thus expresses the expansion rate of the universe in terms of

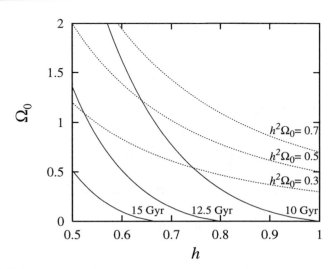

Figure 17.1: The age of the universe as functions of the density parameter Ω_0 and Hubble parameter, assuming the universe was matter dominated throughout. The dashed lines indicate contours of equal total density, which is proportional to $h^2\Omega_0$.

its the energy density and plays a crucial role in the study of various cosmological situations.

Solution of Eq. (17.9) gives the dependence of the scale factor a with time. Using Eq. (17.6) for a radiation dominated (RD) universe, we get

$$t \propto a^2 . \tag{17.11}$$

Similarly, using Eq. (17.7), we obtain

$$t \propto a^{3/2} \tag{17.12}$$

for matter dominated (MD) universe. The *cosmological red-shift* z for any era is defined by the relation $a_0/a = 1 + z$, where a_0 is the scale factor of the universe now. Therefore,

$$t = \begin{cases} \frac{1}{2}(1+z)^{-2}H_0^{-1} & \text{(for RD)} \\ \frac{2}{3}(1+z)^{-3/2}H_0^{-1} & \text{(for MD)} \end{cases} \tag{17.13}$$

where H_0 the Hubble expansion rate now. In the rest of this chapter, we will always use the subscript 0 to indicate the present era.

Exercise 17.1   *Eqs. (17.11) and (17.12) are valid only when $k = 0$. In general,*

$$\frac{k}{a_0^2} = (\Omega_0 - 1)H_0^2 , \qquad (17.14)$$

where Ω_0 is defined by

$$\Omega_0 H_0^2 = \frac{8\pi G}{3}\rho_0 , \qquad (17.15)$$

ρ_0 being the present energy density of the universe. Using Eq. (17.8), show that the present age of the universe can be written as

$$t_0 = \frac{1}{H_0} \int_0^1 dx \left[1 - \Omega_0 + \frac{\Omega_0}{x^{n-2}} \right]^{-\frac{1}{2}} , \qquad (17.16)$$

where $\rho(t) \propto [a(t)]^{-n}$, i.e., $n = 3$ for an MD universe and $n = 4$ for an RD universe. Using this, prove Eq. (17.13).

Exercise 17.2   *The cosmological constant Λ is defined by the presence of a term $\Lambda g^{\mu\nu}$ on the right side of Eq. (17.2). In Eq. (17.8), this modifies the right side to $\frac{1}{3}[8\pi G\rho(t) + \Lambda]$. Show that, if the total energy density has contributions from matter, radiation and the Λ-term, the age formula is*

$$t_0 = \frac{1}{H_0} \int_0^1 dx \left[1 - \Omega_0 + \frac{\Omega_{mat}}{x} + \frac{\Omega_{rad}}{x^2} + \Omega_\Lambda x^2 \right]^{-\frac{1}{2}} , \qquad (17.17)$$

where $\Omega_\Lambda = \Lambda/3H_0^2$.

The age of the universe plays an important role in constraining neutrino properties, as we will see later.

The next important piece of information we need in our study of cosmology is the energy density and number density of different species of particles — relativistic or non-relativistic, fermions or bosons. We give this below:

- Relativistic gas:

$$\rho = \begin{cases} (\pi^2/30)gT^4 & \text{Bose} \\ \frac{7}{8}(\pi^2/30)gT^4 & \text{Fermi}, \end{cases} \qquad (17.18)$$

$$n = \begin{cases} (\zeta(3)/\pi^2)gT^3 & \text{Bose} \\ \frac{3}{4}(\zeta(3)/\pi^2)gT^3 & \text{Fermi}. \end{cases} \qquad (17.19)$$

- Non-relativistic gas (Bose and Fermi):

$$\rho = mn \,,$$

$$n = g \left(\frac{mT}{2\pi}\right)^{3/2} \exp\left(\frac{\mu - m}{T}\right). \qquad (17.20)$$

In the above expressions, g denotes the number of spin states, μ is the chemical potential, and $\zeta(3) \simeq 1.202$ is a Riemann zeta function. Total energy density is of course the sum of the individual energy densities. From Eq. (17.6) and Eq. (17.18), we get that aT is constant for relativistic particles. Since the k-term is negligible in the early universe anyway, we can then rewrite Eq. (17.9) as

$$\left(\frac{\dot{T}}{T}\right)^2 = \frac{8\pi G}{3}\frac{\pi^2}{30}g_\star T^4 \qquad (17.21)$$

where $g_\star$ is the effective number of degrees of freedom contributing to the energy density. Using $G = 1/M_{\rm Pl}^2$, we get solving Eq. (17.21)

$$t = 0.3g_\star^{-1/2}\frac{M_{\rm Pl}}{T^2} \simeq g_\star^{-1/2}\left(\frac{1\,{\rm MeV}}{T}\right)^2 \cdot 1\,{\rm s}\,. \qquad (17.22)$$

Exercise 17.3 *Show that*

$$\int_0^\infty \frac{E^N dE}{e^{\beta E} + 1} = \left(1 - \frac{1}{2^N}\right)\int_0^\infty \frac{E^N dE}{e^{\beta E} - 1}\,. \qquad (17.23)$$

Use this to derive that the number and energy densities of a massless Fermi gas are smaller by the factors 3/4 and 7/8 respectively compared to a massless Bose gas with equal number of spin degrees of freedom, as shown in Eqs. (17.18) and (17.19).

17.2 Neutrino decoupling

In giving the energy density and the particle number densities we have assumed that all the particles are in thermal equilibrium. It is therefore appropriate to describe at this stage the condition for equilibrium in an expanding universe. In a gaseous ensemble of atoms or molecules, thermal equilibrium is said to be attained when the state of the system (number of particles at a given energy level) does not change with

Table 17.1: Milestones in the history of the universe.

Age of the universe (approximately)	Temperature (in K)	Major event
10^{-43} s	10^{32}	Gravitation becomes negligible
10^{-43}—10^{-35} s	10^{32}—10^{28}	Period of GUT inflation
10^{-35}—10^{-12} s	10^{28}—10^{16}	Generation of baryon asymmetry
10^{-12} s	10^{16}	Electroweak phase transition
10^{-4} s	10^{12}	Quark-hadron transition, μ^+-μ^- annihilations
1 s	10^{10}	Neutrinos decouple, nucleosynthesis begins
10 s	5×10^9	e^+-e^- annihilations
10^5 yr	4000	Atoms form; radiation decouples from matter
2×10^5—10^9 yr		Galaxy formation
$(10$—$18) \times 10^9$ yr	2.7	Now

time. In an expanding universe, the temperature is constantly changing. Therefore, to be in equilibrium particles must adjust there energy faster than the time it takes to change the temperature. For this to be achievable, the interaction rate of the particles must be faster than the expansion rate of the universe, i.e.,

$$\Gamma_{\text{int}}(t) > H(t). \tag{17.24}$$

If this condition fails to hold at some era during the evolution of the universe for some particular species of particles, those particles are said to *decouple* or *freeze-out*, i.e., they fall out of thermal equilibrium.

As an application of this, consider the interaction of the neutrinos. Neutrinos interact only via weak interaction. At energies small compared to the W-mass, the cross section of these interactions is of order $G_F^2 E^2$, where for relativistic particles the energy E is of order of the temperature T. The number density for relativistic particles is of order

T^3. Therefore, a typical neutrino interaction rate is given by:

$$\Gamma_{\text{int}}^{\nu} \simeq \langle \sigma_{\text{wk}} n_{\nu} v \rangle \simeq G_F^2 T^5 \,. \tag{17.25}$$

On the other hand, from Eq. (17.21), we get

$$H(t) \simeq 1.66 g_{\star}^{1/2} \frac{T^2}{M_{\text{Pl}}} \,. \tag{17.26}$$

Thus, $\Gamma_{\text{int}}^{\nu}$ falls faster than $H(t)$ as the temperature of the universe decreases. The equilibrium condition in Eq. (17.24) then implies that, below a temperature T_D when the interaction rates are smaller than the expansion rates, the neutrinos are no more in thermal equilibrium. This temperature is given by a solution of the equation

$$G_F^2 T_D^5 = g_{\star}^{1/2} \frac{T_D^2}{M_{\text{Pl}}} \,, \tag{17.27}$$

i.e.,

$$T_D = \left(\frac{g_{\star}^{1/2}}{G_F^2 M_{\text{Pl}}} \right)^{1/3} \,. \tag{17.28}$$

Using $M_{\text{Pl}} = 1.2 \times 10^{19} \, \text{GeV}$ and $G_F \simeq 10^{-5} \, \text{GeV}^{-2}$, we obtain $T_D \simeq 1 \, \text{MeV}$. In practical terms, it means that there is (almost) no neutrino scattering below a temperature of 1 MeV or after the universe is older than about 1 s. Neutrinos expand as a free gas, with their temperature going down like $1/a$.

In Table 17.1, we present some milestones in the history of the universe. Of course, many entries in the column denoting the age of the universe are expectations rather than established times.

17.3   Nucleosynthesis and the number of neutrino species

Nucleosynthesis [2] is the first step in the formation of heavy elements out of neutrons and protons in the universe. As we see from Table 17.1, below the quark-hadron transition temperature $T_H \simeq 100 \, \text{MeV}$, the universe consists of a hot soup of neutrons, protons, pions etc. These particles are in equilibrium via the strong, electromagnetic and weak interactions. Since the protons and neutrons are already non-relativistic, their number densities are given by Eq. (17.20). For $T \gg 1\,\mathrm{MeV}$, the weak interaction reactions

$$n \rightarrow pe\hat{\nu}_e, \quad \nu_e n \rightarrow pe^-, \quad e^+ n \rightarrow p\hat{\nu}_e \qquad (17.29)$$

are much faster than the expansion rate of the universe (Eq. (17.26)). As a result, we have $\mu_p = \mu_n$ for the case of no lepton degeneracy, i.e., when the number of neutrinos equal that of antineutrinos, and the number of electrons equal that of positrons. The nuclear binding energies are of the order of several MeVs – e.g., $2.2\,\mathrm{MeV}$ for $^2\mathrm{H}$, $8.5\,\mathrm{MeV}$ for $^3\mathrm{H}$, $7.7\,\mathrm{MeV}$ for $^3\mathrm{He}$, $28.2\,\mathrm{MeV}$ for $^4\mathrm{He}$. So, unless the temperature of the universe is below an MeV, the energetic photons in the tail of the thermal distribution will wipe out any nuclei manufactured out of neutrons and protons. The ratio of protons and neutrons in the universe is fixed by weak interactions. We saw that neutrinos decouple below $T_D \simeq 1\,\mathrm{MeV}$. So at $t \simeq 1\,\mathrm{s}$, we have

$$\frac{n_n}{n_p} = \exp\left(\frac{m_p - m_n}{1\,\mathrm{MeV}}\right) \simeq \frac{1}{6}. \qquad (17.30)$$

Beyond this point, the ratio of neutron to proton remains frozen except for small changes due to neutron beta decay.

$$\frac{n_n}{n_p} \simeq \frac{1}{6} e^{-t/\tau}. \qquad (17.31)$$

Roughly speaking the $^4\mathrm{He}$ formation takes place soon after this. If we assume that all the neutrons combine to form $^4\mathrm{He}$, then the final number of $^4\mathrm{He}$ nuclei is given by $n_n/2$. The $^4\mathrm{He}$ abundance parameter Y is then calculated to be

$$Y \equiv \frac{\text{Mass of }^4\mathrm{He}}{\text{Total mass}} = \frac{\frac{1}{2}n_n \times 4}{n_p + n_n} = \frac{2n_n/n_p}{1 + n_n/n_p}. \qquad (17.32)$$

If we use $n_n/n_p = 1/6$ as in Eq. (17.30), we obtain $Y = 2/7$. Since beta decay reduces the value of n_n/n_p slightly to $1/7$, one actually gets $Y \simeq 0.25$. A crucial assumption going into calculations is the value of the neutrino decoupling temperature T_D.

If there are more species of neutrinos they will contribute to the energy density, thereby increasing the value of the decoupling temperature. The higher the decoupling temperature, the higher the neutron to proton ratio and from that one would get higher Y. Thus, qualitatively it is clear that the observed ^{4}He abundance would imply a restriction on the number of neutrino species. It is worth noting here that, if there is a right-handed neutrino ν_R which can convert to ν_L via some interaction that is in equilibrium at the epoch of nucleosynthesis (such as a large magnetic moment), that ν_R will count as an additional particle species in the energy density and its number will be restricted. Similar arguments will apply to massless bosons such as a triplet Majoron.

For a quantitative expression of the dependence of Y on the number of additional species of neutrinos or some other kind of particles, we see that,

$$Y = \frac{2\exp(-\Delta m/T_D)}{1 + \exp(-\Delta m/T_D)} \tag{17.33}$$

where Δm is the neutron proton mass difference. Thus

$$\Delta Y = \frac{1}{2}Y(2 - Y)\ln\left(\frac{2 - Y}{Y}\right)\cdot(\Delta T_D/T_D). \tag{17.34}$$

By differentiating the expression for T_D in Eq. (17.28), we get

$$\frac{\Delta T_D}{T_D} = \frac{1}{6}\frac{\Delta g_\star}{g_\star}. \tag{17.35}$$

For three generation of neutrinos, we have

$$\begin{aligned}
g_\star &= g_\gamma + \frac{7}{8}\times\left(3g_\nu + 3g_{\bar\nu} + g_{e^-} + g_{e^+}\right)\\
&= 2 + \frac{7}{8}\times(3 + 3 + 2 + 2) = 10.75.
\end{aligned} \tag{17.36}$$

Notice that the spin degeneracy has been set equal to one for each neutrino, but two for the electron or the positron. This is because what is important here is the number of interacting species. The right handed neutrinos either do not exist or they have little interaction with other kind of matter, so we neglect them. If the number of neutrino species is not 3 but $3 + \Delta N_\nu$, we would get $\Delta g_\star = (7/4)\Delta N_\nu$. Using Eq. (17.34) and Eq. (17.35), we thus get $\Delta Y \simeq 0.01\Delta N_\nu$.

There are also other factors that control Y. These include, for example, the beta decay lifetime of the neutron, the initial baryon to photon ratio etc. For instance, a shorter neutron halflife reduces the Y value and so does a smaller baryon to photon ratio since the rate of the nuclear reaction responsible for ^{4}He production depends on the initial n_B/n_γ. All these factors have been carefully analyzed [3, 4] to lead to $\Delta N_\nu \leq 0.4$ from a fit to observed abundances of ^{4}He, deuterium, ^{3}He and ^{7}Li. The observed abundances for these are:

$$
\begin{aligned}
0.22 < \quad & Y \quad < 0.26 \\
10^{-5} \leq \quad & (D/H) \\
& [D + {}^3He]/H \leq 8 \times 10^{-5} \\
10^{-10} \leq \quad & (^7Li/H) \quad \leq 2 \times 10^{-10}\,.
\end{aligned}
\tag{17.37}
$$

The dependence of Y on the various variables are

$$
Y = 0.226 + 0.012\Delta N_\nu + 0.010 \log \eta_{10} + 0.012(\tau_n - 887\,\mathrm{s})\,,
\tag{17.38}
$$

where $\Delta N_\nu = N_\nu - 3$, $\eta_{10} = 10^{10} n_B/n_\gamma$ and τ_n is the neutron lifetime. In obtaining the bound $\Delta N_\nu \leq 0.4$, the latest measurement of neutron lifetime [5] is used, which gives $\tau_n = 887.0 \pm 2.0$ s, which implies a half-life of $\tau_n \ln 2 = 10.25$ min.

17.4 Constraints on stable neutrino properties

If neutrinos are stable, their energy density is constrained by cosmological considerations. The basic idea is simple and obvious: the energy density of neutrinos must not be larger than the total energy density of the universe, ρ_0. It is customary to represent ρ_0 in units of the critical density:

$$
\rho_0 \equiv \Omega_0 \rho_c\,,
\tag{17.39}
$$

where

$$
\rho_c = \frac{3H_0^2}{8\pi G}\,.
\tag{17.40}
$$

The value of H_0 is usually parameterized in the form

$$H_0 = 100h \, \text{km} \, \text{s}^{-1} \, \text{Mpc}^{-1} \,. \tag{17.41}$$

Substituting this in Eq. (17.40), we find

$$\rho_\nu < \rho_0 = 10^4 h^2 \Omega_0 \, \text{eV} / \, \text{cm}^3 \,. \tag{17.42}$$

Neither h nor Ω_0 is very well-known. All modern observations indicate that $\frac{1}{2} \le h \le 1$. From Fig. 17.1, it appears that if we accept that the age of the universe is larger than $10^{10} \, \text{yr}$, one obtains

$$h^2 \Omega_0 < 0.5 \,. \tag{17.43}$$

In fact, studies on globular clusters etc indicates that the age is probably closer to $15 \times 10^9 \, \text{yr}$, in which case the constraint is much stronger. In any case, we will use Eq. (17.43) in what follows.

17.4.1 Bound on the degeneracy of massless neutrinos

Let the neutrinos have a chemical potential μ. Consider, for simplicity, the temperature of the neutrino gas to be zero. In this case, the Fermi distribution function is unity for all energies less than or equal to μ, and zero for higher energies. The energy density of the neutrinos is then given by

$$\rho_\nu = \int_{|\boldsymbol{p}|=0}^{\mu} \frac{d^3\boldsymbol{p}}{(2\pi)^3} |\boldsymbol{p}| = \frac{\mu^4}{8\pi^2} \,. \tag{17.44}$$

Using Eq. (17.42), then, we obtain [6]

$$\mu < (h^2 \Omega_0)^{1/4} 8.8 \times 10^{-3} \, \text{eV} < 7.4 \times 10^{-3} \, \text{eV} \,. \tag{17.45}$$

17.4.2 Bound on light neutrino masses

One can also use Eq. (17.43) to put an upper limit on the masses of the light stable neutrinos [7]. To derive this limit, we have to find out the number density of neutrinos in the present universe. This depends on whether the neutrinos were relativistic or non-relativistic at the time of decoupling. For now, let us assume the former possibility, which implies $m_\nu \ll 1\,\text{MeV}$. Then, at the time of decoupling $t \simeq 1\,\text{s}$,

$$n_\nu(T_D) = \frac{3}{4}n_\gamma(T_D)\,. \tag{17.46}$$

The ratio of neutrino density to photon density remains constant since that time until $T \leq \frac{1}{2}\,\text{MeV}$ when e^+e^- annihilation to photons increased the photon temperature. To see how much this increase in temperature was, we recall that before and after the e^+e^- annihilation, the entropy remains same. The entropy density of a relativistic gas is given by

$$s = \frac{\rho + p}{T} = \frac{4}{3}\frac{\rho}{T} = \frac{2\pi^2}{45}g_\star T^3\,. \tag{17.47}$$

But before e^+e^- annihilation, $g_\star = 2 + 2 \times (7/4) = 11/2$ coming from γ, e^- and e^+ whereas afterwards the contribution is just from the photons, so $g_\star = 2$. Entropy conservation therefore implies

$$g_{\star<}T_<^3 \quad = \quad g_{\star>}T_>^3\,, \tag{17.48}$$

which gives

$$T_> \quad = \quad (11/4)^{1/3}T_<\,. \tag{17.49}$$

The T_ν however remained same as before. Therefore, in the present universe,

$$T_\gamma^0 = (11/4)^{1/3}T_\nu^0\,. \tag{17.50}$$

Using $T_\gamma^0 = 2.7\,\text{K}$, we get $T_\nu^0 = 1.9\,\text{K}$. Using Eq. (17.19), we then get the present number density of a Dirac neutrino to be

$$n_\nu = \frac{3}{4}\left(\zeta(3)/\pi^2\right) \times 2 \times (1.9\,\text{K})^3 = 109/\,\text{cm}^3\,, \tag{17.51}$$

using $1\,\text{K} = 4.3\,\text{cm}^{-1}$. Since the neutrino energy density $\rho_\nu = m_\nu n_\nu$ must be less than ρ_0, we obtain [7]

$$m_\nu \leq 92h^2\Omega_0\,\text{eV} < 46\,\text{eV}\,, \tag{17.52}$$

using Eqs. (17.42) and (17.43). This is a very stringent constraint indeed.

17.4.3 Bound on heavy stable neutrino masses

If neutrino masses are much bigger than $1\,\mathrm{MeV}$, the neutrinos become non-relativistic while in equilibrium. Therefore, their number density falls exponentially with temperature as $\exp(-m/T)$. In such a case, when they decouple, their number density is already so small that their contribution to mass density need not exceed the total allowable density [8]. The reduction in the number density of heavy neutrinos occurs via its annihilation to $f\widehat{f}$ where f is a lighter fermion. Let us parameterize the matrix element of the effective four-Fermion interaction as

$$\mathcal{L}_{\mathrm{eff}} = \frac{4G_F}{\sqrt{2}}\left[\overline{u}_H\gamma_\lambda\left(\frac{a-b\gamma_5}{2}\right)v_H\right]\cdot\left[\overline{u}_f\gamma^\lambda\left(\frac{C_f+C'_f\gamma_5}{2}\right)v_f\right],\tag{17.53}$$

where the subscripts H and f denote spinors for the heavy neutrinos and for the fermions f. The annihilation cross-section can then be easily calculated to give

$$\sigma_{\mathrm{ann}}v = \frac{1}{4\pi}G_F^2 E^2\sqrt{1-z_f^2}\times$$
$$\left[(a^2+b^2)(C_f^2+C_f'^2)\{1+\frac{1}{3}(1-\gamma^{-2})(1-z_f^2)\}\right.$$
$$+(a^2+b^2)(C_f^2-C_f'^2)z_f^2(1-\frac{1}{2}\gamma^{-2})$$
$$+(a^2-b^2)(C_f^2+C_f'^2)\gamma^{-2}(1-\frac{1}{2}z_f^2)$$
$$\left.+(a^2-b^2)(C_f^2-C_f'^2)\gamma^{-2}z_f^2\right]\tag{17.54}$$

where E is the energy of the incoming neutrinos in the center of mass frame, $z_f = m_f/E$ and $\gamma = E/m_\nu$. In $SU(2)_L\times U(1)_Y$ model, we can put $a=1$, $b=1$ for a Dirac neutrino, and C_f and C'_f are given by $I_{3L}-2Q\sin^2\theta_W$ and I_{3L} respectively. For Majorana neutrinos described in the models of Ch. 5, the matrix element of the vector current vanishes, i.e., $a=0$. The matrix element of the axial current can be evaluated in the way shown in §11.1.2, and the result is $b=2$. If ν_H participates in interactions beyond the $SU(2)_L\times U(1)_Y$ model, C and C' may have new contributions.

To calculate the neutrino decoupling temperature, we must solve the following equation for the number density n_ν:

$$\frac{dn_\nu}{dt} = -3\frac{\dot{a}}{a}n_\nu - \langle\sigma_{\mathrm{ann}}v\rangle\left(n_\nu^2 - n_{\mathrm{eq}}^2\right).\tag{17.55}$$

On the right hand side, the first term denotes change of the number density due to the overall expansion of the universe and the second term due to annihilation. We now define new dimensionless variables which are convenient for the problem at hand:

$$x \equiv m_\nu/T\,, \qquad y \equiv \frac{n_\nu}{s}\,, \tag{17.56}$$

where $s = (4\pi^2/45)T^3$ is the photon entropy density. Using Eq. (17.22), we can write

$$\frac{dx}{dt} = \frac{dx}{dT} \cdot \frac{dT}{dt} = xH(T)\,, \tag{17.57}$$

where $H(T)$ is the Hubble parameter at the era when the temperature is T, as given in Eq. (17.26). With these variables, we can rewrite Eq. (17.55) as

$$
\begin{aligned}
\frac{dy}{dx} &= -\frac{s\,\langle\sigma_{\mathrm{ann}}v\rangle}{xH(T)}(y^2 - y_{\mathrm{eq}}^2) \\
&= -\frac{y_{\mathrm{eq}}}{x}\cdot\frac{\Gamma_{\mathrm{ann}}}{H(T)}\left[\left(\frac{y}{y_{\mathrm{eq}}}\right)^2 - 1\right]\,,
\end{aligned}
\tag{17.58}
$$

where in the last step we have introduced $\Gamma_{\mathrm{ann}} \equiv n_{\mathrm{eq}}\langle\sigma_{\mathrm{ann}}v\rangle$.

From this equation, it is easily seen that if the condition $\Gamma_{\mathrm{ann}} \ll H(T)$ is satisfied, y cannot change noticeably. We have assumed that we are talking of heavy neutrinos which decouple when they are non-relativistic. Using Eq. (17.54) for a Dirac neutrino and neglecting the masses of the product particle, we obtain $\sigma_{\mathrm{ann}}v \approx G_F^2 m_\nu^2/2\pi$ for each channel of annihilation. We thus write $N_{\mathrm{ann}}G_F^2 m_\nu^2/2\pi$ for the total annihilation cross section, where N_{ann} is the number of channels open for annihilation, whose value will be specified later. As for n_{eq}, we can use Eq. (17.20) with $g = 2$, assuming $\mu = 0$. Then,

$$\frac{\Gamma_{\mathrm{ann}}}{H(T)} = 1.2 \times 10^{-2} g_\star^{-1/2} N_{\mathrm{ann}}G_F^2 m_\nu^3 M_{\mathrm{Pl}} x^{1/2} e^{-x}\,, \tag{17.59}$$

which shows that it vanishes as x becomes large, so that y does not change. Physically, it means that the neutrinos become so rare that after a while, they cannot find each other, so that annihilation effectively stops. We can thus say that the neutrinos decouple when $\Gamma_{\mathrm{ann}}/H \sim 1$.

From Eq. (17.59), we thus see that x_D, the value of x at decoupling, will be the solution of the equation

$$x_D^{1/2} e^{-x_D} = \frac{82 g_*^{1/2}}{N_{\text{ann}}} \left(G_F^2 m_\nu^3 M_{\text{Pl}} \right)^{-1} , \qquad (17.60)$$

or

$$x_D - \frac{1}{2} \ln x_D = \ln \left[\frac{N_{\text{ann}}}{82 g_*^{1/2}} G_F^2 m_\nu^3 M_{\text{Pl}} \right] . \qquad (17.61)$$

To obtain a rough analytic solution to this equation, we can neglect the log term on the left side to a first approximation, since x_D is large by our assumption. This gives a zeroth order solution

$$x_D^{(0)} = \ln \left[\frac{N_{\text{ann}}}{82 g_*^{1/2}} G_F^2 m_\nu^3 M_{\text{Pl}} \right] . \qquad (17.62)$$

Putting this back into Eq. (17.61), we can obtain a better estimate:

$$x_D \approx \ln \left[\frac{N_{\text{ann}}}{82 g_*^{1/2}} G_F^2 m_\nu^3 M_{\text{Pl}} \right] + \frac{1}{2} \ln \ln \left[\frac{N_{\text{ann}}}{82 g_*^{1/2}} G_F^2 m_\nu^3 M_{\text{Pl}} \right] . \quad (17.63)$$

Above a few hundred MeV, we should take the quark-gluon degrees of freedom into account to estimate g_*. Thus we get, for $m_s < T_D < m_\tau$,

$$\begin{aligned} g_* &= g_\gamma + 8 g_g \\ &\quad + \frac{7}{8} \times \left(g_{\nu_e} + g_{\nu_\mu} + g_{\nu_\tau} + g_e + g_\mu + 3[g_u + g_d + g_s] \right) \\ &= 59.75 , \end{aligned} \qquad (17.64)$$

where we have put $g_\nu = 2$ for each species of neutrinos, and $g = 4$ for all charged fermions, and the factor of 3 in front of the quarks denote the color degeneracy. Also, $N_{\text{ann}} \approx 10$ for the range of temperature we are looking at. Putting these values, we obtain

$$\ln \left[\frac{N_{\text{ann}}}{82 g_*^{1/2}} G_F^2 m_\nu^3 M_{\text{Pl}} \right] = 17 + 3 \ln \left(\frac{m_\nu}{1 \, \text{GeV}} \right) . \qquad (17.65)$$

And then the second term on the right side of Eq. (17.63) can be written as

$$\begin{aligned} &\frac{1}{2} \ln \left[17 \times \left\{ 1 + \frac{3}{17} \ln \left(\frac{m_\nu}{1 \, \text{GeV}} \right) \right\} \right] \\ &= 1.4 + 0.09 \ln \left(\frac{m_\nu}{1 \, \text{GeV}} \right) , \end{aligned} \qquad (17.66)$$

where in the last step we have assumed that $m_\nu \gg 1\,\text{GeV}$, so that $3\ln(m_\nu/1\,\text{GeV}) \ll 17$. Putting these numerical estimates back in Eq. (17.63), we thus obtain

$$x_D \approx 18.4 + 3.09\ln(m_\nu/1\,\text{GeV}) \,. \tag{17.67}$$

This implies for instance that, a 2 GeV neutrino will decouple roughly at 93 MeV.

By definition, the value of y does not change after the decoupling time, so the density goes down only due to volume expansion, leading to the present density

$$n_\nu(T_0) \;=\; n_\nu(T_D) \times \left(\frac{T_0}{T_D}\right)^3 \,. \tag{17.68}$$

Using Eq. (17.20), we thus obtain

$$\begin{aligned}
n_\nu(T_0) &= \frac{2}{(2\pi)^{3/2}} x_D^{3/2} e^{-x_D} T_0^3 \\
&= 8 \times 10^{-6} x_D \left(\frac{1\,\text{GeV}}{m_\nu}\right)^3 \text{cm}^{-3} \,,
\end{aligned} \tag{17.69}$$

using Eq. (17.60). The contribution of these relic neutrinos to the density parameter Ω_0 is then easily calculated, using the value of the critical density from Eq. (17.40):

$$\Omega_\nu(T_0) \;=\; \frac{m_\nu n_\nu(T_0)}{10^4 h^2 \,\text{eV}/\text{cm}^3} \tag{17.70}$$

which gives, using Eq. (17.69) and inserting the solution for x_D from Eq. (17.67),

$$h^2 \Omega_\nu(T_0) \;=\; 14.7 \left[1 + 0.16\ln\left(\frac{m_\nu}{1\,\text{GeV}}\right)\right] \left(\frac{1\,\text{GeV}}{m_\nu}\right)^2 \,. \tag{17.71}$$

Using Eqs. (17.42) and (17.43) now, we get $m_\nu \geq 5.5\,\text{GeV}$. We have presented a rather rough derivation of the lower bound to illustrate the physics. A more careful evaluation gives $m_\nu \geq 2\,\text{GeV}$ for Dirac neutrinos [8], where $h^2\Omega_0 < 1$ was assumed for the calculations. This is known as the Lee-Weinberg bound.

For Majorana neutrinos, looking back at Eq. (17.54), we find that for $z_f = 0$, the cross section is proportional to the square of the velocity of the incoming neutrinos. Since the neutrinos are non-relativistic, this means that the cross-sections are much smaller. As a result, the annihilation process is less efficient, and one needs to go to higher values of mass in order to obtain acceptable energy density. Detailed calculations show that $m_\nu \geq 6\,\text{GeV}$ for Majorana neutrinos [9]. From this, we see that for stable neutrinos, the cosmologically forbidden mass range is

$$92h^2\Omega_0 \text{ eV} \leq m_\nu \leq \frac{2}{\sqrt{h^2\Omega_0}}\,\text{GeV}\,. \qquad (17.72)$$

Exercise 17.4 *Use the thermal average velocity of a non-relativistic particle, $\sqrt{8T/\pi m}$, to estimate the annihilation cross-section for a Majorana neutrino, assuming $z_f = 0$. Using this, follow the steps given in the text to estimate the energy density of relic Majorana neutrinos in the present universe and hence find the minimum allowed mass for such a neutrino.*

17.5 Constraints on heavy unstable neutrinos

So far, we have been discussing stable neutrinos. However, in most gauge models for massive neutrinos, we expect the heavier neutrinos to be unstable. It is therefore important to derive constraints on the masses of the unstable heavy neutrinos. Obviously this will depend on the lifetime of the neutrino. For example, if an unstable neutrino has a lifetime bigger than $10^{17}\,$s (the present age of the universe), for our purposes it is stable and the mass bounds derived in the previous section will apply. On the other hand, if a neutrino has a lifetime less than the age of the universe the shorter its lifetime, the heavier it can be for the following reason. When a neutrino decays, its decay products (typically, lighter neutrinos, photon or Majoron) are all relativistic and will therefore undergo red-shift as the universe expands. If they are red-shifted so much that their energy is small enough to contribute less than the critical energy density at present, those masses of heavy neutrinos are acceptable. But whether their energy red-shifts enough depends both on the mass of the parent heavy neutrino (ν_H) and the time of its decay. Thus, in this case, we will get a correlated bound between m_{ν_H} and τ_{ν_H}. Further ramifications to this bound arise from considerations of what the decay products do to the universe.

To get this bound [10], we first consider the case $m_{\nu_H} < 1\,\text{MeV}$. Thus, the ν_H are relativistic when they decouple when $t \sim 1\,$s. Their

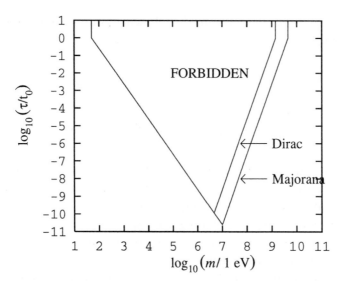

Figure 17.2: Bounds on the lifetimes on unstable neutrinos from the energy density of the universe.

number density for later times is given by Eq. (17.68).

To obtain a simple estimate [11], we assume that all the neutrinos decay at the era $t \sim \tau$, where τ is the lifetime. The decay products share the total energy m_{ν_H} at the time of decay. Since the decay products are in general much lighter, they are relativistic. Their energy makes the universe radiation dominated. In Eq. (17.11), we saw that in an RD universe, the scale parameter varies as the square root of time. Therefore, the energy of the decay products red-shift according to $(\tau/t_0)^{1/2}$. The contribution to the present energy density from ν_H-decay can be written as:

$$ n_{\nu_H}(T_0) \cdot m_{\nu_H} \left(\frac{\tau}{t_0} \right)^{1/2} . \tag{17.73} $$

Using the number density from Eq. (17.51), we thus obtain

$$ m_{\nu_H} \left(\frac{\tau}{t_0} \right)^{1/2} \leq 92 h^2 \Omega_0 \, \text{eV} . \tag{17.74} $$

This bound was used in §6.2.2 to put constraints on the left-right symmetric model.

A similar treatment leads to the following bounds for neutrinos heavier than an MeV [12]:

$$\left(\frac{1\,\text{GeV}}{m_{\nu_H}}\right)^2 \left(\frac{\tau}{t_0}\right)^{1/2} \leq 0.5h^2 \quad \text{Dirac}$$

$$\left(\frac{1\,\text{GeV}}{m_{\nu_H}}\right)^2 \left(\frac{\tau}{t_0}\right)^{1/2} \leq 0.05h^2 \quad \text{Majorana.} \qquad (17.75)$$

These limits are shown in Fig. 17.2. These limits, as discussed, come from energy density arguments and are valid irrespective of the nature of the decay products. If it happens that the final state contains a photon or charged leptons, stronger bounds may apply, as discussed in §17.6. But such particles in the final state do not necessarily arise since neutrinos can decay to invisible particles.

Two types of invisible decays have been implied by particle physics models: i) $\nu_H \rightarrow 3\nu$ [16] and ii) $\nu_H \rightarrow \nu_e + J$ [17], where J is the Majoron. However, since these decay modes generally tend to fill the universe with relativistic particles, they may interfere with some mechanisms for galaxy formation. The reason for this is that galaxy formation requires gravitation density fluctuations to grow. The presence of free streaming relativistic particles will generally impede this process. This would imply that even the lifetimes for invisible decays must be short enough so that the relativistic final states have enough time to degrade their energy through red-shifting. These kind of bounds, however, do not apply, when the galaxy formation is triggered by superconductive strings. In the later case, the masses and lifetimes are subject only to the general mass density bounds.

17.6 Limits for radiative neutrino decays

If one of the decay products of neutrino decay is an easily detectable particle such as $e^\pm$ or photon, there exists other constraints on the mass and lifetime [13]. There exist a large variety of measurements of the photon background from $\lambda \sim 10^3\,\text{m}$ (radio waves) to $\lambda \sim 10^{-24}\,\text{m}$ (x-rays), as shown in Fig. 17.3. These can be used for the purpose of getting detailed limits [14]. The limits arise only if the neutrinos matter dominate and vary depending on the various ranges of lifetimes. For

light neutrinos, the conditions for matter dominance is $T/m_{\nu_H} \leq 0.1$
and for heavy neutrinos, it is $T/m_{\nu_H} \leq 6 \times 10^{-8}(m_{\nu_H}/1\,\text{GeV})^3$.

- $\tau \gtrsim 3 \times 10^{17}\,\text{s}$

If the lifetime of a heavy neutrino is bigger than the age of the universe,
then its decay is continuously contributing to the photon background
and roughly speaking, the differential photon flux will be given by

$$F_\gamma \simeq \frac{n_\nu t_0}{4\pi\tau} \simeq \frac{10^{29}\,\text{s}}{\tau}\ \text{cm}^{-2}\,\text{s}^{-1}\,\text{sr}^{-1}\,. \tag{17.76}$$

Observations indicate that, for photons of energy E_γ,

$$F_\gamma \lesssim \frac{1\,\text{MeV}}{E_\gamma}\,\text{cm}^{-2}\,\text{s}^{-1}\,\text{sr}^{-1}\,. \tag{17.77}$$

We expect $E_\gamma \simeq m_\nu/2$. This gives

$$\tau \gtrsim \left(\frac{m_\nu}{1\,\text{eV}}\right)\cdot 10^{23}\,\text{s}\,, \tag{17.78}$$

using the number density of neutrinos from Eq. (17.51).

Exercise 17.5 *Show that, if the decay time τ for $X \to Y + \gamma$ is much
larger than t_0, the differential flux of photons from the cosmological decay
of X per unit solid angle in the sky is given by*

$$\frac{dF_\gamma}{dE} = \frac{cn_0}{4\pi\tau}\int_1^\infty dy\,y\,\delta(yE - E_\gamma)\cdot\left(-\frac{dt}{dy}\right)\,, \tag{17.79}$$

*where n_0 is the present number density of the X particles, E_γ is the
energy of the photon in the rest frame of the decaying X, and $y = 1 + z$,
z being the cosmological red-shift. For the path of light in a matter-
dominated universe, show that*

$$\frac{dt}{dy} = -\frac{1}{H_0 y^2}\left(1 - \Omega_0 + \Omega_0 y\right)^{-1/2}\,, \tag{17.80}$$

so that

$$\frac{dF_\gamma}{dE} = \frac{cn_0}{4\pi H_0 \tau E_\gamma}\left(1 - \Omega_0 + \Omega_0\frac{E_\gamma}{E}\right)^{-1/2}\,. \tag{17.81}$$

For neutrinos in the GeV range, we have to use Eq. (17.69) for n_ν.
This gives the theoretical expectation of

$$F_\gamma \simeq \frac{10^{23}\,\text{cm}^{-2}\,\text{s}^{-1}\,\text{sr}^{-1}}{(m_{\nu_H}/1\,\text{GeV})^3\,(\tau_\nu/1\,\text{s})}\,. \tag{17.82}$$

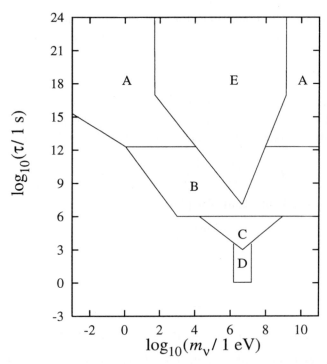

Figure 17.3: The regions marked A, B, C and D are not allowed if the decay products contain photons. The region marked E is ruled out by the mass density of the universe, as shown in Fig. 17.2. (Adapted from Ref. [12])

Using Eq. (17.77) once again, we derive

$$\tau \geq \frac{5 \times 10^{25} \, \text{s}}{(m_\nu / 1 \, \text{GeV})^2} \, . \tag{17.83}$$

As lifetimes becomes shorter, other constraints apply.

- $2 \times 10^{12} \, \text{s} \leq \tau \leq 3 \times 10^{17} \, \text{s}$

The lower limit here corresponds to recombination time, after which matter becomes transparent to radiation. So, any photon from neutrino decay would survive. If $m_\nu \leq T$, the decay products will contribute an energy density which is less than the existing radiation energy density and will escape detection. If however, $m_\nu \geq T$, these photons are detectable. No evidence exists for any such additional energy density.

Therefore, if $m_\nu \leq T$ at the time of decay, this range of lifetimes is forbidden.

- $10^6\,\text{s} \leq \tau \leq t_{\text{rec}}$

Here the lower limit corresponds to the thermalization time which means that if energetic photons appear before this time, they scatter and rescatter against electrons and thermalize. On the other hand, if the neutrinos decay after this time with $m_\nu > T$, it is likely to cause a distortion of the cosmic blackbody spectrum.

- $3\,\text{min} \leq \tau \leq 10^6\,\text{s}$

Here the lower limit corresponds to the time when nucleosynthesis ends. The point here is that if neutrinos matter dominate the universe before $10^6\,\text{s}$ and decay before they dump extra photons into the universe thereby reducing $\eta \equiv n_B/n_\gamma$ from its value at nucleosynthesis. But the value required for successful completion of nucleosynthesis is between (2 to 6) $\times 10^{-10}$. The presently inferred value of η is also about 10^{-9}—10^{-10}. Therefore, dumping of such extra photons would not be acceptable ruling out this range of lifetimes for a neutrino that matter dominates in this regime.

- $1\,\text{s} \leq \tau \leq 3\,\text{min}$

This range of lifetimes corresponds to the duration of nucleosynthesis. If neutrinos matter dominate the universe during this phase, in Eq. (17.24) describing the equilibrium condition, we will have new contributions in the right-hand side leading to higher than desired neutrino decoupling temperature T_D and thus, higher value of the Helium abundance parameter Y. Therefore [15], for $m_\nu \simeq 1$—$10\,\text{MeV}$ (required to matter dominate), $\tau \leq 1\,\text{s}$. There are also constraints from photo-fusion of deuterium by photons produced in radiative neutrino decays if m_ν is in the MeV range [13].

On the whole, we see that severe constraints must be satisfied by mass and lifetimes of unstable neutrinos, if the final state involves a photon or e^+e^-.

17.7 Limits on neutrino properties from nucleosynthesis

In §17.3, we saw that the most recent analysis of nucleosynthesis implies that numbers of extra neutrino species beyond the three known (i.e. ν_e, ν_μ and ν_τ) is less than 0.4. This result has important implications for the properties of neutrinos beyond the standard model, as we discuss now. It must however be pointed out that, these bounds are sensitively dependent on $\Delta N_\nu < 0.4$. For instance, if the upper bound is 0.8, all the bounds disappear.

17.7.1 Limit on interaction of right-handed neutrinos

As a paradigm, we discuss the limit on the strengths of possible new interactions of the right-handed neutrino. We will focus on the neutral current interactions of ν_R mediated by a heavy Z-boson such as the one present in the left-right symmetric model and charged Higgs interactions discussed in connection with the magnetic moment of the neutrino.

The general argument goes as follows. If the new interaction is strong enough that the scattering of ν_R with electrons, protons or neutrons can keep the ν_R in equilibrium at $t = 1\,$s, then the ν_R will contribute to the energy density of the universe like an extra species of neutrino implying $\Delta N_\nu = 1$ in disagreement with nucleosynthesis data. On the other hand, if the ν_R decouples at a higher temperature $T_R > 1\,$MeV, then after decoupling, the ν_R's undergo free expansion. The density relative to photon density, ρ_{ν_R}/ρ_γ remains unchanged with cosmological evolution unless the temperature T_R so high that it exceeds the muon mass. In the later case, $\mu^+\mu^-$ annihilation will increase photon temperature while T_R remains same. The effect of ν_R on the cosmological expansion is then reduced by a factor $[g_\star(T_\nu)/g_\star(T_R)]^{1/3}$ where $g_\star(T_\nu)$ are the number of degrees of freedom of particles contributing to energy density at the neutrino decoupling temperature T_ν whereas $g_\star(T_R)$ denotes the same quantity at T_R.

The nucleosynthesis constraint then implies that

$$\left[\frac{g_\star(T_\nu)}{g_\star(T_R)}\right]^{4/3} \leq 0.4\,, \tag{17.84}$$

which gives $g_\star(T_R) \geq 21.5$. From Table 17.2, we thus infer $T_R \geq$

Table 17.2: [
]Effective number of spin degrees of freedom for different temperatures.

Temperature range	Contributing particles			$g_\star$
	Gauge bosons	Fermions	Scalars	
$1\,\mathrm{MeV} < T < m_\mu$	γ	$\nu_e, \nu_\mu, \nu_\tau, e$		$\frac{43}{4}$
$m_\mu \leq T \leq m_\pi$	γ	above and μ		$\frac{57}{4}$
$m_\pi \leq T \leq T_{\mathrm{QCD}}$	γ	above	$\pi^\pm, \pi^0$	$\frac{69}{4}$
$T_{\mathrm{QCD}} \leq T \lesssim 1\,\mathrm{GeV}$	γ, 8 gluons	above, u, d, s		$\frac{261}{4}$

200—400 MeV, above the deconfinement temperature in QCD.

To infer the interaction strength from this, we denote the cross section of the fastest process involving ν_R and any other fermion by σ and demand that the interaction rate is comparable to the expansion rate given in Eq. (17.26). Let us apply this discussion to the left-right model assuming $m_{\nu_R} \leq 1\,\mathrm{MeV}$. Using Eq. (6.24) for the ν_R-e neutral current coupling via the Z' exchange, we obtain

$$\sigma \simeq \frac{1}{30} G_F^2 \epsilon^2 E^2 \,, \tag{17.85}$$

where $\epsilon \equiv (M_Z/M_{Z'})^2$. Using $\langle E^2 \rangle \simeq 9T^2$ for a thermalized gas, we get $\langle \sigma n v \rangle \simeq G_F^2 \epsilon^2 T^5$. This leads to

$$T_R \simeq \left[\frac{g_\star^{1/2}}{\epsilon^2 G_F^2 M_{\mathrm{Pl}}} \right]^{1/3} . \tag{17.86}$$

Imposing $T_R \geq 200\,\mathrm{MeV}$ then gives $M_{Z'} \geq 36 M_Z$. One can obtain similar bounds for Higgs mediated interactions as well. If all three neutrinos are of Dirac type and $m_{\nu_\tau} < 1\,\mathrm{MeV}$, then we get $g_\star(T_R) > 50$, which still implies $T_R > 200\,\mathrm{MeV}$ as before. So, no new constraint emerges for this case.

17.7.2 Neutrino mass

The gauge interactions distinguish between the left and the right *chiralities* of the neutrino, ν_L and ν_R. If neutrinos have mass, these are not the same as the helicity eigenstates ν_- and ν_+:

$$\nu_- \approx \nu_L + \frac{m}{E}\nu_R \,, \qquad \nu_+ \approx \nu_R - \frac{m}{E}\nu_L \,. \tag{17.87}$$

Thus, if the mass is large enough, even the ν_+ can thermalize owing to the small admixture of ν_L in it [27]. This can violate the bound of Eq. (17.84). The cross sections of reactions with only one ν_R in the initial or final states is now given by

$$\sigma \sim (m/E)^2 G_F^2 E^2 = G_F^2 m^2 \,. \tag{17.88}$$

So, in place of Eq. (17.27), we now obtain

$$G_F^2 m_\nu^2 T_+^3 \approx \sqrt{g_*(T_+)} \times T_+^2/M_P \,, \tag{17.89}$$

Using $g_*(T_+) \geq 21.5$ and $T_+ \geq 200$ MeV as before, we get

$$m_\nu \leq 160 \text{ keV} \,. \tag{17.90}$$

This bound is useless for the electron neutrino because the direct experimental bounds are many orders of magnitude smaller. In the light of the mass bound given in Eq. (17.52), this might seem irrelevant for ν_μ and ν_τ as well. But this is not so, because here we have *not* assumed the neutrinos to be surviving to the present era. They need to be present only at the time of nucleosynthesis.

More detailed calculations, taking all different modes of scattering into account, produces [28] the upper limits of 480 keV for ν_μ and 740 keV for ν_τ.

17.7.3   Neutrino magnetic moment

The magnetic moment interaction changes chirality. Thus, a magnetic moment is bounded for the same reason the mass is [29]. The magnetic moment interaction implies a neutrino effective coupling with the photon. Through this, it can interact with the electron. The cross-section of $\nu_L e \leftrightarrow \nu_R e$ mediated by a photon is given by

$$\sigma \sim \alpha^2 \mu_\nu^2 \,. \tag{17.91}$$

Using this in place of Eq. (17.88) and following the same steps, we now obtain

$$\mu_\nu \leq 8 \times 10^{-11} \mu_B \,, \tag{17.92}$$

where μ_B is the Bohr magneton.

17.8 Neutrinos and dark matter in the universe

Earlier in this chapter, we mentioned that the total energy density of the universe, ρ, has an upper bound of ρ_c from various considerations. Inflationary models predict that ρ should in fact equal ρ_c. Luminous baryonic matter in stars account for $\frac{1}{10}\rho_c$ at most. Thus, if one believes the inflationary models, an immediate conclusion is that 90% or more energy of the universe is in the form of non-luminous or dark matter. Even without the inflationary models, there are various indications of dark matter [19] in the universe, as we briefly discuss in this section.

If the neutrinos are massless, their energy density is comparable to that of the photons and hence negligible compared to baryons. If they are massive and stable (or $\tau \gg t_0$ if unstable), we showed in §17.4 that considerations of the total energy density of the universe constrain their masses to be less than $92h^2\Omega_0$ eV or greater than about 2 GeV. If the masses are close to these extreme values, neutrino mass can be the dominant source of energy density of the universe. The possibility of the existence of a neutrino barely heavier than 2 GeV has been ruled out by measurements of the decay width of the Z boson. In this section, we consider the implications of the other possibility, i. e., of $m_\nu \simeq 92h^2\Omega_0$ eV, on the dark matter problem.

17.8.1 Galactic halos and neutrinos

A galaxy is composed of stars moving under the spell of mutual attraction. If one plots the speed of these stars against the distance of the stars from the galactic center, the resulting plot is called the *rotation curve* of the galaxy. In Fig. 17.4, we show some such rotation curves for various types of galaxies.

Galaxies have a central bright region which extend typically upto a distance of 10 kpc from the center. Almost all the mass of a galaxy is concentrated in this region. Beyond that, there are a few stars, upto a distance of several tens of kpc, which are still gravitationally bound to the galaxy. The outer parts of the rotation curves are obtained by observations on these lone stars and the diffused Hydrogen gas in these parts. However, the curves show that the velocities are roughly constant for stars as far as 40 kpc for some galaxies. This presents a puzzle for

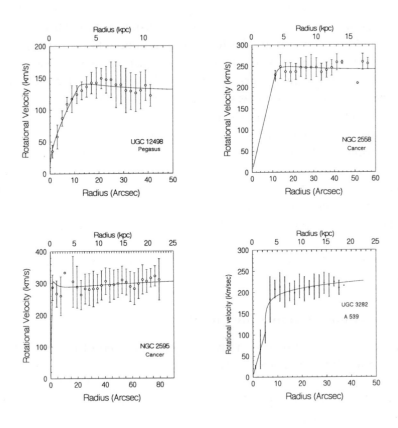

Figure 17.4: Rotation curves of various galaxies. (Courtesy R. D. Prabhu.)

the reason described below.

Consider, for example, a particle of mass m, rotating in a spherically symmetric mass distribution with an orbit of radius r. It speed v is given by

$$\frac{GM(r)m}{r^2} = \frac{mv^2}{r},$$ (17.93)

where $M(r)$ is the total mass within a sphere of radius r. Thus

$$v^2 = \frac{GM(r)}{r}.$$ (17.94)

If luminous matter constitutes all matter in a galaxy, $M(r)$ hardly increases for $r > 10\,\text{kpc}$. Thus, from Eq. (17.94), we would expect v to go down as $r^{-1/2}$. On the contrary, the rotation curves indicate an almost constant v, implying $M(r) \propto r$. In other words, there is indication that galactic matter extends to radii several times larger than the radius of luminous matter. It is not known what constitutes this dark halo of galaxies.

A measure of this halo could be obtained from the *mass to light ratio* M/L, where M is the mass determined from the rotation curves using Eq. (17.94) and L is the absolute luminosity. For typical galaxies, one obtains

$$(M/L)_{\text{gal}} \sim 10h(M/L)_\odot,$$ (17.95)

where the subscript $\odot$ stands as usual for the sun, and h parameterizes the expansion rate of the universe through Eq. (17.41). The halo becomes more prominent if one considers bound objects larger than galaxies. For example, in small groups of galaxies, one typically obtains

$$(M/L)_{\text{small group}} \sim 100h(M/L)_\odot.$$ (17.96)

This means that bulk of the matter is dark.

We know that a particle can be bound by the gravitational pull of a body of mass M and radius R if its velocity is less than the escape velocity

$$v_{\text{esc}} = \sqrt{\frac{2GM}{R}}.$$ (17.97)

For typical galaxies with $M \sim 10^{11} M_{\odot}$ and $R \sim 10\,\text{kpc}$, we get $v_{\text{esc}} \sim 10^{-3}$. If neutrinos have mass, they become non-relativistic at some era of cosmological evolution and can be bound to the galaxies, thereby providing the non-luminous halo.

For this to happen, there is a strong bound to be satisfied [21]. It comes from the fact that the neutrinos are fermions, and therefore there cannot be more than one neutrino in unit volume of phase space. Since the maximum velocity is v_{esc}, the maximum momentum is is $m_{\nu} v_{\text{esc}}$ for a neutrino of mass m_{ν}. Within a sphere of radius R, the phase space volume is thus given by

$$\frac{4\pi}{3} R^3 \cdot \frac{4\pi}{3} (m_{\nu} v_{\text{esc}})^3 , \qquad (17.98)$$

which also gives the maximum number of neutrinos in this volume. The total mass of these neutrinos is then bounded by

$$M < \left(\frac{4\pi}{3}\right)^2 m_{\nu}^4 \, (2GMR)^{3/2} , \qquad (17.99)$$

using the expression for the escape velocity. Thus

$$m_{\nu} \gtrsim \left(G^3 R^3 M\right)^{-1/8} . \qquad (17.100)$$

As we saw earlier, non-luminous matter constitutes bulk of the mass. Hence, a good approximation for M in this equation could be the total mass of the galaxy, which is typically or order $10^{11} M_{\odot}$. Using $R \simeq 10\,\text{kpc}$, we obtain

$$m_{\nu} \gtrsim 10\,\text{eV} . \qquad (17.101)$$

This lower bound, though marginally acceptable for ν_e, is well below the experimental upper bounds on ν_{μ} and ν_{τ} masses shown in Eq. (3.1). Therefore, ν_{μ} or ν_{τ} can easily form the halo of a galaxy.

While the above comments are true for typical galaxies, some problems arise [22] if one considers a special class of galaxies which are smaller ($R \sim 0.1\,\text{kpc}$) and lighter ($M \sim 10^7 M_{\odot}$). For these dwarf galaxies, Eq. (17.100) gives $m_{\nu} > 500\,\text{eV}$. Although this produces no conflict with laboratory bounds on $m_{\nu_{\mu}}$ and $m_{\nu_{\tau}}$, it violates the cosmological bounds on stable neutrino masses.

17.8.2 Galaxy formation and neutrinos

In the big bang model, the universe is assumed to be homogeneous and isotropic. Observations of the microwave background indicate that this is probably true at large scale. However, the assumption of homogeneity is certainly not true at smaller scales where we see stars and galaxies or clusters of galaxies, with vast empty space in between. One can then ask how these various structures were formed if the early universe was a homogeneous soup of particles.

The standard approach to attack this problem is as follows. One assumes that the early universe had small density fluctuations at all scales. The origin of these fluctuations are undefined in the standard big bang model, but can be explained in inflationary models. Once the fluctuations are there, they might either grow or die, depending on the size of the fluctuation and the dynamics of the universe. If they happen to grow, we get some structures in the universe.

Let us see the implications of such analysis on a baryon dominated universe. For $T \gtrsim 1\,\mathrm{eV}$, atoms cannot form, as indicated in Table 17.1. In that era, protons and electrons are in thermal equilibrium with photons, and their interaction hinders growth of fluctuations at galactic scales. At the era when atoms form (called the *recombination era*), the density fluctuation of the photons must be related to that of the baryons. Observations of the microwave background radiation indicates $(\delta\rho/\rho)_\gamma \lesssim 10^{-4}$. Starting from this value at the recombination era, the baryonic perturbation can grow. However, the universe is matter dominated in this era, and perturbations grow as the scale factor a in a matter dominated universe. Since $z \sim 10^3$ at the recombination era, starting from the value of order 10^{-4} at that era, the density perturbation can grow only upto $\sim 10^{-1}$. However, since the intergalactic space is nearly empty, $\delta\rho/\rho \simeq 1$ in the present universe. This discrepancy clearly indicates that non-baryonic matter plays a very important role in the dynamical evolution of the universe.

If massive neutrinos dominate the energy density of the universe, the scenario of structure formation is quite different. As long as the neutrinos are relativistic, no fluctuations can grow [23]. They can *free-stream* and wash out perturbations at all scales within the horizon (which has a radius t at time t). This free-streaming stops at the era $T \sim m_\nu$ when the neutrinos become non-relativistic. In order to survive the free-streaming, a fluctuation in neutrino density must be larger in linear dimension than d_ν, which is the horizon size at the era $T \sim m_\nu$. Using Eq. (17.22), we see that $d_\nu \sim M_{\rm Pl}/m_\nu^2$. Since the energy density at that era is of order T^4, i.e. m_ν^4, the minimum mass that can survive freestreaming [23] is given by $M_J \sim d_\nu^3 m_\nu^4 \sim M_{\rm Pl}^3/m_\nu^2$. For m_ν in the range of tens of eV, this gives $M_J \sim 10^{16} M_\odot$. This is the typical mass of a supercluster of galaxies. Thus, superclusters would be the first structures to form in a neutrino dominated universe.

Galaxies typically have masses of order $10^{11} M_\odot$. They can form subsequently, when the supercluster size structures evolve to develop gravitational instabilities [24]. Numerical simulations, however, indicate that it takes a large time for these instabilities to grow, so that the galaxies can form only at time when [25] $z < 2$. However, various ways of measuring the ages of galaxies indicate that galaxies were formed at $z > 3$. Mainly because of this discrepancy, the possibility of a universe dominated by massive neutrinos is disfavored today.

There are of course, some assumptions going into the last statement. One of them is that the contribution to the energy density of the universe from any particles other than neutrinos is negligible. This may not be true. Neutrinos may be only one major component in a complicated distribution of energy density. The other assumption is that the dominant interaction of neutrinos is that mediated by the weak gauge bosons. This may not be true. There can be Higgs boson mediated interations between neutrinos which are stronger than the weak gauge interactions [26]. In this case also, the neutrinos can be viable candidates for dark matter.

While the present thinking tends to disfavor massive neutrinos as the sole contributor to the dark matter of the universe, a popular scenario that agrees with the present data on structure on various scales is the so-called "cocktail" picture where 70% of the matter in the universe is believed to be in the form of cold dark matter (CDM), 20% in the form of hot dark matter (HDM) and 10% in the form of baryons. The massive neutrinos are of course the obvious candidates for HDM in which case taking Hubble parameter to be $50\,{\rm km\ s^{-1}\,Mpc^{-1}}$, leads to a neutrino mass of 4-5 eV. In this model, the cosmological constant is assumed to be zero. This picture satisfies the prediction of inflationary models that $\Omega_0 = 1$. If one tries to fit this scenario with various informations from the solar and atmospheric neutrino data, one can then conclude that the total mass of $4-5$ eV required by the structure data is divided equally between ν_μ and ν_τ or even between all three known species [30] leading to a scenario of nearly degenerate neutrinos.

Structure formation can produce important bounds for unstable neutrinos. The point is that, the decay products are in general much lighter than the parent particle, and so they are relativistic at the time the decay occurs. So they can make the universe RD. In the RD era, density fluctuations cannot grow, so structures cannot form easily, as discussed earlier. If the decay products have some mass m_{dk}, this RD phase will end when their energies will red-shift to a value around m_{dk}, i.e., at a time t_{NR} given by

$$m_\nu \left(\frac{\tau_\nu}{t_{NR}} \right)^{1/2} \approx m_{dk} \,. \tag{17.102}$$

After t_{NR}, the universe is matter dominated (MD) again. In an MD era, the time-evolution of the scale factor is given in Eq. (17.12), which enables us to write

$$1 + z_{NR} \equiv \frac{a_0}{a_{NR}} = \left(\frac{t_0}{t_{NR}} \right)^{2/3} \,. \tag{17.103}$$

Now require [31, 32] that t_{NR} must have been sufficiently early, say earlier than some specific time t_*, so that structures can form between then and now. This means $z_{NR} > z_*$. Then

$$t_{NR} \leq \frac{t_0}{(1 + z_*)^{3/2}} \,, \tag{17.104}$$

or, using Eq. (17.102),

$$m_\nu \left(\frac{\tau_\nu}{t_0} \right)^{1/2} \leq \frac{m_{dk}}{(1 + z_*)^{3/4}} \,. \tag{17.105}$$

To use it in a specific situation, consider the possibility of $\nu_\tau \rightarrow 3\nu_\mu$, where these ν_μ's are stable. But in this case, it should be emphasized that the bound on the mass of ν_μ is much stronger than that given in Eq. (17.52). This is because the number density of the ν_μ's in the present universe not only has a contribution from the primordial neutrinos, but also from the decay of ν_τ. Since each ν_τ contributes three ν_μ's, the total number density of ν_μ is 4 times the number given in Eq. (17.51). Thus, the mass of these neutrinos is bounded by [11]

$$m_\nu < 23 h^2 \Omega_0 \text{ eV} . \tag{17.106}$$

As for z_*, one can use a conservative estimate of $z_* = 5$ since quasars have been observed at such red-shifts. Then Eq. (17.105) simplifies to

$$m_\nu \left(\frac{\tau_\nu}{t_0}\right)^{1/2} \leq 6 h^2 \Omega_0 \text{ eV} < 3 \text{ eV} , \tag{17.107}$$

using Eq. (17.43) at the last step.

Part V

Epilogue

Chapter 18

Summary and outlook of neutrino physics

18.1   Introduction

In this book, we have given a comprehensive summary of the theory, phenomenology as well as cosmology of a massive neutrino. In doing so, we have painstakingly avoided the interesting details about some of the present experimental results which have not been conclusively established and hence we have refrained from engaging in extensive discussion about their possible theoretical implications. This is not meant to reflect any kind of opinion on the part of the authors with respect to their validity — but rather that neutrino experiments are extremely difficult and one must wait for careful reconfirmation of any experimental result. Recent examples of exciting results which failed this "reconfirmation" criterion are the indications of a 17 keV neutrino, the high y-anomaly, possible indication of a neutrino mass in the tritium endpoint beta spectrum etc. On the other hand, current neutrino research by its very nature deals with any and all of the results in print since they always lead to new insight. A book however has a different goal and needs to contain discussions and results which are likely to have long term validity while at the same time not ignoring the current excitements in a field. This requires a very delicate balance. To solve this problem we have added this chapter to present an overview of the present puzzles in neutrino physics suggested by all available data and a glimpse into their implications for physics beyond the standard model. We would like the readers to understand that there are ongoing and planned experiments which will either confirm or refute the experimental results summarized in this chapter. This could then lead to revision of the consequent theoretical ideas.

One such result already alluded to in Ch. 9 is the possible end point anomaly in the electron spectrum in the tritium beta decay. Most physicists believe that this is due to lack of our understanding of the atomic effects in the actual material that contains the tritium nucleus. If this effect is however genuine, unconventional interpretations in terms of either a tachyonic neutrino [1] or a neutrino cloud [2] "hovering" around the earth due to some new long range forces [3] can be given. We will not dwell on this issue any further in this chapter. Instead we will concentrate on the experimental results that indicate a nonzero neutrino mass in what follows.

As already mentioned earlier in the book, there are several different observations involving neutrinos which receive a plausible and satisfactory explanation if the neutrinos are massive. First is the well-known solar neutrino deficit [4], observed by four different experiments [5]. Second is the deficit of muon neutrinos relative to electron neutrinos produced in the atmosphere, as measured by three experiments [6]. Third is the reported evidence for $\hat{\nu}_\mu$ to $\hat{\nu}_e$ oscillation from the Los Alamos Liquid Scintillation Neutrino Detector (LSND) experiment [8]. Finally, there is the likely need for a neutrino component of the dark matter of the universe to understand the structure and density on all distance scales [12]. Since the highly successful standard model of particle physics predicts zero mass for all the neutrinos, confirmation of any one of the above observations by ongoing and future experiments will already be a major step towards decoding the nature of new physics beyond the standard model. If however all the above findings are substantiated in future, one can reasonably expect the nature of this new physics to fall into only a very few categories. One case involves the mixing among all three neutrinos and another requires the existence of a new ultra-light sterile neutrino. Several speculative scenarios which explain the lightness of the sterile neutrino are presented including one that involves a shadow universe which has identical particle and force content as the familiar universe but with the weak scale being somewhat higher.

18.1.1 Solar neutrino deficit

For massive neutrinos which can oscillate from one species to another, the solar electron neutrino observations [5] can be understood if the neutrino mass differences and mixing angles fall into one of the following ranges [4], where the Mikheyev-Smirnov-Wolfenstein (MSW) mechanism is included:

a) small-angle MSW $\quad\begin{cases} \Delta_{ei} \sim 5 \times 10^{-6} - 10^{-5} \mathrm{eV}^2, \\ \sin^2 2\theta_{ei} \sim 7 \times 10^{-3} \end{cases}$

b) large-angle MSW $\quad\begin{cases} \Delta_{ei} \sim 9 \times 10^{-6} \, \mathrm{eV}^2, \\ \sin^2 2\theta_{ei} \sim 0.6, \end{cases}$ $\qquad$ (18.1)

If the solar ν_e's oscillate into sterile neutrinos, the MSW effect is different from the ν_e to ν_μ case and the large angle solution is no more allowed. The above results are based on the approximation that only two of the neutrino species are involved in the oscillation.

18.1.2 Atmospheric neutrino deficit

The second set of experiments indicating non-zero neutrino masses and mixings has to do with the missing atmospheric muon neutrinos. As already discussed in Ch. 10, observations of $\mu^\pm$ and $e^\pm$ indicate a far lower value for ν_μ and $\hat{\nu}_\mu$ than suggested by naïve counting arguments which imply that $N(\nu_\mu + \hat{\nu}_\mu) = 2N(\nu_e + \hat{\nu}_e)$ [6]. If one assumes that the oscillation of ν_μ to ν_τ provides an explanation of these results, then to fits to both the sub-GeV and multi-GeV data require that [7]

$$\Delta_{\mu\tau} \approx 0.025 \text{ to } 0.005 \, \mathrm{eV}^2, \ \sin^2 2\theta_{\mu\tau} \approx .6 \text{ to } 1. \qquad (18.2)$$

18.1.3 Results from the LSND experiment

Recently, the LSND collaboration has published the results of their search for $\hat{\nu}_\mu$ to $\hat{\nu}_e$ oscillation using the liquid scintillation detector at Los Alamos. Combining their results which indicate a positive result with the negative results by the E776 group and the Bugey reactor data, one can conclude that a mass difference squared between the ν_e and the ν_μ lies between

$$0.27 \, \mathrm{eV}^2 \leq \Delta \leq 2.3 \, \mathrm{eV}^2, \qquad (18.3)$$

with points at 6 and 10 eV^2 also perhaps allowed. The LSND group has also recently announced that they have evidence for oscillation in the ν_μ to ν_e channel with roughly similar mass difference and mixing angles [9]. The recent results from a CCFR collaboration [10] and the NOMAD [11] seem to rule out the high mass difference square region, although these also remain to be confirmed.

18.1.4 Hot dark matter

There is increasing evidence that more than 90% of the mass in the universe must be detectable so far only by its gravitational effects. This dark matter is likely to be a mix of $\sim$ 20% of particles which were relativistic at the time of freeze-out from equilibrium in the early universe (hot dark matter) and $\sim$ 70% of particles which were non-relativistic (cold dark matter). Such a mixture gives the best fit of any available model to the structure and density of the universe on all distance scales, such as the anisotropy of the microwave background, galaxy-galaxy angular correlations, velocity fields on large and small scales, correlations of galaxy clusters, etc. A very plausible candidate for hot dark matter is one or more species of neutrinos with total mass of $m_{\nu_H} = 92h^2 F_H \Omega = 4.8$ eV, if $h = 0.5$ (the Hubble parameter in units of 100 km·s^{-1}·Mpc^{-1}, as defined in Eq. (17.41)), $F_H = 0.3$ (the fraction of dark matter which is hot), and $\Omega = 1$ (the ratio of density of the universe to closure density).

It is usually assumed that the ν_τ would supply the hot dark matter. However, if the atmospheric ν_μ deficit is due to $\nu_\mu \to \nu_\tau$, the ν_τ alone cannot be the hot dark matter, since the ν_μ and ν_τ need to be closer to each other in mass. It is interesting that instead of a single $\sim$ 4.8 eV neutrino, sharing that $\sim$ 4.8 eV between two or among three neutrino species provides a better fit to the universe structure and particularly a better understanding of the variation of matter density with distance scale [12].

It is worth noting that an equally popular picture adopts the hypothesis that there is a large cosmological constant ($\Omega_\Lambda = .8$ or so) in a low density baryon plus CDM universe to make up $\Omega = 1$ [13]. This has been inspired by reported large values ($h = 0.7\text{-}0.8$ or so) of the Hubble parameter from several observations [14] which have hard time fitting the age of the universe (e.g. from globular clusters) with $\Omega = 1$ without a cosmological constant. There are however other observations that give a lower value for h ($h \simeq .5$). The final verdict on the dark matter picture of the universe will therefore have to wait. It is nevertheless heartening that there is a compelling case for a neutrino mass in the eV range from structure formation in the universe.

In understanding the detailed implications of these data for physics beyond the standard model, one must also take into account other constraints on neutrinos, from nucleosynthesis, the Heidelberg-Moscow [16] $\beta\beta_{0\nu}$ experiment searching for the Majorana mass of the neutrino using enriched ^{76}Ge and the synthesis of heavy elements supposedly by the rapid neutron capture process (the so-called *r-process*) around supernovae [17].

18.1.5 Other constraints

(i) While the Z width limits the number of weakly interacting neutrino species to three, the nucleosynthesis limit [15] of about 3.3 on the number of light neutrinos is more useful here, since it is independent of the neutrino interactions. Invoking a fourth neutrino, ν_s, which is sterile, meaning it does not have the usual weak interaction, must be done with parameters such that it will not lead to overproduction of light elements in the early universe. For example, the atmospheric ν_μ problem cannot be explained by $\nu_\mu \to \nu_s$, since $\sin^2 2\theta_{\mu s} \approx 0.5$ is too large for the $\Delta_{\mu s}$ involved, and that ν_s would have been brought into equilibrium in the early universe. On the other hand, the solar ν_e puzzle can be explained by $\nu_e \to \nu_s$ for either the small-angle MSW or the vacuum oscillation solutions, but not for the less favored large-angle MSW solution.

(ii) The Heidelberg-Moscow ^{76}Ge experiment [16] has been conducting a high precision search for neutrinoless double beta decay for the past several years and have at present set the most stringent upper limits on the effective Majorana mass of the neutrino: $\langle m_\nu \rangle \leq 0.56$ eV.

(iii) It has been pointed out that in minimal model with three massive neutrinos, supernova r-processes provide a very stringent constraint on the neutrino mixings for eV mass range or higher. The origin of this constraint can be understood as follows. Inside the supernova, the MSW phenomenon enhances the conversion of the muon neutrinos (which have higher energy) to electron neutrinos if the mass difference square $\Delta \geq 4$ eV^2 while leaving the $\hat{\nu}_\mu$'s unaffected. The newly born high energy ν_e's deplete the neutron content of the supernova environment via the reaction $\nu_e + n \to e^- + p$. This reduction of the neutron content slows down the r-process making it difficult to understand the heavy element abundance of the present universe. This result crucially hinges on the assumption that $m_{\nu_\mu} \geq m_{\nu_e}$ and that there are neutrinos that ν_μ mixes with. In fact in the presence of sterile neutrinos, its mixing with ν_μ can lead to MSW enhancement of ν_μ to ν_s conversion deeper in the supernova providing a way out of this constraint [18].

18.2 Neutrino mass textures consistent with data

In discussing the neutrino mass textures in this section, we will assume that all the neutrinos are Majorana particles, since it is easier to understand the smallness of Majorana masses of neutrinos within the framework of grand unified theories. Before going to a detailed discussion of the allowed mass matrices, let us note two generic requirements for the allowed mass matrices dictated by the data:*(i) at least two neutrinos must be degenerate in mass; and (ii) there is a very compelling case for the existence of a sterile neutrino in the present data.*

18.2.1 Are neutrinos degenerate?

If only two of the above hints (either solar and atmospheric data or solar and HDM) are taken seriously, then one can maintain a hierarchical picture for neutrino masses i.e. $m_{\nu_e} \ll m_{\nu_\mu} \ll m_{\nu_\tau}$. Such a pattern emerges very naturally in one class of the see-saw models (see below). However, if we take any three of the above four hints for neutrino masses, then we must have at least two neutrinos degenerate [20]. To see the case for a sterile neutrino, let us first note that it is not easy to write down a neutrino mass matrix within the three generation picture that can accommodate all the above observations as well as constraints. The main obstacle comes from the conflict between the LSND data and the MSW resolution of the solar neutrino puzzle. The first requires that the $\Delta_{\nu_e\nu_\mu}$ is in the eV range which is much larger than the mass difference required to explain the solar neutrino puzzle. If we ignore the LSND data, the solar, atmospheric data and the HDM neutrino can be accommodated in a three neutrino scenario by assuming that all three neutrinos are degenerate in mass [20]. One possibility [19] has been advocated recently using a variant of this, to accommodate the LSND results in this picture provided the LSND Δ is chosen to be around $.3\,\mathrm{eV}^2$. First point to note is that since the solar neutrino puzzle requires that $\Delta_{e\mu} \simeq 10^{-5}\,\mathrm{eV}^2$, to understand the LSND results in this scenario, one must use the complete three neutrino oscillation keeping all mixing angles [22]. This requires first that ν_e-ν_τ mixing angle is not too small. Secondly, we must have $\Delta_{\mu\tau}$ be $\approx .3\ eV^2$. Thus the oscillation frequency is determined by ν_e-ν_τ mass difference. The main problem for this scenario comes from the atmospheric neutrino data, since the original analysis of the Kamiokande sub-GeV and the multi-GeV data by the Kamiokande group excludes $\Delta \geq .1\,\mathrm{eV}^2$ at 90% confidence level (c.l.). While at its face a value of $m_{\nu_e} \simeq 1.6$ eV may appear to be in conflict with the neutrinoless double beta decay limit [16], one can hide under the uncertainties of nuclear matrix element calculations which typically could be as much as a factor of $2-3$. As the precision in $\beta\beta_{0\nu}$ search improves further (say to the level of 0.1 eV), nuclear matrix element uncertainties cannot come to the rescue and this mass texture will then be ruled out. One can write the neutrino Majorana mass matrix for this case as follows:

$$M = \begin{pmatrix} m + \delta_1 s_1^2 & -\delta_1 c_1 c_2 s_1 & -\delta_1 c_1 s_1 s_2 \\ -\delta_1 c_1 c_2 s_1 & m + \delta_1 c_1^2 c_2^2 + \delta_2 s_2^2 & \delta_1 c_1^2 c_2 s_2 - \delta_2 c_2 s_2 \\ -\delta_1 c_1 s_1 s_2 & \delta_1 c_1^2 s_2 c_2 - \delta_2 c_2 s_2 & m + \delta_1 c_1^2 s_2^2 + \delta_2 c_2^2 \end{pmatrix}, \quad (18.4)$$

where $c_i = \cos\theta_i$ and $s_i = \sin\theta_i$, $m = 1.6$ eV; $\delta_1 \simeq 1.5 \times 10^{-5}$ eV; $\delta_2 \simeq .1$ eV; $s_1 \simeq 0.05$; and $s_2 \simeq 0.4$ for the small-angle MSW solution.

18.2.2 The need for a sterile neutrino

We thus see that if the above scenario is ruled out, for instance by the tightening of the double beta decay limit on the Majorana mass of ν_e or by the atmospheric neutrino data, then the only way to understand all neutrino results will be to assume the existence of an additional neutrino species which in view of the LEP data must not couple (or couple extremely weakly) to the Z-boson. We will call this the sterile neutrino. The picture then would be as follows [20, 21]: the solar neutrino puzzle is explained by the ν_e-ν_s oscillation; atmospheric neutrino data would be explained by the ν_μ-ν_τ oscillation. The LSND data would set the overall scale for the masses of ν_μ and ν_τ (which are nearly degenerate) and if this scale is around 2 to 3 eV as is allowed by the data [8], then the $\nu_{\mu,\tau}$ would constitute the hot dark matter of the universe. The mass matrix in this case, in the basis $(\nu_s, \nu_e, \nu_\mu, \nu_\tau)$, would be

$$M = \begin{pmatrix} \mu_1 & \mu_3 & \epsilon_{11} & \epsilon_{12} \\ \mu_3 & \mu_2 & \epsilon_{21} & \epsilon_{22} \\ \epsilon_{11} & \epsilon_{21} & m & \delta/2 \\ \epsilon_{12} & \epsilon_{22} & \delta/2 & m+\delta \end{pmatrix}. \tag{18.5}$$

For simplicity, we set the $\epsilon_{12} = \epsilon_{22} = 0$ and $\mu_2 \ll \mu_1 \simeq 10^{-3}\,\text{eV}$. The ϵ_{11} term is responsible for the ν_e-ν_μ oscillation that can explain the LSND data. The apparent problem for such a scenario comes from the supernova r-process nucleosynthesis. But it has been argued [18] that in such a scenario, the ν_μ can oscillate into the ν_s at a smaller proto-neutron star radius before it reaches the radius where ν_μ to ν_e MSW transition occurs. This may enable one to evade the r-process bound for $\Delta_{e\mu} \geq 4\,\text{eV}^2$. Clearly the crucial test of the sterile neutrino scenario will come when SNO collaboration obtains their results for neutral current scattering of solar neutrinos. One would expect that $\Phi_{\text{cc}} = \Phi_{\text{nc}}$ if the ν_e oscillation to ν_s is responsible for the solar neutrino deficit. There should be no signal in $\beta\beta_{0\nu}$ search. Precision measurement of the energy distribution in charged current scattering of solar neutrinos at super-Kamiokande can also shed light on this issue.

Before proceeding to the discussion of the theoretical implications of the mass textures outlined above, we want to note that if the atmospheric neutrino data is excluded but LSND, HDM and solar neutrino constraints are kept, a theoretical explanation for them can be found also with an inverted mass texture [23] for neutrinos where the $m_{\nu_e} \simeq m_{\nu_\tau} \simeq 2.4\,\text{eV} \gg m_{\nu_\mu}$ and which does not invoke the sterile neutrino. This texture is consistent with the supernova r-process constraints and uses the $\nu_e \to \nu_\tau$ large angle MSW solution to explain the solar neutrino data. This could therefore be tested once super-Kamiokande results for the neutrino energy spectrum as well as the data on day-night variation is in.

18.3 Implications for higher unification and two types of see-saw mechanism

In this section we address the question of what implications the small nonzero neutrino masses and in particular any of the scenarios discussed above have for the nature for the nature of new physics beyond the standard model. To start with let us remind the reader that in the standard model the presence of an exact global $B - L$ symmetry combined with the absence of the right handed neutrino leads to zero mass for all neutrinos. As has already been emphasized, the simplest way to generate a nonzero neutrino mass is to add three right handed neutrinos N_i, one per generation. It is easy to see that as soon as the N_i are included, the maximal anomaly-free gaugeable symmetry becomes $SU(2)_L \times SU(2)_R \times U(1)_{B-L} \times SU(3)_c$ which can eventually lead to an $SO(10)$ grand unification of fermions. As has been shown in Ch. 6, this class of models provide the most natural framework for describing the neutrino masses [24, 25]. One obtains the see-saw matrix for the neutrinos [26, 24, 25]:

$$M = \begin{pmatrix} 0 & m_f \\ m_f & M_N \end{pmatrix} \tag{18.6}$$

The diagonalization matrix leads to the generic formula for neutrino masses

$$m_{\nu_i} \simeq \frac{(m_f)^2}{M_{N_i}} \tag{18.7}$$

This formula has two interesting implications. The first is that the neutrinos, which are now necessarily Majorana fermions, have masses which are suppressed compared to the masses of the charged fermions of the corresponding generation. Secondly, the neutrino masses show a generationwise hierarchical pattern linked to the square of the masses of the charged fermions of the corresponding generation (i.e. $m_{\nu_e} \ll m_{\nu_\mu} \ll m_{\nu_\tau}$). We will call this the type-I see-saw formula.

It was pointed out in [25] that when the spontaneous symmetry breaking of the left-right model (or its $SO(10)$ grand unified version) is carefully analyzed, one actually gets a modified neutrino mass matrix given by

$$\begin{pmatrix} \frac{\lambda v_W^2}{v_R} & m_f \\ m_f & M_N \end{pmatrix} \tag{18.8}$$

Diagonalization of the above mass matrix leads to what we call the type-II see-saw formula for the neutrino masses:

$$m_{\nu_i} \simeq \frac{\lambda v_W^2}{v_R} - \frac{m_f^2}{M_{N_i}} \qquad (18.9)$$

Note that the first term in the above formula is practically generation independent. Therefore, the neutrino mass pattern in this case is not hierarchical and could lead to a nearly degenerate spectrum as has been advocated in the previous section.

There are conditions under which the type-II see-saw formula reduces to a type-I see-saw formula [27]: for instance when the discrete parity symmetry of the left-right or $SO(10)$ models is broken at a scale higher than the $SU(2)_R \times U(1)_{B-L}$ gauge symmetry, then the first term in the type-II see-saw formula is replaced by $\lambda v_W v_R^2 / M_P^2$ (M_P being the scale of discrete parity breaking), which can clearly be arranged to yield the hierarchical mass pattern. Another class of models where the type-I see-saw can emerge are some supersymmetric models with restricted Higgs representations.

A very interesting point worth emphasizing here is that if we look at the typical masses needed to solve the solar as well as the atmospheric neutrino puzzles and use the see-saw formula to find the scale of $B - L$ symmetry breaking, we find that $v_R \approx 10^{12} - 10^{13}$ GeV. It may be more than a mere coincidence that the $B - L$ breaking scale of $v_R \sim 10^{13}$ GeV emerges naturally from constraints of $\sin^2 \theta_W$ and α_s in non-supersymmetric $SO(10)$ grand-unified theories [30], as well as supersymmetric $SO(10)$ [28] theories. In the least it enhances the reason for an $SO(10)$ scenario. It is however possible to construct TeV scale right handed neutrino scenarios [29] where the suppression of the neutrino mass originates from the fact that the Dirac masses are radiatively induced. To summarize this section, it is reasonable to conclude that evidence for a small neutrino mass would indicate the existence of a local $B - L$ symmetry in nature and perhaps even a left-right symmetry, which will be a major new dimension to our understanding of particle physics. Secondly, the generic class of grand unified models where the see-saw mechanism (both type-I and type-II) is naturally implemented are based on the $SO(10)$ GUT group with the type-I see-saw leading to a hierarchical pattern for neutrino masses whereas the type-II leads to a near degenerate pattern. In the next section, we explore whether definite predictions can be made for the neutrino masses and mixing angles in this class of models.

18.4 Predictions from minimal SO(10) grand unification models

If there are only three light neutrinos, it is both economical and elegant to work within simple $SO(10)$ grand unified models. While simple electroweak gauge theories without additional symmetries do not have the capability to predict fermion masses, the assumption of grand unification improves this record somewhat (e.g. the prediction of b-quark mass in $SU(5)$). In the minimal $SO(10)$ models, the neutrino Dirac mass and the up-quark mass matrices become equal since they both arise from the Yukawa couplings of the fermions (which belong to the **16** dim representations) to the Higgs boson in the **10**-dim representation, thereby reducing the number of free parameters. This raises the possibility for a prediction of the neutrino masses in these models. The problem however is that the Majorana mass of the N_i arises from the couplings of the fermions to the $\overline{\textbf{126}}$ dimensional Higgs bosons. Since these couplings are arbitrary, in general no specific predictions can be made. It was however pointed out [31] that in the minimal $SO(10)$ models, the standard model doublets arise from an admixture of the $SU(2)_L$ doublets in **10** and $\overline{\textbf{126}}$ dimensional Higgs bosons. Therefore, the $\overline{\textbf{126}}$ Yukawa couplings (as well as those to **10**) get predicted in terms of the quark, lepton masses and their mixings. This model (which is a realization of the type-I see-saw mechanism) therefore leads to numerical predictions for the neutrino masses and mixings. The reader is referred to the original papers for the detailed predictions for the non-supersymmetric [31, 32] as well as supersymmetric versions [33] of the model. There are actually six solutions depending on the relative signs of the various quark masses. Here we simply want to note that there are predictions in both versions that can accommodate the small angle MSW solution to the solar neutrino puzzle but not the atmospheric nor the LSND nor the HDM neutrino.

There is another class of $SO(10)$ models [34] where additional symmetries are imposed to fix the heavy Majorana mass matrix for the right handed neutrinos and different popular quark mass textures are used for the neutrino Dirac masses. They also implement the type-I see-saw formula and give generic predictions that can accommodate only the small angle MSW solutions to the solar neutrino puzzle.

Finally a different class $SO(10)$ models were studied [35] where the type-II see-saw mechanism was implemented. Using an additional S_4-permutation symmetry on the fermions and the Higgs bosons, it was possible to obtain a realization of the degenerate neutrino mass mixing angle predictions that can solve both the solar as well as the atmospheric neutrino problems.

18.5 Beyond grand unification: into the shadow universe

Once we admit the possibility of light sterile neutrinos, one needs to go beyond simple grand unified models to understand why the sterile neutrino is so light. The reason for this is that the sterile neutrino by definition is an $SU(2)_L \times U(1)_Y$ singlet and therefore is allowed to have an arbitrary mass unless there are some compelling new symmetries that keep it massless. Attempts have been made using additional $U(1)$'s and supersymmetry [36] etc to achieve this goal. But it is perhaps fair to say that there are no compelling motivations for such symmetries. To circumvent such arguments, it was conjectured [37] that there is an exact duplication of the standard model in both the gauge as well as the fermion content i.e. an extra G'_{standard} with $Q', u^{c\prime}, d^{c\prime}, L', e^{c\prime}$ etc. (this adds a new sector to the world of elementary particles, which will be called the shadow sector). It is then clear that we have three extra neutrinos which do not interact with the Z-boson. We further assume that the only interactions that connect the known and the shadow sector is the gravitational interaction.

Within this framework, it is easy to understand that the shadow neutrinos (which will be the sterile neutrinos) are massless in the renormalizable theory for exactly the same reason that the ordinary neutrinos are (i.e. the existence of a $B' - L'$ symmetry in the shadow standard model sector). We may assume that there is a "shadow" see-saw mechanism which operates exactly the same way to give tiny masses to the shadow neutrinos. The next question is how do the shadow (or sterile) neutrinos acquire small masses and mix with the known neutrinos? Here we use the existing lore that all global symmetries are broken by Planck scale effects. It was already pointed out [38] that one can write

Planck scale induced operators such as LH_uLH_u/M_P, $LH_uL'H'_u/M_P$ and $L'H'_uL'H'_u/M_P$ which violate both $B - L$ as well as $B' - L'$ symmetries and after electroweak symmetry breaking in both the sectors lead to ν-ν' mixing. If we now make the additional assumption that $v'_W \simeq 30v_W$, the resulting ν_e-ν'_e mass matrix gives a solution to the solar neutrino puzzle with small mixing angles. When this idea is combined with the postulate that there exists an $L_e + L_\mu - L_\tau$ symmetry (instead of the overall $B - L$ symmetry) that is broken by Planck scale effects, we come up with a neutrino mass matrix that explains all neutrino puzzles using the four neutrino mass texture [20].

The next interesting feature of these models is that if the $m_{\nu_\mu} \simeq m_{\nu_\tau} \simeq 2$ eV or so, then the $m_{\nu'_\mu} \simeq m_{\nu'_\tau} \simeq 2$ keV. Thus $\nu'_{\mu,\tau}$ can qualify as warm (or cool) dark matter of the universe, a possibility which does not appear to have been ruled out present cosmological observations. Such models have many interesting implications for cosmology [39], which we will not go into here.

18.6   Conclusions

The solar, atmospheric and LSND neutrino data, along with a need for some hot dark matter, if all are due to neutrino mass, there are two very profound implications: (i) at least two neutrinos must be degenerate in mass, a feature nor shared by charged fermions and not expected in the minimal SO(10) type models; (ii) there is a very good possibility that there is need for a sterile neutrino. Here we have considered the various implications of these conclusions for physics beyond the standard model, such as the simple $SO(10)$ scenarios and conclude that one needs to go beyond such simple models if all the present indications neutrino mass are correct. We then outline the recent suggestion [37] that a scenario with a shadow universe with identical gauge and fermion structure (but with an asymmetric weak scale) can explain all the neutrino puzzles without the need for any other ingredients.

References

The following abbreviations have been used for names of journals:

(APPh) Astroparticle Physics, (A&A) Astronomy and Astrophysics, (ApJ) Astrophysical Journal, (ApJL) Astrophysical Journal Letters, (ApSp) Astrophysics and Space Science, (ARAA) Annual Reviews of Astronomy and Astrophysics, (ARNP) Annual Reviews of Nuclear and Particle Sciences, (ComNPPh) Comments on Nuclear and Particle Physics, (EuPL) Europhysics Letters, (FPh) Fortschritte der Physik, (JETP) Journal of Experimental and Theoretical Physics, (JETPL) Journal of Experimental and Theoretical Physics Letters, (ModPL) Modern Physics Letters, (MNRS) Monthly Notices of the Royal Society, (NuIns) Nuclear Instruments and Methods, (NPh) Nuclear Physics, (NuCim) Nuovo Cimento, (PhRep) Physics Reports, (PL) Physics Letters, (PR) Physical Review, (PRL) Physical Review Letters, (PPNPh) Progress in Particle and Nuclear Physics, (PTP) Progress of Theoretical Physics, (RepPrPh) Reports on Progress in Physics, (RMP) Reviews of Modern Physics, (SJNP) Soviet Journal of Nuclear Physics, (Usp) Soviet Physics Uspekhi, (ZPh) Zeitschrift fur Physik.

References to Chapter 1

1. References to early literature on the subject can be obtained in K Winter (ed): *Neutrino Physics* (Cambridge University Press 1991).
2. For an English translation of the text of this letter, see Pauli's article in Ref. [1].
3. E Fermi: Ricercha Scient. 2 (1933) 12; ZPh 88 (1934) 161.
4. E C G Sudarshan, R E Marshak: in *Proc. Padua-Venice conf. on mesons and recently discovered particles* (1957);
   R P Feynman, M Gell-Mann: PR 109 (1958) 193.
5. Particle Data Group: *Review of Particle Physics*, PR D54 (1996) 1.
6. For a review, see L Littenberg: Proceedings of *1989 Lepton-Photon Symposium*, ed. R Taylor, World Scientific (1990).
7. R N Mohapatra, J Subba Rao, R E Marshak: PR 171 (1968) 1502;
   B L Ioffe, E P Shabalin: SJNP 6 (1967) 328;

See R E Marshak, Riazuddin, C P Ryan: *Theory of Weak Interactions in Particle Physics* (John Wiley 1969).

8. P Higgs: PL 12 (1964) 132; PRL 13 (1964) 508;

   F Englert, R Brout: PRL 13 (1964) 321;

   G S Guralnik, C R Hagen, T W B Kibble: PRL 13 (1964) 585;

   T W Kibble: PR 155 (1967) 1554.

9. There exist a number of books on gauge theories dealing with the idea of spontaneous breakdown and their applications to weak interactions;

   J C Taylor: *Gauge Theories of Weak Interactions* (Cambridge Univ. Press 1976);

   C Quigg: *Gauge Theories of Weak and Electromagnetic Interactions* (Benjamin Cummings 1983);

   T P Cheng, L F Li: *Gauge Theories of Weak Interactions* (Oxford University Press 1985);

   R N Mohapatra: *Unification and Supersymmetry* (Springer-Verlag, 2nd edition 1991);

   G G Ross: *Grandunified Theories* (Benjamin Cummings 1985);

   L B Okun: *Leptons and Quarks* (North Holland 1984);

   P Frampton: *Gauge Field Theories* (Benjamin Cummings 1987);

   P Renton: *Electroweak Interactions* (Cambridge University Press 1990);

   J F Donoghue, B R Holstein, E Golowich: *Dynamics of the Standard Model* (Cambridge University Press 1992);

   W Greiner, B Müller: *Gauge theory of Weak Interaction* (Springer-Verlag, 2nd edition 1996);

   P D B Collins, A D Martin, E J Squires: *Particle Physics and Cosmology* (Wiley Interscience c1989).

10. G 't Hooft: NPh B33 (1971) 173; NPh B35 (1971) 167.

11. S L Adler: PR 177 (1969) 2426;

    J S Bell, R Jackiw: NuCim A60 (1967) 47.

References to Chapter 2

1. S L Glashow: NPh 22 (1961) 579.

2. A Salam, J C Ward: PL 13 (1964) 168.

3. S Weinberg: PRL 19 (1967) 1264.

4. A Salam: in *Elementary Particle Theory*, ed. N Svartholm (Almquist and Forlag 1968).

5. S L Glashow, J Iliopoulos, L Maiani: PR D2 (1970) 1285.

6. G 't Hooft: NPh B33 (1971) 173; B35 (1971) 167.

7. D Gross, F Wilczek: PRL 30 (1973) 1343;
   H D Politzer: PRL 30 (1973) 1346.

8. A Sirlin: PR D22 (1980) 971; PR D29 (1984) 89;
   W Marciano, A Sirlin: PR D22 (1980) 2695;
   D Kennedy, B W Lynn: NPh B322 (1989) 1;
   D Y Bardin, S Riemann, T Riemann: ZPh C32 (1986) 121;
   F Jegerlehney: ZPh C32 (1986) 425;
   For extensive references, see *Radiative Corrections in $SU(2)_L \times U(1)$ Model*, ed. B W Lynn and J F Wheater, (World Scientific 1984);
   W Hollik: FPh 38 (1990) 165.

9. P Langacker, J Erler: in *Review of Particle Physics*, PR D54 (1996) 1.

10. F J Hasert et al: PL B46 (1973) 121.

11. A Benvenuti et al: PRL 32 (1974) 800.

12. For a review see K Winter: *Neutrinos* ed. H V Klapdor (Springer-Verlag) p. 35.

13. U Amaldi et al: PR D36 (1987) 1385.

14. G Costa et al: NPh B297 (1988) 244.

15. For reviews, see D Schaile, in *Precision Tests of the Standard Electroweak Model*, ed. P Langacker (World Scientific, Singapore, 1995); A Blondel, *ibid.*

16. K Abe et al PRL 73 (1994) 25.

17. F Abe et al: PRL 73 (1994) 225.

18. R C Allen et al: PRL 55 (1988) 2401;
    D Krakauer: Univ. of Maryland PhD. Thesis (1988).

19. F Reines, H S Gurr, H W Sobel: PRL 37 (1976) 315.

20. H Faissner et al: PRL 41 (1978) 213;
    N Armenise et al: PL B86 (1979) 225;
    A M Crops et al: PRL 41 (1978) 6;
    R H Heisterberg et al: PRL 44 (1980) 635;
    F Bergsma et al: PL B147 (1984) 481;
    L A Ahrens et al: PRL 51 (1984) 1514;
    J Blietschau et al: NPh B114 (1976) 189.

21. For details, see, e.g., the book by Renton, Ref. [9] of Ch. 1.

22. G Degrassi, A Sirlin, W J Marciano: PR D39 (1989) 287;
    For experimental studies, see R Allen et al: PRD 43 (1991) 1.

References to Chapter 3

1. Most material introduced in this chapter will be discussed throughout the book. Detailed references will be given in later chapters. For additional references and recent reviews of status of neutrinos in gauge theories, see

   P Langacker, in *Neutrinos* ed. H Klapdor, (Springer-Verlag 1988) p. 71;

   R N Mohapatra: in *Neutrinos* ed. H Klapdor, (Springer-Verlag 1988) p. 117;

   B Kayser, F Gibrat-Debu, F Perrier: *The Physics of Massive Neutrinos*, (World Scientific 1989);

   F Boehm, P Vogel: *Physics of Massive Neutrinos*, (Cambridge University Press 1987);

   P H Frampton, P Vogel: PhRep 82 (1982) 339;

   S M Bilenky, S T Petcov: RMP 59 (1987) 671.

   C W Kim, A Pevsner: *Neutrinos in Physics and Astrophysics*, (Harwood Academic Publishers 1993);

   M Fukugita, A Suzuki (eds.): Physics and Astrophysics of Neutrinos (Springer-Verlag, 1994);

   G Gelmini, E Roulet: RepPrPh 58 (1995) 1207.

2. N G Deshpande: Oregon preprint OITS-107 (1979);

   C Q Geng, R E Marshak: PR D39 (1989) 693;

   K S Babu, R N Mohapatra: PRL 63 (1989) 938;

   R Foot, G C Joshi, H Lew, R R Volkas: ModPL A5 (1990) 95;

   S Rudaz: PR D41 (1990) 2619;

   E Golowich, P B Pal: PR D41 (1990) 3537.

References to Chapter 4

1. E Majorana: NuCim 14 (1937) 171.

2. For other expositions of the properties of Majorana spinors, see e.g., K M Case: PR 107 (1957) 307.

3. Discussion of this section follows the following articles. The notation has been modified in places so that the expressions are independent of the representation of the γ-matrices.

   B Kayser, A Goldhaber: PR D28 (1983) 2341;

   B Kayser: PR D30 (1984) 1023.

4. At the time Majorana wrote his paper, violation of C was not known, so a Majorana spinor could be described as a C eigenstate. The

importance of $\mathcal{CPT}$ was first emphasized, to our knowledge, in the following paper:

J F Nieves: PR D26 (1982) 3152.

5. Y B Zeldovich: Doklady 86 (1952) 505;
   E J Konopinski, H M Mahmoud: PR 92 (1953) 1045.

References to Chapter 5

1. T P Cheng, L F Li: PR D22 (1980) 2860.
2. A Zee: Phys. Lett B93 (1980) 389.
3. L Wolfenstein: NPh B175 (1980) 93.
4. K S Babu: PL B203 (1988) 132.
5. Y Chikashige, R N Mohapatra, R D Peccei: PL B98 (1981) 265, PRL 45 (1980) 1926.
6. G B Gelmini, M Roncadelli: PL B99 (1981) 411.
7. C S Aulakh, R N Mohapatra: PL B119 (1983) 136.
8. S Bertolini, A Santamaria: NPh B310 (1988) 714.
9. D A Bryman, E T H Clifford: PRL 57 (1986) 2787.
10. V Barger, W Y Keung, S Pakvasa: PR D25 (1982) 907.
11. See, e.g., D D Clayton: *Principles of stellar evolution and nucleosynthesis* (Univ. of Chicago Press, 2nd edition 1983).
12. D A Dicus, E W Kolb, V L Teplitz, R Wagoner: PR D18 (1978) 1829;
    M Fukugita, S Watamura, M Yoshimura: PR D26 (1982) 1840;
    A Pantziris, K Kang: PR D33 (1986) 3509;
    R Chanda, J F Nieves, P B Pal: PR D37 (1988) 2714.
13. K Choi, C W Kim: PR D37 (1988) 3225.
14. J Schechter, J W F Valle: PR D25 (1982) 774;
15. J M Cline, K Kainulainen, S Paban: PL B319 (1993) 513.
16. R N Mohapatra, P B Pal: PR D38 (1988) 2226.
17. H Georgi, S L Glashow, S Nussinov: NPh B193 (1981) 297.
18. D Chang, W Y Keung, P B Pal: PRL 61 (1988) 2420.

References to Chapter 6

1. J C Pati, A Salam: PR D10 (1974) 275;
   R N Mohapatra, J C Pati: PR D11 (1975) 566, 2558;
   G Senjanović, R N Mohapatra: PR D12 (1975) 1502.

2. R N Mohapatra, R E Marshak: PL B91 (1980) 222;
   A Davidson: PR D20 (1979) 776.
3. R N Mohapatra, R E Marshak: PRL 44 (1980) 1316.
4. For a detailed review of neutron-antineutron oscillation, see R N Mohapatra, Proceedings of the Workshop on *Fundamental Physics with slow neutrons*, ed. D Dubbers, NuIns A284 (1989) 1.
5. H Georgi, S Weinberg: PR D17 (1978) 275.
6. R N Mohapatra, D P Sidhu: PRL 38 (1977) 667;
   R N Mohapatra, F E Paige, D P Sidhu: PR D17 (1978) 2462.
7. R N Mohapatra, G Senjanović: PRL 44 (1980) 912: PR D23 (1981) 165.
8. L Durkin, P Langacker: PL B166 (1986) 436;
   P Langacker, M Luo: PR D45 (1992) 278.
9. V Barger, J Hewett, T Rizzo: PR D42 (1990) 152.
10. F Abe et al: PR D51 (1995) 949.
11. G Altarelli, R Cassalbuoni, D Dominici, F Feruglio, R Gatto: NPh B342 (1990) 15.
12. M A B Beg, R Budny, R N Mohapatra, A Sirlin: PRL 38 (1978) 1252.
13. A Jodidio et al: PR D34 (1986) 1967.
14. The bounds are based on:

    a) A Jodidio et al: Ref. [13];
    b) F T Calaprice et al: PRL 35 (1975) 1566; B Holstein, S Treiman: PR D16 (1977) 2369;
    c) J van Klinken: NPh 75 (1966) 145;
    d) J Peoples: Ph. D. thesis (Columbia University), Nevis Cyclotron Laboratory report 147 (1966);
    e) R N Mohapatra: PR D34 (1986) 909;
    f) G Beall et al: Ref. [15];
    g) V A Wichers, T R Hageman, J van Klinken, H W Wilschut, D Atkinson: PRL 58 (1987) 1821;
    h) J Deutsch: Proceedings of the workshop on *Breaking of Fundamental Symmetries in Nuclei*, ed. J Ginocchio and S P Rosen (World Scientific 1989);
    i) CDF Colloboration, F Abe et al: PRL 74 (1995) 2900.

For additional references, see :
P Herczeg: PR D34 (1986) 3449;
A S Carnoy, J Deutsch, B R Holstein: PR D38 (1988) 1636;

J F Donoghue, B R Holstein: PL B113 (1982) 382;
I I Bigi, J M Frère: PL B110 (1982) 255.

15. G Beall, M Bander, A Soni: PRL 48 (1982) 848.

16. R N Mohapatra, G Senjanović, M Trahn: Phys. Rev. D28 (1983) 546;
G Ecker, W Grimus, H Neufeld: NPh B229 (1983) 421;
F J Gilman, M H Reno: PR D29 (1984) 937.

17. D Chang, J Basecq, L F Li, P B Pal: PR D30 (1984) 1601;
W S Hou, A Soni: PR D32 (1985) 163;
J Basecq, L F Li, P B Pal: PR D32 (1985) 175.

18. I I Bigi, J M Frère: PL B129 (1983) 469;
G Ecker, W Grimus: NPh B258 (1985) 328.

19. R Barbieri, R N Mohapatra: PR D39 (1989) 1229;
G Raffelt, D Seckel: PRL 60 (1988) 1793.

20. K Kanaya: PTP 64 (1980) 2278.

21. M Gell-Mann, P Ramond, R Slansky: in *Supergravity*, ed. P van Nieuwenhuizen and D Z Freedman (North Holland 1979);
T Yanagida: in Proceedings of *Workshop on Unified Theory and Baryon number in the Universe*, ed. O Sawada and A Sugamoto (KEK 1979).

22. S S Gershtein, Y B Zeldovich: JETP Lett. 4 (1966) 120;
R Cowsik, J Mclelland: PRL 29 (1972) 699.

23. D A Dicus, E W Kolb, V L Teplitz: PR D18 (1978) 1819.

24. U Chattopadhyay, P B Pal: PR D34 (1986) 3444.

25. M Roncadelli, G Senjanović: PL 107B (1983) 59.

26. P B Pal: NPh B227 (1983) 237.

27. P Herczeg, R N Mohapatra: PRL 69 (1992) 2475.

28. S Willis et al: PRL 44 (1980) 522.

29. K Jungman: Talk at the *Beyond 97 conference in Ringberg Castle*, June 8–12, 1997 (to appear in the proceedings).

30. A Dar, S Dado: PRL 59 (1987) 2368.

31. A Boesgard, G Steigman: ARAA 23 (1985) 318.

32. R N Mohapatra, S Nussinov: PR D39 (1989) 1378.

33. W Keung, G Senjanović: PRL 50 (1983) 1427;
B Kayser, N G Deshpande, J Gunion: *Proceedings of 3rd Telemark miniconference on neutrino mass and low energy weak interactions*, p 221.

34. D Chang, R N Mohapatra: PR D32 (1985) 1248.

35. D Chang, R N Mohapatra, M K Parida: PRL 52 (1984) 1072.
36. A Kumar, R N Mohapatra: PL B150 (1985) 191;
    R N Mohapatra, P B Pal: PR D38 (1988) 2226.
37. R N Mohapatra: PL B201 (1988) 517;
    B S Balakrishna, R N Mohapatra: PL B216 (1989) 349.
38. K S Babu, X G He: ModPL A4 (1989) 61.
39. D Chang, R N Mohapatra: PRL 58 (1987) 1600.
40. S Rajpoot: PL B191 (1987) 122;
    A Davidson, K C Wali: PRL 59 (1987) 393;
    K S Babu, R N Mohapatra: PRL 62 (1989) 1079.
41. B S Balakrishna: PRL 60 (1988) 1602;
    B S Balakrishna, A Kagan, R N Mohapatra: PL B205 (1988) 345.

References to Chapter 7

1. R N Mohapatra: *Unification and Supersymmetry* (Springer-Verlag, 2nd edition 1991);
   G G Ross: *Grandunified Theories* (Benjamin-Cummings 1985).
2. H Georgi, S L Glashow: PRL 32 (1974) 438.
3. G B Gelmini, M Roncadelli: PL B99 (1981) 411.
4. Mark II collaboration, G S Abrams et al: PRL 63 (1989) 2173;
   CDF collaboration, F Abe et al: PRL 63 (1989) 720;
   ALEPH collaboration, D Buskulic et al: ZPh C62 (1994) 539;
   L3 collaboration, M Acciarri et al: ZPh C62(1994) 551;
   OPAL collaboration, R Akers et al: ZPh C61 (1994) 19;
   DELPHI collaboration, P Abreu et al: NPh B418 (1994) 403.
5. For ways to construct the spinor and other low dimensional representations of $SO(10)$, see:
   R N Mohapatra, B Sakita: PR D21 (1980) 1062;
   F Wilczek, A Zee: PR D25 (1982) 553.
6. D Chang, R N Mohapatra, M K Parida: PRL 52 (1984) 1072.
7. D Chang, R N Mohapatra, M K Parida: PR D30 (1984) 1052;
   D Chang, R N Mohapatra, J Gipson, R E Marshak, M K Parida: PR D31 (1985) 1718;
8. N G Deshpande, E Keith, P B Pal: PR D46 (1992) 2261;
   R N Mohapatra, M K Parida: PR D47 (1993) 264.

9. M Gell-Mann, P Ramond, R Slansky: in *Supergravity*, ed. P van Nieuwenhuizen and D Z Freedman (North Holland 1979);
   T Yanagida: in Proceedings of *Workshop on Unified Theory and Baryon number in the Universe*, ed. O Sawada, A Sugamoto (KEK 1979);
   R N Mohapatra, G Senjanović: PRL 44 (1980) 912, PR D23 (1981) 165.

10. J F Wilkerson et al: PRL 58 (1987) 2023;
    H Kawakami et al: PL B187 (1987) 198.

11. F Abe et al: PRL 73 (1994) 225.

12. R N Mohapatra, P B Pal: PR D38 (1988) 2226.

13. R N Mohapatra, G Senjanović: ZPh C17 (1983) 53;
    R Holman, G Lazarides, Q Shafi: PR D27 (1983) 995;

14. J E Kim: PRL 43 (1979) 103;
    M Dine, W Fischler, M Srednicki: PL B104 (1981) 199;
    A Zhitnitskii: SJNP 31 (1980) 260;
    M Shifman, A Vainstein, V Zakharov: NPh B166 (1980) 493.

15. E Witten: PL B91 (1980) 81.

16. D Chang, R N Mohapatra: PR D32 (1985) 1248.

17. K S Babu, R N Mohapatra: PRL 70 (1993) 2845;
    D-G Lee, R N Mohapatra: PR D51 (1995) 1353.

18. S Dimopoulos, L Hall, S Raby: PR D47 (1993) 3697;
    Y Achiman, T Greiner: NP B443 (1995) 3;
    G Amelino-Camelia, O Pisanti, L Rosa: NP (Proc. Supp.) B43 (1995) 86.

19. F Gursey, P Sikivie, P Ramond: PL B60 (1976) 117;
    F Gursey, M Serdaroglu: NuCim A65 (1981) 337;
    Y Achiman, B Stech: PL B77 (1977) 389;
    Q Shafi: PL B79 (1978) 301;
    For a review and detailed references, see J L Hewett, T G Rizzo: PhRep 183 (1989) 193.

20. J Breit, B Ovrut, G Segrè: PL B158 (1985) 33;
    A Sen: PRL 55 (1985) 33.

21. R N Mohapatra, J W F Valle: PR D34 (1986) 1642.

References to Chapter 8

1. J Bagger, J Wess: *Supersymmetry and Supergravity* (Princeton University Press 1983);

P C West: *Introduction to Supersymmetry and Supergravity* (World Scientific, 2nd edition 1990).

2. R N Mohapatra: *Unification and Supersymmetry*, Springer-Verlag, Second edition (1991);

3. H Haber, G Kane: PhRep 117 (1984) 76;

  H P Nilles: PhRep 110 (1984) 1.

4. M Grisaru, M Rocek, W Siegel: NPh B159 (1979) 429.

5. *Supersymmetry-96: Theoretical Perspectives and Experimental Outlook*, ed. R N Mohapatra, A Rasin, NP (Proc. Supp.) B52A (1997).

6. R Arnowitt, A Chamsheddine, P Nath: PRL 49 (1982) 970;

  R Barbieri, S Ferrara, C Savoy: PL B119 (1982) 343.

7. M Dine, A Nelson: PR D48 (1993) 1277;

  L Alvarez-gaume, M Claudson, M Wise: NPh B207 (1982) 96.

8. P Binetruy, E Dudas: PL B389 (1996) 503;

  G Dvali, A Pomarol: PRL 77 (1996) 3728;

  R N Mohapatra, A Riotto: PR D55 (1997) 1138.

9. R N Mohapatra: PR D34 (1986) 3457;

  R Kuchimanchi, R N Mohapatra: PR D48 (1993) 4352.

10. C S Aulakh, R N Mohapatra, PL B119 (1982) 136.

11. V Barger, G F Giudice, M Y Han: PR D40 (1989) 2987;

  For a recent review, see G Bhattacharyya: in Ref. [5].

12. S Giddings, A Strominger: NPh B307 (1988) 854.

13. M Francis, M Frank, C S Kalman: PR D43 (1991) 2369;

  K Huitu, J Malaampi, M Raidal: NPh B420 (1994) 449;

  R Kuchimanchi, R N Mohapatra: PRL 75 (1995) 3989.

References to Chapter 9

1. R G H Robertson, D A Knapp: ARNP 38 (1988) 185.

2. F Boehm, P Vogel: *Physics of Massive Neutrinos* (Cambridge Univ Press 1987).

3. V A Lubimov, E G Novikov, V Z Nozik, E F Tretyakov, V S Kosik: PL B94 (1980) 266.

4. V A Lyubimov, E G Novikov, V Z Nozik, E F Tretyakov, V S Kosik, N F Myásoedov: JETP 54 (1981) 616.

5. K E Bergkvist: PL B159 (1988) 408;

  C L Bennett, A B McDonald, P T Springer, T E Chupp, M L Tate: PR C31 (1985) 197.

6. M Fritschi et al: PL B173 (1986) 485; PL B287 (1992) 381.

7. J F Wilkerson: in Proceedings of *Salt Lake City Meeting*, ed. C DeTar, J Ball (World Scientific 1987).

8. J F Wilkerson et al: PRL 58 (1987) 2023.

9. H Kawakami et al: PL B256 (1987) 105.

10. Ch. Weinheimer et al: PL B300 (1993) 210.

11. W Stoeffl, D Decman: PRL 75 (1995) 3237.

12. A I Belesev et al: PL B350 (1995) 263.

13. G J Stephenson Jr., T Goldman: hep-ph/9309308; R N Mohapatra, S Nussinov: PL B395 (1997) 63.

14. A Chodos, V A Kostelecky, R Potting, E Gates: ModPL A7 (1992) 467.

15. For a review and references on this episode, see: F E Wietfeldt, E B Norman: PhRep 273 (1996) 149.

16. Particle Data Group: Review of Particle properties, PR D54 (1996) 1.

17. A Assamagan et al PR D53 (1996) 6065.

18. D Buskulic et al: PL B349 (1995) 585.

19. A de Rújula: NPh B188 (1981) 414; C L Bennett et al: PL B107 (1981) 19.

20. P T Springer, C L Bennett, P A Baisden: PR A35 (1987) 679.

21. N G Deshpande, G Eilam: PRL 53 (1984) 2289.

22. J F Nieves, P B Pal: PR D32 (1985) 1849.

References to Chapter 10

1. B Pontecorvo: JETP 6 (1958) 429; V Gribov, B Pontecorvo: PL B28 (1969) 493.

2. S M Bilenky, B Pontecorvo: PhRep 41 (1978) 225.

3. C Athanasopoulos et al: PRL 77 (1996) 3082.

4. S M Bilenky, S T Petcov: RMP 59 (1987) 671.

5. G S Vidyakin et al: JETP 66 (1987) 243.

6. P Nemethy et al: PR D23 (1981) 262.

7. C Angelini et al: PL B179 (1986) 307.

8. N Ushida et al: PRL 57 (1986) 2897.

9. A E Asratyan et al: PL B105 (1981) 301.

10. F Dydak et al: PL B134 (1984) 281; I E Stockdale et al: ZPh C27 (1985) 53.

11. L Borodovsky et al: PRL 68 (1992) 274.

12. B Zeitneitz et al: PPNPh 32 (1994) 351.
13. S Nussinov: PL B63 (1976) 201;
    B Kayser: PR D24 (1981) 110;
    C Giunti, C W Kim, W Lam: PR D44 (1991) 3635.
14. K S Hirata et al (Kam-II Collaboration): PL B280 (1992) 146.
15. R Becker-Szendy et. al (IMB Collaboration): PR D46 (1992) 3720;
    D Casper et al: PRL 66 (1991) 2561.
16. M C Goodman (Soudan2 Collaboration): Talk at the Neutrino 94 Conference, NP (Proc. Supp.) B38 (1995) 337.
17. C Berger et al (FREJUS Collaboration): PL B245 (1990) 305;
    M Aglietta et al (NUSEX Collaboration): EuPL 8 (1991) 611.
18. T K Gaisser, T Stanev, G Barr: PR D38 (1988) 85;
    W Frati, T Gaisser, A K Mann, T Stanev, PR D48 (1993) 1140.
19. Y. Totsuka, Talk at the Lepton-Photon meeting, Hamburg (1997).
20. J Learned, S Pakvasa, T Weiler: PL B207 (1988) 79;
    V Barger, K Whisnant: PL B209 (1988) 36;
    K Hidaka, M Honda, S Midorikawa: PRL 61 (1988) 1537.
21. Y Fukuda et al: PL B335 (1994) 237;
    O Yasuda, H Minakata: hep-ph/9602386.
22. Y Suzuki: Invited talk at the Erice Neutrino Workshop, 1997.
23. J A Frieman, H E Haber, K Freese: PL B200 (1988) 115.
24. L Wolfenstein: PR D17 (1978) 2369.
25. S P Mikheyev, A Y Smirnov: NuCim C9 (1986) 17.
26. For reviews and detailed references, see, e.g.,
    T K Kuo, J Pantaleone: RMP 61 (1989) 937;
    P B Pal: IJMP A7 (1992) 5387.
27. S P Rosen, J M Gelb: PR D34 (1986) 969;
    T K Kuo, J Pantaleone: PRL 57 (1986) 1805, PR D35 (1987) 3432, PL B198 (1987) 406;
    C W Kim, S Nussinov, W K Sze: PL B184 (1987) 403;
    C W Kim, W K Sze: PR D35 (1987) 1404;
    A Bottino, J Ingham, C W Kim: PR D39 (1989) 909;
    S T Petcov, S Toshev: PL B187 (1987) 120;
    S T Petcov: PL B214 (1988) 259;
    S P Mikheyev, A Y Smirnov: PL B200 (1988) 560;
    H W Zaglauer, K H Schwarzer: PL B198 (1987) 556, ZPh C40 (1988) 273;

A Baldini, G F Giudice: PL B186 (1987) 211.

References to Chapter 11

1. J F Nieves: PR D26 (1982) 3152.
2. B Kayser, A Goldhaber: PR D28 (1983) 2341.
3. B Kayser: PR D30 (1984) 1023.
4. P B Pal, L Wolfenstein: PR D26 (1982) 766.
5. B W Lee, R E Shrock: PR D16 (1977) 1444.
6. K Fujikawa, R E Shrock: PRL 45 (1980) 963.
7. W Marciano, A Sanda: PL B67 (1977) 303.
8. S T Petcov: SJNP 25 (1977) 340; [Erratum ibid, p. 698].
9. S L Glashow, J Iliopoulos, L Maiani: PR D2 (1970) 1285.
10. J Schecter, J W F Valle: PR D24 (1981) 1883 [Erratum 25 (1982) 283].
11. S T Petcov: PL B115 (1982) 401.
12. U Chattopadhyay, P B Pal: PR D34 (1986) 3444.
13. K Enqvist, J Maalampi, A Masiero: PL B222 (1989) 453;
    K S Babu, R N Mohapatra: PRL 64 (1990) 9.
14. J Maalampi, K Mursula, M Roos: PRL 56 (1986) 1031.
15. M A B Beg, W J Marciano, M Ruderman: PR D17 (1978) 1395;
    ʲ R E Shrock: NPh B206 (1982) 359;
    M J Duncan, J A Grifols, A Mendez, S UmaSankar: PL B191 (1987) 304;
    J Liu: PR D35 (1987) 3447.
16. M Voloshin, M Vysotskii, L B Okun: SJNP 44 (1986) 544;
    R Cisneros: ApSp 10 (1971) 87.
17. M Fukugita, T Yanagida: PRL 58 (1987) 1807;
    K S Babu, V S Mathur: PL B196 (1987) 218;
    A Zee: PL B161 (1985) 141.
18. R N Mohapatra: PL B201 (1988) 517; Proceedings of the 1988 INS symposium on Neutrino masses and related topics, eds. S Kato and T Oshima, (World Scientific, 1988) p. 69.
19. M B Voloshin: SJNP 48 (1988) 512.
20. R Barbieri, R N Mohapatra: PL B218 (1989) 225;
    N G Deshpande, P B Pal: PR D45 (1992) 3183.
21. K S Babu, R N Mohapatra: PRL 63 (1989) 228.
22. K S Babu, R N Mohapatra: PRL 64 (1990) 1705;

G Ecker, W Grimus, H Neufeld: PL B232 (1990) 217;

D Chang, W Y Keung, G Senjanović: PR D42 (1990) 1599;

N Marcus, M Leurer: PL B237 (1990) 81.

23. For reviews and references, see, e.g.

P B Pal: IJMP A7 (1992) 5387.

24. H Georgi, L Randall: PL B244 (1990) 196.

25. S M Barr, E M Freire, A Zee: PRL 65 (1990) 2626.

References to Chapter 12

1. For a recent update on the various aspects of the double beta decay phenomena, see *Double-Beta Decay and Related Topics*, ed. H. Klapdor-Kleingrothaus and S. Stoica, (World Scientific 1995).

2. S R Elliot, A A Hahn, M K Moe: PRL 59 (1987) 2020.

3. M Gunther et al: Ref. [1], page 455.

4. T Kirsten et al: Proceedings of the *International Symposium on Nuclear Beta Decay and Neutrino*, ed. T Kotani et al, (World Scientific 1986) p. 81.

5. O Manuel: Proc. of *International Symposium on Nuclear Beta Decay and Neutrino*, ed. T Kotani et al, (World Scientific 1986), p. 71.

6. Y Chikashige, R N Mohapatra, R D Peccei: PL B98 (1981) 265; PRL 45 (1980) 1926.

7. Heidelberg-Moscow Collaboration: A Balysh et al: PL B356 (1995) 450; this surpasses the earlier limit by more than an order of magnitude;

D O Caldwell et al: PR D33 (1986) 2737; PRL 59 (1987) 419.

8. J Hellmig et al: Ref. [1], p 130; this is almost an order of magnitude better than the earlier limit set in F T Avignone et al: PR C34 (1986) 666.

9. H Ejiri et al: NPh A448 (1986) 271.

10. E Bellotti et al: NuCim A95 (1986) 1.

11. For extensive reviews, see:

F T Avignone, R L Brodzinski: PPNPh 21 (1988) 99;

D Caldwell: NuIns A264 (1988) 106.

12. For excellent reviews of double beta decay, see:

H Primakoff, S P Rosen: ARNP 31 (1981) 145;

M Doi, T Kotani, E Takasugi: PTP Suppl. 83 (1985) 1;

W Haxton, G J Stephenson: PPNPh 12 (1984) 409;

J D Vergados: PhRep 133 (1986) 1;

K Muto, H V Klapdor: in *Neutrinos*, ed. H Klapdor (Springer-

Verlag 1988), p. 183;

M G Shehepkin: Usp 27 (1984) 555;

T Tomoda: RepPrPh 54 (1991) 53;

H. Klapdor-Kleingrothaus: in Ref. [1], p. 1;

H. Klapdor-Kleingrothaus and A. Staudt, *Non-Accelerator Particle Physics* (IOP Pub., Philadelphia 1995).

13. M Doi, T Kotani, H Nishiura, E Takasugi: PL B103 (1981) 219; PTP 66 (1981) 1739, 1765;

E Takasugi: private communication (1988).

14. H Georgi, S L Glashow, S Nussinov: NPh B193 (1981) 297.

15. R N Mohapatra, E Takasugi: PL B211 (1988) 192.

16. S P Rosen: *Neutrino '88*, ed. J Schneps et al, (World Scientific 1988), p. 78.

17. For a lucid discussion of this see, B Kayser et al: *Massive Neutrinos*, (World Scientific 1989).

18. W C Haxton, G J Stephenson, Jr., D Strottman: PRL 47 (1981) 153, PR D26 (1982) 1805;

K Muto, H V Klapdor: PL B201 (1988) 420;

T Tomoda, A Faessler, K W Schmid, F Grummer: NPh A452 (1986) 591;

T Tomoda, A Faessler: PL B199 (1987) 475;

J D Vergados: NPh B218 (1983) 109;

P Vogel, M Zirnbauer: PRL 57 (1986) 3148.

19. L Wolfenstein: PL B107 (1981) 77.

20. D Chang, P B Pal: PR D26 (1982) 3113;

D Choudhury, U Sarkar: PR D41 (1990) 1591.

21. A Halprin, P Minkowski, H Primakoff, S P Rosen: PR D13 (1976) 2567.

22. R N Mohapatra, G Senjanović: PR D23 (1981) 165.

23. G B Gelmini, M Roncadelli: PL B99 (1981) 411.

24. R N Mohapatra, J D Vergados: PRL 47 (1981) 1713.

25. L Wolfenstein: PR D26 (1982) 2507.

26. W C Haxton, S P Rosen, G J Stephenson Jr.: PR D26 (1982) 1805.

27. C Piccioto, M Zahir: PR D26 (1982) 2320.

28. R N Mohapatra: PR D34 (1986) 909.

29. R N Mohapatra: PR D34 (1986) 3457;

J D Vergados: PL B184 (1987) 55;

R Kuchimanchi, R N Mohapatra: PR D48 (1993) 4352.

30. M Hirsch, H Klapdor-Kleingrothaus, S Kovalenko: PRL 75 (1995) 17.

382 SUMMARY AND OUTLOOK OF NEUTRINO PHYSICS

31. K S Babu, R N Mohapatra: PRL 75 (1995) 2276.
32. M Hirsch, H Klapdor-Kleingrothaus, S Kovalenko: PR D54 (1996) 4207.
33. C S Aulakh, R N Mohapatra: PL B119 (1983) 136.
34. S Bertolini, A Santamaria: NPh B310 (1988) 714.
35. For a comprehensive discussion of general scalar emission in neutrinoless double beta decay, see C Burgess, P Bamert, R N Mohapatra: NPh B449 (1995) 25.
   The present experimental situation with respect to these processes are summarized in J Hellmig et al: Ref. [1], p. 130.
36. J W F Valle: private communication.
37. J Schecter, J W F Valle: PR D25 (1982) 2951.

References to Chapter 13

1. S T Petcov: SJNP 25 (1977) 340; [Erratum ibid, p. 698].
2. W J Marciano, A I Sanda: PL B67 (1977) 303.
3. S M Bilenky, S T Petcov, B Pontecorvo: PL B67 (1977) 309.
4. Particle Data Group: Review of Particle Properties, PR D54 (1996) 1.
5. T P Cheng, L F Li: PRL 45 (1980) 1908.
6. B W Lee, R E Shrock: PR D16 (1977) 1444.
7. H Georgi, S L Glashow, S Nussinov: NPh B193 (1981) 297.
8. P B Pal: NPh B227 (1983) 237.
9. G Feinberg, S Weinberg: PR 123 (1961) 1439.
10. G A Beer et al: PRL 57 (1986) 671;
    B Ni et al: PRL 59 (1987) 2716;
    T Huber et al: PRL 61 (1989) 2189;
    V A Gordeev et al: JETPL 57 (1993) 270.
11. K Jungman: Talk at the *Beyond 97 conference in Ringberg Castle*, June 8–12, 1997 (to appear in the proceedings).
12. A Halprin: PRL 48 (1982) 1313.
13. R N Mohapatra, P B Pal: PL B179 (1986) 105.
14. P Herczeg, R N Mohapatra: PRL 69 (1992) 2475.
15. R N Mohapatra: in *Proceedings of Eighth Workshop on Grandunification*, ed. K C Wali (World Scientific 1987);
    D Chang, W Y Keung: PRL 62 (1989) 2583;
    W S Hou, G G Wong: PR D53 (1996) 1537;
    M L Schwarz: PR D40 (1989) 1521.

16. R N Mohapatra: ZPh C56 (1992) 117;
    A Halprin, A Masiero: PR D48 (1993) 2987.
17. J C Pati, A Salam: PR D10 (1974) 275;
    R N Cahn, H Harari: NPh B176 (1980) 135;
    S Dimopoulos, S Raby, G L Kane: NPh B182 (1981) 77;
    N G Deshpande, R J Johnson: PR D27 (1983) 1193.
18. M Kobayashi, K Maskawa: PTP 49 (1973) 652.
19. S M Bilenky, J Hosek, S T Petcov: PL B94 (1980) 495.
20. J Schecter, J W F Valle: PR D22 (1980) 2227.
21. M Doi, T Kotani, H Nishiura, K Okuda, E Takasugi: PL B102 (1981) 323.
22. J Schecter, J W F Valle: PR D23 (1981) 1666.
23. For a review, see W Bernreuther, M Suzuki: RMP 63 (1991) 313.
24. E P Shabalin: SJNP 28 (1978) 75.
25. J F Nieves, D Chang, P B Pal: PR D33 (1986) 3324.
26. G Beall, A Soni: PRL 47 (1981) 552.

References to Chapter 14

1. H A Weldon: PR D26 (1982) 2789.
2. J F Nieves: PR D40 (1989) 866.
3. L Dolan, R Jackiw: PR D9 (1974) 3320.
4. S Weinberg: PR D9 (1974) 3357.
5. For detailed discussion on the formalism and extensive references, see, e.g., N P Landsman, C G van Weert: PhRep 145 (1987) 141.
6. J F Nieves: PR D42 (1990) 4123.
7. D Nötzold, G Raffelt: NPh B307 (1988) 924.
8. P B Pal, T N Pham: PR D40 (1989) 259.
9. K Enqvist, K Kainulainen, J Maalampi: NPh B349 (1991) 754.
10. J C D'Olivo, J F Nieves, M Torres: PR D46 (1992) 1172.
11. J F Nieves, P B Pal: PR D40 (1989) 1693.
12. V B Semikoz, Y A Smorodinskiĭ: JETP 68 (1989) 20.
13. V N Oraevsky, A Y Plakhov, V B Semikoz, Y A Smorodinskiĭ: JETP 66 (1987) 890; Erratum JETP 68 (1989) 1309.
14. J C D'Olivo, J F Nieves, P B Pal: PR D40 (1989) 3679. Note that the formulas in this reference assume $e < 0$. Here, we present the formulas with the convention $e > 0$.

15. T Altherr, K Kainulainen: PL B262 (1991) 79.

16. J F Nieves, P B Pal: PR D39 (1989) 652.

17. V N Oraevsky, V B Semikoz: SJNP 42 (1985) 446; Physica A142 (1987) 135.

18. J F Nieves, P B Pal: PR D49 (1994) 1398.

19. J C D'Olivo, J F Nieves, P B Pal: PRL 64 (1990) 1088.

20. C Giunti, C W Kim, W P Lam: PR D43 (1991) 164.

21. J F Nieves, P B Pal: PR D56 (1997) 365.

22. S Esposito, G Capone: ZPh C70 (1996) 55;
    J C D'Olivo, J F Nieves: PL B383 (1996) 87;
    P Elmfors, D Grosso, G Raffelt: NPh B479 (1996) 3.

23. A Kusenko, G Segrè: PRL 77 (1996) 4872.

24. R F Sawyer: PR D46 (1992) 1180.

25. J C D'Olivo, J F Nieves, P B Pal: PL B365 (1996) 178.

26. W Grimus, H Neufeld: PL B315 (1993) 129.

27. C Giunti, C W Kim, U W Lee, W P Lam: PR D45 (1992) 1557.

28. Z G Berezhiani, M I Vysotsky: PL B199 (1987) 281.

References to Chapter 15

1. An excellent discussion of the solar neutrino physics is given by J N Bahcall: *Neutrino Astrophysics* (Cambridge Univ. Press 1989).

2. D D Clayton: *Principles of Stellar Evolution and Nucleosynthesis* (Univ. of Chicago Press, 2nd edition 1983).

3. H Bethe: PR 55 (1939) 434.

4. J N Bahcall, M H Pinsonneault: RMP 67 (1995) 781.

5. A Cummings, W Haxton: PRL 77 (1996) 4286;
   A Dar, G Shaviv: ApJ 468 (1996) 933.

6. R Davis Jr.: PRL 12 (1964) 303.

7. J N Bahcall, R Ulrich: RMP 60 (1988) 297.

8. S Turck-Chièze, S Cahen, M Cassé, C Doom: ApJ 335 (1988) 415.

9. J N Bahcall, M H Pinsonneault: RMP 64 (1992) 885.
   For other calculations of the solar neutrino flux, see
       S Turck-Chiéze, I Lopes: ApJ 408 (1993) 347;
       A Dar, G Shaviv: NPh (supp) B38 (1995) 81;
       C Proffitt: ApJ 425 (1994) 849.

10. V A Kuzmin: JETP 22 (1966) 1051.

11. V V Kuzminov, A Pomansky, V L Chihladze: NuIns A271 (1988) 257.

12. W Haxton: PRL 60 (1988) 768.

13. K S Hirata et al: PRL 58 (1987) 1490.

14. B T Cleveland et al: in *Neutrino '94*, Proceedings of the 16th International conference on Neutrino physics and astrophysics, edited by A Dar, G Eilam, M Gronau, appeared in NPh B (Proc. Suppl.) 38 (1995) 47.

15. Y Suzuki: in *Neutrino '94*, Proceedings of the 16th International conference on Neutrino physics and astrophysics, edited by A Dar, G Eilam, M Gronau, appeared in NPh B (Proc. Suppl.) 38 (1995) 54.

16. J N Abdurashitov et al: PL B328 (1994) 234.

17. P Anselmann et al: PL B327 (1994) 377; PL B 357 (1995) 237.

18. D R O Morrison: Particle World 3 (1992) 30.

19. J N Bahcall: PL B238 (1994) 276.

    V Castellani, S Degl'Innocenti, G Fiorentini: A&A 271 (1993) 601;

    S Degl'Innocenti, G Fiorentini, A Lissia: NPh Proc. Suppl. B43 (1995) 66;

    N Hata, S A Bludman, P Langacker: PR D49 (1994) 3622;

    W Kwong, S P Rosen: PRL 73 (1994) 369.

20. V Castellani, S Degl'Innocenti, G Fiorentini, M. Lissia, B Ricci: PR D50 (1994) 4749;

    V Berezinsky, G Fiorentini, M Lissia: PL B365 (1996) 185;

    V Berezinsky: ComNPPh 21 (1994) 249.

21. N Hata, P Langacker: hep-ph/9705339.

22. J N Bahcall, P I Krastev: PR D53 (1996) 4211;

23. P I Krastev, S T Petcov: PR D53 (1996) 1665;

    S T Petcov: NP (Supp) B43 (1995) 12;

    Z Berezhiani, A Rossi: PL B367 (1996) 219.

24. P B Pal: IJMP A7 (1992) 5387.

25. L Wolfenstein: PR D17 (1978) 2369.

26. S P Mikheyev, A Yu Smirnov: NuCim C9 (1986) 17.

27. V Barger, R J N Phillips, K Whisnant: PR D34 (1986) 980.

28. H Bethe: PRL 56 (1986) 1305.

29. W C Haxton: PRL 57 (1986) 1271, PR D35 (1987) 2352;

    S J Parke: PRL 57 (1986) 1275.

30. P Pizzochero: PR D36 (1987) 2293.

31. S T Petcov: PL B200 (1988) 373;

    T K Kuo, J Pantaleone: PR D39 (1989) 1930;

M M Guzzo, J Bellandi, V M Aquino: PR D49 (1994) 1404.

32. S P Rosen, J Gelb: PR D34 (1986) 969;
    E W Kolb, M S Turner, T P Walker: PL B175 (1986) 478.

33. V Barger, N Deshpande, P B Pal, R J N Phillips, K Whisnant: PR D43 (1991) R1759.

34. S A Bludman, N Hata, D C Kennedy, P G Langacker: PR D47 (1993) 2220;
    P I Krastev, S T Petcov: NP B449 (1995) 605.

35. J Bahcall, N Cabibbo, A Yahil: PRL 28 (1972) 316.

36. Y Chikashige, R N Mohaptra, R D Peccei: PL B98 (1981) 265.

37. J Bahcall, S T Petcov, S Toshev, J W F Valle: PL B181 (1986) 369.

38. J A Frieman, H E Haber, K Freese: PL B200 (1988) 115.

39. A Cisneros: ApSp 10 (1971) 87.

40. M Voloshin, M Vysotskii, L B Okun: JETP 64 (1986) 446; SJNP 44 (1986) 440.

41. A V Kyuldjiev: NPh B243 (1984) 387.

42. D Krakauer et al: PL B252 (1990) 177.

43. R Barbieri, G Fiorentini: NPh B304 (1988) 909.

44. M Fukugita, S Yazaki: PR D36 (1987) 3817.

45. Charm II Experiment, G Geiregat et al: PL B232 (1989) 539.

46. K Abe et al: PRL 58 (1987) 636.

47. R C Allen et al: in Neutrino '88, ed. J Schneps et al (World Scientific 1989).

48. F Reines, H S Gurr, H W Sobel: PRL 37 (1977) 315.

49. J Morgan: PL B102 (1981) 247.
    G Raffelt: PRL 64 (1990) 2856.

50. R Barbieri, R N Mohapatra: PRL 61 (1988) 27;
    J Lattimer, D Cooperstein: PRL 61 (1988) 23;
    D Nötzold: PR D38 (1988) 1658;
    For a qualitative discussion, see I Goldman, G Alexander, S Nussinov, Y Aharonov: PRL 60 (1988) 1789.

51. E Akhmedov, M Y Khlopov: ModPL A3 (1988) 451;
    C S Lim, W Marciano: PR D37 (1988) 1368;
    E Akhmedov: PL B213 (1988) 64; PL B 257 (1991) 163.

52. K S Babu, R N Mohapartra, I Z Rothstein: PR D44 (1991) 2265;
    E Akhmedov, A Lanza, S Petcov: PL B303 (1993) 85.

53. K S Babu, R N Mohapatra: PRL 63 (1989) 228; PR D42 (1990) 3778.

54. Various models give rise to a large neutrino magnetic moment. For a review, see Ref. [24].

55. H H Chen: PRL 55 (1985) 1524;
W F Davidson, P Depommier, G T Ewan,, H B Mak: Proceedings of ICOBAN '84 (Park City, Utah) p. 273.

56. R S Raghavan, S Pakvasa, T Brown: PRL 57 (1986) 1801.

57. A Halprin, C N Leung: PRL 67 (1991) 1833.

58. A Dar, A Mann: Nature 325 (1987) 790;

59. D N Spergel, W H Press: ApJ 294 (1985) 663;
L M Krauss, K Freese, D N Spergel, W H Press: ApJ 299 (1985) 1001.
G Gelmini, L Hall, M Lin: NPh B281 (1987) 726;
S Raby, G West: NPh B292 (1987) 79.

References to Chapter 16

1. For a recent survey, see G Raffelt: *Stars as laboratories for Fundamental Physics* (Chicago University Press 1996).

2. For a detailed theoretical description of the neutrino signal from supernova collapse, see
A Burrows, J Lattimer: ApJ 307 (1986) 178, ApJ 318 (1987) L63;
R Mayle, J R Wilson, D Schramm: ApJ 318 (1987) 288.

3. M Aglietta et al: EuPL 3 (1987) 1321.

4. E N Alekseev et al: PL B205 (1988) 209.

5. K Hirata et al: PRL 58 (1987) 1490.

6. R Bionta et al: PRL 58 (1987) 1494.

7. W Arnett, J Rosner: PRL 58 (1987) 1906;
J N Bahcall, S L Glashow: Nature 326 (1987) 476;
E Kolb, A Stebbins, M Turner: PR D35 (1987) 3598 [Erratum 36 (1987) 3820].

8. K Gaemers, R Gandhi, J Lattimer: PR D40 (1989) 309;
A Burrows, R Gandhi: PL B246 (1990) 149.

9. R Barbieri, R N Mohapatra: PRL 61 (1988) 27;
J Lattimer, J Cooperstein: PRL 61 (1988) 24;
D Nötzold: PR D38 (1988) 1658;
I Goldman, G Alexander, S Nussinov, Y Aharonov: PRL 60 (1988) 1789;
for recent works, see A Goyal, S R Dutta, S Raichoudhury: PL B346 (1995) 312;
S Mohanty, M K Samal: PRL 77 (1996) 806;

P Elmfors, K Enquist, G Raffelt, G Sigl: hep-ph/9703214.

10. R Barbieri, R N Mohapatra, T Yanagida, PL B213 (1988) 69.
11. L B Okun: Preprint ITEF-88-079 (1988);
    M Voloshin: SJNP 48 (1988) 512.
12. G Barbiellini, G Cocconi: Nature 329 (1987) 21.
13. J Bahcall, D Speigel, W Press: "SN1987A in The Large Magellanic Cloud", Proceedings of the *Fourth George Mason Astronomy Workshop*, Fairfax, Virginia, ed. M Kafatos (Cambridge Univ. Press) p. 172.
14. R Barbieri, R N Mohapatra: PR D39 (1989) 1229;
    G Raffelt, D Seckel: PRL 60 (1988) 1793.
15. E Chupp, C Reppin, W Vestrand: PRL 62 (1989) 505.
16. F von Feilitzsch, L Oberauer: PL B200 (1988) 580;
    E Kolb, M Turner: PRL 62 (1989) 509.
17. R Cowsik: PRL 39 (1977) 784.
18. R N Mohapatra, S Nussinov: PR D51 (1995) 3843;
    J Soares, L Wolfenstein: PR D40 (1989) 3666.
19. R N Mohapatra, S Nussinov, X Zhang: PR D49 (1994) 3434;
    L Oberauer, G Raffelt, G Sigl: APPh 1 (1993) 377;
    R Cowsik, D Schramm, R Hopflich: PL B218 (1989) 91.
20. Y-Z Qian, G M Fuller, G J Mathews, R W Mayle, J R Wilson, S E Woosley: PRL 71 (1993) 1965;
    G M Fuller, Y-Z Qian: Seattle preprint, 1994.
21. L Krauss, S B Tremaine: PRL 60 (1988) 176;
    M J Longo: PRL 60 (1988) 173.
22. E Kolb, D Schramm, M Turner: *Neutrino Physics*, ed. K Winter (Cambridge Univ. Press, 1989).

References to Chapter 17

1. There exist several excellent books and review articles on Cosmology. Some examples are:
   P J E Peebles: *Physical Cosmology* (Princeton Univ. Press 1971);
   S Weinberg: *Gravitation and Cosmology* (Wiley 1972);
   S Weinberg: *The first three minutes* (Basic Books 1977);
   C W Misner, K S Thorne, J A Wheeler: *Gravitation* (Freeman 1973);
   J R Primack: Lectures presented at the *International School of Physics "Enrico Fermi"* (Verena, 1984);

E W Kolb: *Proceedings of the 1986 Theoretical Advanced Studies Institute*, (Santa Cruz, 1986);

E W Kolb, M Turner: *The Early Universe* (Addition-Wesley 1990).

G Börner: *The early universe* (Springer-Verlag 1988).

2. R A Alpher, J W Follin, R C Herman: PR 92 (1953) 1347;

R V Wagoner, W A Fowler, F Hoyle: ApJ 148 (1967) 3.

3. J Yang, M S Turner, G Steigman, D N Schramm, K Olive: ApJ 281 (1984) 493;

For a recent review, see S Sarkar: RepPrPh 59 (1996) 1493.

4. K A Olive, D N Schramm, G Steigman, T P Walker: PL B236 (1990) 454.

5. W Mampe et al: PRL 63 (1989) 593.

6. See, e.g., Weinberg's *Gravitation and cosmology* in [1].

7. S S Gershtein, Y B Zeldovich: JETP Lett. 4 (1966) 120;

R Cowsik, J Mclelland: PRL 29 (1972) 669.

8. B W Lee, S Weinberg: PRL 39 (1977) 165;

P Hut: PL B69 (1977) 85;

K Sato, M Kobayashi: PTP 58 (1977) 1775;

M I Vysotskii, A D Dolgov, Y B Zeldovich: JETP Lett. 26 (1977) 188.

9. E W Kolb, K A Olive: PR D33 (1986) 1202 [Erratum ibid 34 (1986) 2531].

10. D A Dicus, E W Kolb, V L Teplitz: PRL 39 (1977) 169; ApJ 221 (1978) 327.

11. P B Pal: NPh B227 (1983) 237.

12. E W Kolb in ref. [1].

13. S Sarkar, A M Cooper: PL B148 (1984) 347;

K Sato, M Kobayashi: PTP 58 (1977) 1775;

J E Gunn, B W Lee, I Lerche, D N Schramm, G Steigman: ApJ 223 (1978) 1015;

R Cowsik: PRL 39 (1977) 784;

D Lindley: MNRS 188 (1979) 15.

14. E W Kolb, D N Schramm, M Turner: in *Neutrino Physics*, ed. K Winter (Cambridge Univ. Press 1989).

15. A M Boesgard, G Steigman: ARAA 23 (1985) 319.

16. M Roncadelli, G Senjanović: PL B107 (1983) 59.

17. Y Chikashige, R N Mohapatra, R D Peccei: PRL 45 (1980) 1926.

18. W Mampe et al: Nucl. Inst. Meth. A284 (1989) 111.

19. For an extensive review of the indication of dark matter in the universe, see e.g. V Trimble: ARAA 25 (1987) 425.
20. J R Primack: ref. [1].
21. S Treimane, J E Gunn: PRL 42 (1979) 407.
22. D N C Lin, S M Faber: ApJL 266 (1983) L21.
23. J R Bond, G Efstathiou, J Silk: PRL 45 (1980) 1980.
24. For a review of structure formation in a neutrino dominated universe, see e.g., S F Shandarin, A G Doroshkevich, Y B Zeldovich: Usp 26 (1983) 46.
25. C Frenk, S D M White, M Davis: ApJ 271 (1983) 417;
J Centrella, A Melott: Nature 305 (1983) 196.
26. P B Pal: PR D30 (1984) 2100;
E D Carlson, L J Hall: PR D40 (1989) 3187;
G Giudice: ModPL A6 (1991) 851.
27. G M Fuller, R A Malaney: PR D43 (1991) 3136.
28. K Enqvist, K Kainulainen, V Semikoz: NPh B374 (1992) 392.
29. J Morgan: PL B102 (1981) 247.
30. D Caldwell, R N Mohapatra: PR D48 (1993) 3259;
J Peltoniemi, J W F Valle: NP B406 (1993) 409.
31. P Hut, S D M White: Nature 310 (1984) 637.
32. G Steigman, M S Turner: NPh B253 (1985) 375.

References to Chapter 18

1. A Chodos, A Hauser, A Kostelecky: PLB 150 (1985) 431;
P Caban, J Rembielinski, K A Smolinski: hep-ph/9707391.
2. G Stephenson, Jr., T Goldman, hep-ph/9309308.
3. R N Mohapatra, S Nussinov: PL B395 (1997) 63.
4. See Ch. 15.
5. R Davis et al: Proceedings of the 21st International Cosmic Ray Conference, Vol. 12, edited by R J Protheroe (Univ. of Adelaide Press, Adelaide, 1990), p. 143;
K S Hirata et al: PR 44 (1991) 2241;
A I Abrazov et al: PRL 67 (1991) 3332;
V N Gavrin in TAUP 93 Workshop, Gran Sasso, Italy, 1993 (unpublished);
P Anselmann et al: PL B285 (1992) 376; B314 (1993) 445.
6. K S Hirata et al: PL B280 (1992) 146;

R Becker-Szendy et al: PR D46 (1992) 3720;

P J Litchfield in International Europhysics Conference on High Energy Physics, Marseille, France, 1993 (unpublished);

7. Y Fukuda et al: PL B335 (1994) 237.

8. C Athanassopoulos et al: PRL 77 (1996) 3082.

9. C Athanassopoulos et al nucl-ex/9706006.

10. A Romosan et al PRL 78 (1997) 2912.

11. NOMAD collaboration, J Altegoer et al, NIM (to appear); cited in K Zuber, Proceedings of the 4th Solar Neutrino Conference, Heidelberg, (to be published).

12. Q Shafi, F Stecker: PRL 53 (1984) 1292;

    for a recent discussion, see K S Babu, R K Schaefer, Q Shafi: PR D53 (1996) 606.

    For a review, see J Primack, astro-ph/9707285.

13. J Ostriker, P Steinhardt: Nature 377 (1995) 600.

14. W L Freedman et al: Nature 371 (1994) 757.

15. For a recent review of nucleosynthesis and its implications for new physics, see S Sarkar: RepPrPh 59 (1996) 1493.

16. H Klapdor-Kleingrothaus: *Double Beta Decay and Related Topics*, ed. H Klapdor-Kleingrothaus and S Stoica, World Scientific, (1995) p. 3;

    A Balysh et al: PL B283 (1992) 32.

17. Y-Z Qian, G M Fuller: PR D52 (1995) 656.

18. J Peltoniemi: hep-ph/9511323.

19. C Y Cardall, G M Fuller: NPh Proc. Suppl. B51 (1996) 259;

    H Minakata, O Yasuda: PR D56 (1997) 1692.

20. D O Caldwell, R N Mohapatra: PR D48 (1993) 3259.

21. J Peltoniemi, J W F Valle: NPh B406 (1993) 409.

22. G Fogli, E Lisi, D Montanino: PR D49 (1994) 3626;

    C Giunti, C W Kim, J D Kim: PL B352 (1995) 357;

    S Goswami, K Kar, A Raychaudhuri: IJMP A12 (1997) 781;

    K S Babu, J C Pati, F Wilczek: PL B359 (1995) 351;

23. G G Raffelt, J Silk: PL B366 (1996) 429;

    D Caldwell, R N Mohapatra: PL B354 (1995) 371.

24. M Gell-Mann, P Ramond, R Slansky in Supergravity, edited by D Freedman et al (1979).

25. R Mohapatra, G Senjanović: PRL 44 (1980) 912; PR D23 (1981) 165.

26. T Yanagida: in KEK Lectures, ed. O Sawada et al (1979).

27. D Chang, R N Mohapatra, J Gipson, R E Marshak and M K Parida: PR D31 (1985) 1718.
28. D G Lee, R N Mohapatra: PR D52 (1995) 4215.
29. K S Babu, R N Mohapatra: PL B267 (1991) 400;
    R N Mohapatra: *Particle Physics and Cosmology at the interface* ed. J C Pati et al, World scientific, (1994), p.273.
30. D Chang, R N Mohapatra: PR D32 (1985) 1248.
31. K S Babu, R N Mohapatra: PRL 70 (1993) 2845.
32. L Lavoura: PR D48 (1993) 5440.
33. D G Lee, R N Mohapatra: PR D51 (1995) 1353.
34. Y Achiman, T Greiner: NPh B443 (1995) 3.
35. D G Lee, R N Mohapatra: PL B329 (1994) 463.
36. E J Chun, A Joshipura, A Smirnov: PL B357 (1995) 608.
37. Z Berezhiani, R N Mohapatra: PR D52 (1995) 6607.
38. R Barbieri, J Ellis, M K Gaillard: PL B90 (1980) 249;
    E Akhmedov, Z Berezhiani, G Senjanović: PRL 69 (1992) 3013.
39. R N Mohapatra, V L Teplitz: Ap J 478 (1997) 29.

Index